Progress in Nonlinear Differential Equations and Their Applications
Volume 52

Editor
Haim Brezis
Université Pierre et Marie Curie
Paris
and
Rutgers University
New Brunswick, N.J.

Partial Differential Equations and Mathematical Physics

In Memory of Jean Leray

Kunihiko Kajitani
Jean Vaillant
Editors

Birkhäuser
Boston • Basel • Berlin

Kunihiko Kajitani
University of Tsukuba
Institute of Mathematics
Ibaraki, 305 Japan

Jean Vaillant
MATHS-Université Paris VI
BC 172, 4 Place Jussieu
Paris Cedex 05
75252 France

Library of Congress Cataloging-in-Publication Data

Partial differential equations and mathematical physics : in memory of Jean Leray /
Kunihiko Kajitani and Jean Vaillant, editors.
   p. cm – (Progress in nonlinear differential equations and their applications ; v. 52)
   Includes bibliographical references.
   ISBN 0-8176-4309-5 (acid-free paper) – ISBN 3-7643-4309-5 (acid-free paper)
   1. Differential equations, Partial. 2. Mathematical physics. I. Leray, Jean 1906-1998 II.
Kajitani, Kunihiko, 1941- III. Vaillant, J. (Jean) IV. Series.

QC20.7.D5 P365 2002
530.15'5353–dc21
                                      2002027973

AMS Subject Classifications: 35M10, 35Q75, 35L15, 58F06, 32S35, 35A10, 35H20, 35A20, 35H20, 35Q35, 35L55, 35J45, 35L30, 35D05, 35D10

Printed on acid-free paper
©2003 Birkhäuser Boston

Birkhäuser

ISBN 0-8176-4309-5   SPIN 1088264
ISBN 3-7643-4309-5

Reformatted from editors' files by TEXniques, Inc., Cambridge, MA
Printed in the United States of America

9 8 7 6 5 4 3 2 1

Birkhäuser Boston • Basel • Berlin
A member of BertelsmannSpringer Science+Business Media GmbH

Contents

Preface

This symposium on Partial Differential Equations and Mathematical Physics in Memory of Jean Leray was organized at Maison Franco-Japonaise in Tokyo on July 2–6, 2001.

The 17 invited research articles in this volume, all written by leading experts in their respective fields, are dedicated to the great French mathematician, Jean Leray. A wide range of topics with significant new results—detailed proofs—are presented in the areas of partial differential equations, complex analysis, and mathematical physics. Key subjects are treated from the mathematical physics viewpoint: nonlinear stability of an expanding universe, the compressible Euler equation, spin groups and the Leray–Maslov index. The Cauchy problem is linked to an intermediate case between effective hyperbolicity and the Levi condition, global Cauchy–Kowalewski theorem in some Gevrey classes, the analytic continuation of the solution, necessary conditions for hyperbolic systems, well posedness in the Gevrey class, uniformly diagonalizable systems and reduced dimension, and monodromy of ramified Cauchy problem. Additional articles examine results on local solvability for a system of partial differential operators, the hypoellipticity of second order operators, differential forms and Hodge theory on analytic spaces, subelliptic operators and sub-Riemannian geometry.

The financial support was obtained from the French Embassy in Japan, The Japanese Ministry of Education and Research, CNRS in Japan, Hironaka's Zaidan and the Institute of Mathematics of Tsukuba University. We want to thank all these organizations for their support.

We also wish to thank all the invited speakers for their inspiring lectures and contributions to this volume.

Kunihiko Kajitani and Jean Vaillant
July 31, 2002

Jean Leray
1906–1998

Dedicated to the memory of Jean Leray

Partial Differential Equations and Mathematical Physics

In Memory of Jean Leray

Differential Forms, Cycles and Hodge Theory on Complex Analytic Spaces

Vincenzo Ancona and Bernard Gaveau

ABSTRACT We summarize our theory of differential forms on complex analytic spaces. This theory is functorial with respect to analytic mappings. We define also dual complexes of chains, and the integration of forms on chains. We construct a mixed Hodge theory on compact algebraic varieties using these forms.

1 Introduction

In a previous paper published in this series of conferences in honor of Jean Leray [AG1], we have extended Leray's theory of residues to any complex analytic space, relatively to any complex subspace. This extension was based on the construction of families of differential forms on a complex space which is described in detail in [AG2, 3]. In this work, we shall use the construction of complexes of differential forms to define a new notion of chains and of integration of forms on these chains, so that Stokes theorem remains valid. This will prove that the singular cohomology is given by the cohomology of the complexes of forms. Moreover, we shall also define, for the case of compact algebraic varieties, a mixed Hodge theory on the cohomology which seems to be identical to the one defined initially by Deligne [D]. Our construction is nevertheless more concrete than Deligne's construction, because it uses neither simplicial schemes, nor *descente cohomologique*. As a consequence, we obtain effortlessly the solutions to certain conjectures. In this work, X will be a reduced complex analytic space, possibly reducible. The constant sheaf of X will be denoted by C_X and if M is a complex analytic manifold $(\mathcal{E}_M^*, d)$ will denote the usual De Rham complex of C^∞ differential forms on M. For any sheaf F, $\Gamma(U, F)$ will be the set of sections of F on U. If $f : X \to Y$ is a mapping and F is a sheaf on X, one defines the direct image $f_* F$ by its sections on any open set $V \subset Y$; $\Gamma(V, f_* F) = \Gamma(f^{-1}(V), F)$.

2   Families of complexes of differential forms. Hypercoverings

Let X be a complex space. We shall associate to X a family $r(X)$ of complexes (Λ_X^*, d) and to any morphism $f : X \to Y$ of complex spaces, a family $r(f)$ of morphisms of complexes, called *pull-back* with the following properties:

(i) (Λ_X^*, d) are fine resolutions of the constant sheaf C_X; namely, all the sheaves Λ_X^* are fine and we have an exact sequence of sheaves

$$0 \to C_X \to \Lambda_X^0 \xrightarrow{d} \Lambda_X^1 \xrightarrow{d} \Lambda_X^2 \to \cdots \to \Lambda_X^p \to \cdots$$

such that if α is a section of Λ_X^{p+1} on a small neighborhood U of x, with $d\alpha = 0$, then there exists a neighborhood $U' \subset U$ of x, and a section β of Λ_X^p on U' with $\alpha = d\beta$. As is well known, a result of Leray implies that the cohomology of X is the cohomology of the global sections of the complex (Λ_X^*, d), namely

$$H^k(X, C) = \frac{\text{Ker } \{d : \Gamma(X, \Lambda_X^k) \to \Gamma(X, \Lambda_X^{k+1})\}}{d\Gamma(X, \Lambda_X^{k-1})}.$$

We shall call an element of Λ_X^p or a section, a p-form on X.

(ii) We have $\Lambda_X^p = 0$ for $p > 2 \dim_C X$.

(iii) For any X and any $\Lambda_X^* \in r(X)$, there exists a smooth open dense subset $U \subset X$ such that $\Lambda_X^*|_U$ is the standard De Rham complex of the manifold U.

(iv) If $f : X \to Y$ is a morphism if Λ_Y^* is an element of $r(Y)$, then there exists an element Λ_X^* of $r(X)$ and a morphism of complexes denoted α,

$$\alpha : \Lambda_Y^* \to f_* \Lambda_X^*,$$

called a pull-back. We shall in general skip the sign f_* of direct image.

(v) If X is a complex manifold, the De Rham complex is an element of $r(X)$ and the De Rham pull back of a morphism $f : X \to Y$ between complex manifolds is an element of $r(f)$.

(vi) If $f : X \to Y, g : Y \to Z$ are morphisms, the composition of corresponding pull-back in $r(f)$ and $r(g)$ is a pull-back in $r(g \circ f)$.

(vii) If $\alpha, \beta : \Lambda_Y^{\cdot} \to \Lambda_X^{\cdot}$ are two pull-backs corresponding to $f : X \to Y$, then $\alpha = \beta$.

It is clear that condition (i), alone, would be fulfilled by any fine resolution of C_X, for example a cochain resolution or an Alexander resolution. But condition (iii) is extremely strong: it says that on an open dense subset contained in the regular points, the resolution $\Lambda_X^{\cdot}$ should be defined by forms.

To specify a complex $(\Lambda_X^{\cdot}, d)$ in $r(X)$, we have to specify the following data.

a) a nowhere dense analytic subset $E \subset X$ which contains the singular points of X, as well as a proper modification

$$
\begin{array}{ccc}
\tilde{E} & \stackrel{i}{\to} & \tilde{X} \\
q\downarrow & & \downarrow p \\
E & \stackrel{j}{\to} & X
\end{array}
\tag{2.1}
$$

where $j : E \to X$ is the inclusion, $\tilde{X}$ is a manifold, $\tilde{E} = p^{-1}(E)$ and $p : \tilde{X} \setminus \tilde{E} \to X \setminus E$ is an isomorphism of complex manifolds.

b) A complex $(\Lambda_E^{\cdot}, d)$ in $r(E)$.

c) Two pull-backs $\varphi : \Lambda_E^* \to \Lambda_{\tilde{E}}^*$ and $\psi : e_{\tilde{X}}^* \to \Lambda_{\tilde{E}}^*$ associated to q and j respectively where $(\Lambda_{\tilde{E}}^*, d)$ is a certain resolution in $r(\tilde{E})$.

Then, the resolution (Λ_X^*, d) is defined as

$$
\Lambda_X^k = p_* \mathcal{E}_{\tilde{X}}^* \oplus i_* \Lambda_E^k \oplus (i \circ q)_* \Lambda_{\tilde{E}}^{k-1}
\tag{2.2}
$$

with a differential

$$
d(\omega \oplus \sigma \oplus \gamma) = (d\omega \oplus d\sigma \oplus (d\gamma + (-1)^k (\psi(\omega) - \varphi(\sigma))))
\tag{2.3}
$$

and the injection $0 \to C_X \to \Lambda_X^0$ given by

$$
c \to (c \oplus c \oplus 0) \in p_* e_{\tilde{X}}^0 \oplus i_* \Lambda_E^0 \oplus 0.
$$

Here we use the obvious convention that

$$
\Lambda_X^k = 0 \text{ whenever } k < 0.
$$

In [AG1], it is proved that given the data E, $\tilde{X}$, $\tilde{E}$ and $\Lambda_E^{\cdot} \in r(E)$, there exists a certain $\Lambda_{\tilde{E}}$ and pull-back φ and ψ as in c) such that the complex (Λ_X^k, d) defined by Eqs (2.2) and (2.3) is a fine resolution of C_X.

The construction proceeds by a rather complicated recursion on the dimensions of the spaces, and the dimensions of spaces related by morphisms. One constructs at the same time, both the complexes and the pull-back.

For any $\Lambda_X^* \in r(X)$, we have also two complexes $\Lambda_E^* \in r(E)$ and $\Lambda_{\tilde{E}}^* \in r(\tilde{E})$ (which are defined or given by recursion). Each of these complexes $\Lambda_E^*, \Lambda_{\tilde{E}}^*$ is also given by a similar construction as in a), b), c) as above and by similar equations as Eqs (2.2)–(2.3). We thus see that for any $\Lambda_X^* \in r(X)$, we will define a certain family $(X_l, h_l)_{l \in L}$ where X_l are complex manifolds, $h_l : X_l \to X$ are morphisms so that

$$
\Lambda_X^k = \oplus_l (h_l)_* \mathcal{E}_{X_l}^{k-q(l)}
$$

where $q(l)$ is a nonnegative integer. Moreover there will be mappings $h_{lm} : X_l \to X_m$ for certain l, m commuting with h_l and h_m when they are defined, such that

$$d(\oplus_l \omega_l) = (\oplus_l (d\omega_l + \sum \varepsilon_{lm}^{(k)} h_{lm}^* \omega_m))$$

where $\omega_l \in \mathcal{E}_{X_l}^{h-q(l)}$ $\varepsilon_{lm}^{(k)} \in \{0, \pm 1\}$ and $q(m) = q(l) - 1$.

Definition. The collection X_l, h_l, h_{lm} is called the hypercovering of X associated to Λ_X^* and $q(l)$ is called the rank of X_l.

Example. Assume that E is a manifold, as well as $\tilde{E}$. Then we take $\Lambda_E^* = \mathcal{E}_E^*$, $\Lambda_{\tilde{E}}^* = \mathcal{E}_{\tilde{E}}^*$ and the pull-backs φ, ψ are the standard pull back between forms associated to q and $i \circ q$ respectively. Skipping the signs of direct image, we write

$$\Lambda_X^k = \mathcal{E}_{\tilde{X}}^k \oplus \mathcal{E}_E^k \oplus \mathcal{E}_{\tilde{E}}^{k-1},$$

$$d(\omega \oplus \sigma \oplus \gamma) = (d\omega \oplus d\sigma \oplus (d\gamma + (-1)^k (j^* \omega - q^* \sigma))).$$

In this case the hypercovering of X is given by $\tilde{X}$, E, $\tilde{E}$ with the natural mappings to X and $q(\tilde{X}) = q(E) = 0$, $q(\tilde{E}) = 1$.

If $f : X \to Y$ is a morphism $\varphi : \Lambda_Y^* \to \Lambda_X^*$ is a pull-back associated to f, one can describe φ using the hypercoverings of X and Y,

$$\Lambda_Y^k = \bigoplus_{b \in B} \mathcal{E}_{Y_b}^{k-q_Y(b)} \to \Lambda_X^k = \bigoplus_{a \in A} \mathcal{E}_{X_a}^{k-q_X(a)},$$

by saying that for any $a \in A$, there exists at most a unique $b(a) \in B$ with $q_Y(b) = q_X(a)$ and a morphism $f_{ab} : X_a \to Y_b$ so that

$$\varphi(\oplus_b \omega_b) = \bigoplus_{a \in A} (f_{ab}^* \omega_b)$$

where we have written for f_{ab}^* the standard De Rham pull-back associated to f_{ab} if f_{ab} is defined and $f_{ab}^* = 0$ if f_{ab} is not defined.

Remark. The main difficulty in constructing pull-back φ is to fulfill the seemingly innocuous sentence that φ *is a morphism of complexes*, namely φ commutes with differentials.

3   Chains, integration and homology

To any $\Lambda_X^* \in r(X)$, we associate a dual cocomplex of *chains* $(S_{X,*}, \partial)$ with the following properties:

(i) $S_{X,k} = 0$ for $k > 2 \dim_{\mathbb{C}} X$ and the sections with compact support of $(S_{X,*}, \partial)$ define a fine coresolution of C_X

$$\cdots \to S_{X,k+1} \to S_{X,k} \to \cdots \to S_{X,1} \to S_{X,0} \to C_X \to 0.$$

(ii) If X is a complex manifold, the cocomplex associated to $\mathcal{E}_X^*$ is the co complex $(S_{X,*}, \partial)$ of subanalytic chains of X (see [B–H], [D–P], [P]).

(iii) If Λ_X^* is given by the data $(\tilde{X}, E, \tilde{E})$ as in Section 2, we have

$$S_{X,*}\big|_{X \setminus E} = S_{X \setminus E,*}.$$

(iv) There is an integration

$$\int_\sigma \omega \qquad \omega \in \Lambda_X^k, \ \sigma \in S_{X,k}$$

where $S_{X,k}$ is associated to Λ_X^k which satisfies Stokes theorem

$$\int_\sigma d\omega = \int_{\partial \sigma} \omega. \tag{3.1}$$

(v) If $f : X \to Y$ is a morphism and $\psi : \Lambda_Y^* \to \Lambda_X^*$ is a pull-back associated to f, there will be a corresponding push-down $\psi' : S_{X,*} \to S_{Y,*}$ such that

$$\int_{\psi'\sigma} \omega = \int_\sigma \psi\omega. \tag{3.2}$$

The construction is easy: given Λ_X^* as in Section 2, Eq (2.2), one associates an $S_{X,*}$ defined by

$$S_{X,k} = p_* s_{\tilde{X},k} \oplus i_* S_{E,k} \oplus (i \circ q)_* S_{\tilde{E},k-1} \tag{3.3}$$

with a differential

$$\partial(\alpha \oplus \beta \oplus \gamma) = ((\partial\alpha + (-1)^k \psi'\gamma) \oplus (\partial\beta - (-1)^k \varphi'\gamma) \oplus \partial\gamma) \tag{3.4}$$

where ψ', φ' are associated to ψ and φ of the definition of Λ_X^*. Here $S_{E,*}$, $S_{\tilde{E},*}$ are defined recursively on the dimension and are associated to Λ_E^* and $\Lambda_{\tilde{E}}^*$. The integration is

$$\int_{\alpha \oplus \beta \oplus \gamma} (\omega \oplus \sigma \oplus \theta) = \int_\alpha \omega + \int_\beta \sigma + \int_\gamma \theta \tag{3.5}$$

where the symbols $\int_\beta \sigma$, $\int_\gamma \theta$ are defined recursively. Then, obviously Stokes theorem is correct, as well as the property (3.2).

Let us comment briefly on this definition. Let X be a complex space. It is known that any singular chain can be realized as a subanalytic chain, namely, any singular chain contains in its homology class, a subanalytic chain, which is essentially such that the support of the subanalytic chain is a subanalytic subset of X, with a certain class of orientation. Moreover when X is a manifold, one can define the integral of any smooth form on X on any subanalytic chain ([He], [D–P], [P]).

Obviously, if (X_l, h_l) is the hypercovering of Λ_X^*,

$$S_{X,k} = \bigoplus_l (h_l)_* S_{X_l,k-q(l)}. \tag{3.6}$$

So ultimately all the integration symbols $\int_\alpha \omega$ are realized by integration of forms on subanalytic chains.

As a consequence, one can prove that the cohomology of X (namely the cohomology of C_X) is indeed the singular cohomology of X and that the cohomology with compact support of X is the dual of the Borel–Moore homology of X. This is indeed formal (but nontrivial) and proceeds along the methods of [W]. The statements are the generalization of the classical De Rham theorems to any complex analytic space (see also [AG3, 4]). Moreover, the cocomplex of subanalytic chains on X, namely, $(S_{X,*}, \partial)$ can be naturally embedded as a subcomplex of $(\mathcal{S}_{X,*}, \partial)$. This is done in the following manner:

1) If N is a subanalytic chain contained in E, one associates $0 \oplus N' \oplus 0$ where N' is obtained by the embedding of N, $s_{E,*} \to S_{E,*}$, which is defined recursively on the dimension.

2) If N has no component in E, one considers $N \setminus E$ which is open dense in N, and take the closure in $\tilde{X}$ of $p^{-1}(N \setminus E)$ (recall that p is an isomorphism $\tilde{X} \setminus \tilde{E} \to X \setminus E$ which is again a subanalytic chain of $\tilde{X}$. Now the subanalytic chain

$$\overline{\partial(p^{-1}(N \setminus E))} \setminus (p^{-1})_* \partial(N \setminus E)$$

is a chain with support in $\tilde{E}$, and the image by q_* of this chain is 0 in E for dimensional and rank reasons. So we associate to N the triplet in $\mathcal{S}_{X,*}$

$$\left(\overline{p^{-1}(N \setminus E)}\right) \oplus (0) \oplus (-1)^k \left[\overline{\partial(p^{-1}(N \setminus E))} \setminus (p^{-1})_* \partial(N \setminus E) \right].$$

The boundary of this chain is easily seen to be

$$\left[(p^{-1})_* \partial(N \setminus E) \right] \oplus (0) \oplus (0)$$

which is also the image in $\mathcal{S}_{X,*}$ of the boundary ∂N, so that the embedding commutes with boundaries. Moreover let $Y_k \subset X$ be a closed complex analytic subspace of complex dimension k. If $Y_k \subset E$, it becomes an element of $\mathcal{S}_{X,*}$ namely $0 \oplus (Y) \oplus 0$.

If Y has no component in E, it becomes the element $\left(\overline{p^{-1}(Y \setminus E)}\right) \oplus 0 \oplus 0$ because $\partial \left(\overline{p^{-1}(Y \setminus E)}\right) \cap \tilde{E} = 0$ for dimensional reasons (the dimension of $\overline{p^{-1}(Y \setminus E)} \cap \tilde{E}$ is $2k - 2$, while the dimension of $\overline{p^{-1}(Y \setminus E)}$ is $2k$).

In particular, let us consider the realization of $\mathcal{S}_{X,2k}$ as in Eq (3.6) using the hypercovering (X_l, h_l) associated to Λ_X^*. We see that any complex analytic subspace Y_k will be represented as a certain element of $\mathcal{S}_{X_l, 2k}$ of the same X_l with a rank $q(l) = 0$. In the embedding of $\mathcal{S}_{X,2k}$ in $S_{X,2k}$, Y_k will have no component in any $\mathcal{S}_{X_r, 2k}$ with $q(r) \geq 1$ in the hypercovering.

4   Filtrations on cohomology

a) Filtration W_m

If X is a complex manifold, we define a trivial filtration on the standard De Rham complex of X by

$$\begin{aligned} W_m \mathcal{E}_X^* &= \mathcal{E}_X^* \quad \text{for } m \geq 0, \\ W_m \mathcal{E}_X^* &= 0 \quad \text{for } m < 0. \end{aligned} \tag{4.1}$$

We define by recursion on the dimension

$$W_m \Lambda_X^k = W_m \mathcal{E}_{\tilde{X}}^k \oplus W_m \Lambda_E^k \oplus W_{m+1} \Lambda_{\tilde{E}}^{k-1}. \tag{4.2}$$

In the language of hypercoverings, one has

$$\begin{aligned} \Lambda_X^k &= \underset{l}{\oplus} (h_l)_* \mathcal{E}_{X_l}^{k-q(l)}, \\ W_m \Lambda_X^k &= \underset{\{l / -q(l) \leq m\}}{\oplus} (h_l)_* \mathcal{E}_{X_l}^{k-q(l)}. \end{aligned} \tag{4.3}$$

W_m is an increasing filtration on Λ_X^* and

$$\begin{aligned} d W_m \Lambda_X^k &\subset W_m \Lambda_X^{k+1}, \\ W_0 \Lambda_X^k &= \Lambda_X^k. \end{aligned}$$

$W_m \Lambda_X^k$ is the set of forms on the spaces X_l of degree $\geq k + m$. The filtration W_m induces a filtration on the cohomology by

$$W_m H^k(X, C) = \frac{\text{Ker } \{d : \Gamma(X, \Lambda_X^k) \to \Gamma(X, \Lambda_X^{k+1})\} \cap W_m \Gamma(X, \Lambda_X^k)}{W_m \Gamma(X, \Lambda_X^k) \cap d\Gamma(X, \Lambda_*^{k-1})}. \tag{4.4}$$

There exists a spectral sequence $E_r^{m,k}(X)$ associated to the filtration W_m which converges towards the graded spaces of the cohomology $\frac{W_m H^k}{W_{m-1} H^k}$. The first term of the spectral sequence is

$$E_1^{m,k} = \underset{\{l / -q(l) \leq m\}}{\oplus} H^{k-m}(X_l, C). \tag{4.5}$$

Remark. We do not use the standard terminology of spectral sequences, but a more natural one, namely $E_r^{m,k}$ is indexed by the cohomological degree k and by the filtered degree m.

One defines the differentials

$$d_r : E_r^{m,k}(X) \to E_r^{m-r,k+1}(X).$$

These differentials d_r can be explicitly expressed by solving standard d equations globally on the spaces X_l of the hypercovering and by the pull-back $h_{ll'}^*$ of the hypercovering.

We see that if X is a complex compact space and Y is a closed complex subspace, then integration of any element of $W_{-1}H^*(X, C)$ on Y gives 0, so that the graded space $\frac{W_0 H^*(X)}{W_{-1}H^*(X)}$ induces linear functionals on the homology classes of analytic subspaces of X.

b) Filtrations F^p, $\bar{F}^q$

If X is a complex manifold, any differential form of degree k on X can be written locally as

$$\omega = \sum_{|I|+|J|=k} \omega_{IJ} dz^I \wedge d\bar{z}^J \qquad (4.6)$$

where $dz^I = dz^{i_1} \wedge \cdots \wedge dz^{i_p}, d\bar{z}^J = d\bar{z}^{j_1} \wedge \cdots \wedge d\bar{z}^{j_q}$ (here $I = \{i_1 < \cdots < i_p\}$, $J = \{j_1 < \cdots < j_q\}$).

We call $F^p \mathcal{E}_X^k$ the subsheaf of $\mathcal{E}_X^k$ which is the set of forms ω given by (4.6) with $|I| \geq p$. We define $\bar{F}^q \mathcal{E}_X^k = \overline{F^q \mathcal{E}_X^k}$: It is the subsheaf of $\mathcal{E}_X^k$ which is the set of forms ω given by (4.6) with $|J| \geq q$.

F^p is a decreasing filtration and

$$dF^p e_X^k \subset F^p \mathcal{E}_X^{k+1}.$$

For any complex space X, we define recursively on the dimension

$$F^p \Lambda_X^k = F^p \mathcal{E}_{\tilde{X}}^k \oplus F^p \Lambda_E^k \oplus F^p \Lambda_{\tilde{E}}^{k-1}$$

so that

$$dF^p \Lambda_X^k \subset F^p \Lambda_X^{k+1}.$$

Obviously, F^p and F^q induce filtrations on the cohomology of X.

5   Mixed Hodge structure

a) The case of a compact kählerian manifold M.

Let M be a compact kählerian manifold. The classical Hodge theory says that the cohomology $H^k(M, C)$ carries a pure Hodge structure. This means that the filtrations F^p and $\bar{F}^q$ induce filtrations (still denoted by F^p and $\bar{F}^q$) on the cohomology and that $H^k(M, C)$ is the direct sum of the graded spaces for F^p (or for $\bar{F}^q$),

$$H^k(M, C) = \bigoplus_{p+q=k} H^{p,q}(M) \qquad (5.1)$$

where

$$H^{p,q}(M) = \frac{F^p H^k}{F^{p+1} H^k} = \frac{\bar{F}^q H^k}{\bar{F}^{q+1} H^k} = F^p \bar{F}^q H^k.$$

We also know that $H^{p,q}(M)$ are the Dolbeault group

$$H^{p,q}(M) = H^q(M, \Omega_M^p)$$

where Ω_M^p is the sheaf of holomorphic p-forms of M (see e.g., [K–M]).

c) compact projective varieties

Let X be a compact projective variety and let (Λ_X^*, d) be a complex in $r(X)$, so that E, $\tilde{E}$ are compact projective varieties and $\tilde{E}$ is a compact projective manifold. The manifolds X_l of the hypercovering of Λ_X^* are all compact projective and so are compact Kählerian.

In particular, each term $E_1^{m,k}(X)$ of the spectral sequence (see Eq (4.5)) has a pure Hodge structure

$$E_1^{m,k} = \bigoplus_{p+q=k-m} (E_1^{m,k})^{p,q},$$

$$(E_1^{m,k})^{p,q} = \bigoplus_{\{l/-q(l)\leq m\}} H^{p,q}(X_l).$$

Theorem. *The filtrations W_m, F^p, $\bar{F}^q$ induce on the cohomology $H^k(X, C)$ of a compact projective variety X, a mixed Hodge structure. This means that the filtrations F^p, $\bar{F}^q$ induce on the graded spaces $\frac{W_m H^k}{W_{m-1} H^k}$ a pure Hodge structure of weight $k + m$, or*

$$\frac{W_m H^k}{W_{m-1} H^k} = \bigoplus_{p+q=k+m} F^p \bar{F}^q \left(\frac{W_m H^k}{W_{m-1} H^k} \right).$$

Moreover this mixed Hodge structure does not depend on the chosen Λ_X^*.

The spectral sequence of the filtration W_m degenerates at $E_2^{m,k}$, that is

$$E_2^{m,k} = \cdots = E_r^{m,k} = \cdots = \frac{W_m H^k}{W_{m-1} H^k} \ (for \ r \geq 2)$$

and all the differentials $d_r = 0$ for $r \geq 2$.

Remark. We conjecture that this mixed Hodge structure is the same as the one of Deligne [D]. Deligne's method uses simplicial schemes and *descente cohomologique*. But the cohomological descent necessitates the introduction of complex manifolds of any dimension, a priori. In our case, all the manifolds X_l of the hypercovering have dimension at most the dimension of X. The fact that the spectral sequence of W_m degenerates at E_2 seems to be unknown. Other proposals for constructing mixed Hodge structures were given in [E] and [GNPP].

6   Ring structure on the cohomology

The cohomology of any topological space carries a ring structure given by the cup product. For a manifold the cup product is induced by the standard exterior product of forms. In our case, we shall define on a complex $\Lambda_X^* \in r(X)$ an exterior product

$$(\omega_1 \oplus \sigma_1 \oplus \gamma_1) \wedge (\omega_2 \oplus \sigma_2 \oplus \gamma_2)$$
$$=((\omega_1 \wedge \omega_2) \oplus (\sigma_1 \wedge \sigma_2) \oplus (\psi(\omega_1) \wedge \gamma_2 + (-1)^{k_2}\gamma_1 \wedge \varphi(\sigma_2))) \qquad (6.1)$$

where $\omega_j \oplus \sigma_j \oplus \gamma_j$ are in $\Lambda_X^{k_j}$ and ψ, φ are the pull-back morphisms used in the definition of Λ_X^* as in Section 2. Here $\omega_1 \wedge \omega_2$ is the usual exterior product in $\mathcal{E}_{\tilde{X}}^*$, $\sigma_1 \wedge \omega_2$ is the exterior product in Λ_E^* defined by recursion on the dimension and $\psi(\omega_1) \wedge \gamma_2$ and $\gamma_1 \wedge \varphi(\sigma_2)$ are exterior products in $\Lambda_{\tilde{E}}^*$ defined also by recursion. The product $\wedge$ is associative. It is possible to prove that it induces the cup product on

$$W_{m_1}H^{k_1}(X) \times W_{m_2}H^{k_2}(X) \to W_{m_1+m_2}H^{k_1+k_2}(X)$$

provided $m_1 + m_2 \geq -2$. Then, we obtain an induced product

$$\frac{W_{m_1}H^{k_1}}{W_{m_1-1}H^{k_1}} \times \frac{W_{m_2}H^{k_2}}{W_{m_2-1}H^{k_2}} \to \frac{W_{m_1+m_2}H^{k_1+k_2}}{W_{m_1+m_2-1}H^{k_1+k_2}}$$

provided $m_1 + m_2 \geq 1$. This product respects the pure Hodge structures on these graded spaces.

References

[AG1]   V. Ancona and B. Gaveau, *La théorie des résidus sur un espace analytique complexe, in Conference Jean Leray*, (M. de Gosson and J. Vaillant, eds.), Kärlskrona, 1999.

[AG2]   V. Ancona and B. Gaveau, *Families of differential forms on complex spaces*, Annali Scuola Norm., Pisa.

[AG3]   V. Ancona and B. Gaveau, Differential forms, integration and Hodge theory on complex analytic spaces, preprint, January 2001.

[AG4]   V. Ancona and B. Gaveau, *The De Rham complex of a reduced space*. In: Contribution to complex analysis and analytic geometry, (H. Skoda and J. Trépreau, eds.), Vieweg, 1994.

[AG5]   V. Ancona and B. Gaveau, *Theorémès de De Rham sur un espace analytique*, Revue Roumaine de Mathématiques pures et appliquées **38** (1993), 579–594.

[B–H]   T. Bloom and M. Herrera, *De Rham cohomology of an analytic space*, Invent. Math. **7** (1969), 275–296.

[D]   P. Deligne, *Théorie de Hodge II et III*, Publi. Math. IHES **40** (1971), 5–58 and **44** (1974), 5–77.

[D–P] P. Dolbeault and J. Poly, *Differential forms with subanalytic singularities, integral cohomology; residues*. In: Proceedings of Symposia in Pure Mathematics Vol. 30 , 255–261, Academic Press, New York.

[E] F. Elzein, *Mixed Hodge structures*, Trans. Am. Math. Soc. **275** (1983), 71–106.

[He] M. Herrera, *Integration on a semi analytic set*, Bull Soc. Math. France **94** (1966), 141–180.

[Hi] H. Hironaka, *Introduction to real-analytic sets and real analytic maps*, Institute di Matematica, Università di Pisa, 1973.

[GNFP] F. Guillen, V. Navarro Aznar, P. Pascual Guainza, and P. Puertas, *Hyperresolutions cubiques et descente cohomologique*. In: Lecture Notes in Math. **1335**, Springer, 1988.

[K–M] K. Kodaira and J. Morrow, *Complex manifolds*, Holt Rinehart, 1975.

[P] J. Poly, *Formule des résidus et intersections de chaînes sous analytiques*, Thèse Poitiers, 1974

[W] A. Weil, *Sur les théorèmes de De Rham*, Comment. Math. Helvetici (1951), 119–145.

Vincenzo Ancona
Dipartimento di Matematica U. Dini
Università degli Studi
67/A Viale Morgagni, Firenze, Italy

Bernard Gaveau
Laboratoire Analyse et Physique Mathématique
14 Avenue Félix Faure
75015 Paris, France
gaveau@ccr.jussieu.fr

On Exact Solutions of Linear PDEs

Richard Beals

ABSTRACT The role of exact fundamental solutions in the study of linear PDEs is illustrated by several examples among equations of mixed type, subelliptic and degenerate elliptic equations, and hyperbolic equations. In particular, we derive exact fundamental solutions for the degenerate hyperbolic operators $\partial_t^2 - t^{2l} \Delta$ in $\mathbb{R}^{2p} \times \mathbb{R}$ for arbitrary l, $p = 1, 2, \ldots$.

1   Introduction

At least since the discovery of exact solutions of the wave equation by D'Alembert and Euler, and of the heat equation by Fourier, the study of PDEs has profited from an interaction between exact solutions and general methods. Both themes are prominent in the work of Leray. After introducing weak solutions, mollifiers, and topological methods in his early papers, he turned in much of his later work to the detailed analysis of fundamental solutions, determined as explicitly as possible.

Much of the work on linear PDEs in the past half century is centered on general problems and methods: elliptic, hyperbolic, and parabolic equations of arbitrary order, general equations with constant coefficients, pseudodifferential operators and Fourier integral operators of very general type, refined regularity theory, propagation of singularities of various strengths, boundary value problems with irregular boundaries .... A student could get the impression that nothing new has been written down in the way of an exact fundamental solution since the Poisson kernel.

We give here a very sketchy account of some post-Poisson constructions of fundamental solutions and related exact formulas, with some motivation and indication of the (actual and potential) consequences for general theory. This account is far from exhaustive, limited to examples from linear theory that have come to this author's attention in recent years. As is well known, discoveries in nonlinear theory related to the KdV equation and other completely integrable evolution equations have provided a wealth of exact solutions of (special) nonlinear equations that have had great impact on the study of those equations and their perturbations.

A common theme among many of the examples to be discussed below is that they do not fall into the standard elliptic/parabolic/hyperbolic trichotomy. These are operators with variable coefficients which have the property, in contrast to the classical operators, that if one freezes the coefficients at a point, one loses the essential character of the operator. Among operators of this kind are opera-

Research supported by NSF Grant DMS-9800605.

tors of mixed type, subelliptic and degenerate elliptic operators, and degenerate hyperbolic equations.

Some operators of mixed type are discussed in §2, in particular some examples from kinetic theory for which exact solutions are known and have played a role in further developments. Also discussed briefly are examples of Euler–Poisson–Darboux or Tricomi type, and work of Leray. The transport operators like those in §2 and subelliptic operators such as the Kohn sub-Laplacian are the two simplest examples of hypoelliptic sums-of-squares operators. Exact fundamental solutions for the Kohn sub-Laplacian, and their ramifications, are discussed in §3. Another class of hypoelliptic sums-of-squares operators are degenerate elliptic operators, discussed in §4. Like subelliptic operators, these have given rise to much study of exact regularity properties. In some cases, exact fundamental solutions and heat kernels are known.

Hyperbolic equations are discussed in §5. The developments in §3 and §4 have implications for this area as well. Exact wave kernels may be constructed by various means from the exact heat kernels or fundamental solutions. We cite some new results in this direction. We also derive exact propagators for the degenerate hyperbolic operators

$$\frac{\partial^2}{\partial t^2} - t^{2l} \sum_{j=1}^m \frac{\partial^2}{\partial x_j{}^2}, \qquad l = 1, 2, \dots, \quad m = 2, 4, \dots .$$

by analytic continuation from fundamental solutions of degenerate elliptic operators.

2   Transport operators; operators of mixed type; sums of squares I

Kinetic theory gives rise to a number of operators of the general type of

$$L = \left(\frac{\partial}{\partial x_1}\right)^2 - bx_1 \frac{\partial}{\partial x_2} - c \frac{\partial}{\partial x_1} \tag{2.1}$$

where $b > 0$ and c are constants. The operator (2.1) was derived by Kolmogorov [43], and he calculated a Green's kernel $G(x, y)$. For kinetic theory one is particular interested in the corresponding time dependent operator $\partial_t - L$, and Kolmogorov appears to have calculated the corresponding "heat kernel" $P(x, y, t)$ with pole at $y \in \mathbb{R}^2$, then integrated with respect to time to obtain $G(x, y)$. The time dependent operator can be put into somewhat simpler form by conjugating by $\exp\{(2cx_1 - c^2 t)/4\}$ in order to eliminate the $c\partial_{x_1}$, and rescaling in x_1 so that $b = 1/2$. With these simplifications, the heat kernel is

$$P(x, y, t) = \frac{c_0}{s^2} \exp\left\{-2\frac{x_1^2 + x_1 y_1 + y_1^2}{s} + 3\frac{x_1 + y_1)(x_2 - y_2}{s^2} - 6\frac{x_2 - y_2)^2}{s^3}\right\}. \tag{2.2}$$

Examination of the Green's kernel $G(x, y)$ obtained by integration in time shows that it is C^∞ off the diagonal $x = y$, a fact to which we return below.

The kinetic operator

$$\frac{\partial}{\partial x_1}\left(\frac{\partial}{\partial x_1} + cx_1\right) - bx_1 \frac{\partial}{\partial x_2} \tag{2.3}$$

was considered by Chandrasekhar [16], who computed the exact heat kernel. The somewhat more complicated version

$$\frac{\partial}{\partial x_1}\left(\frac{\partial}{\partial x_1} + cx_1\right) - bx_1 \frac{\partial}{\partial x_2} + bx_2 \frac{\partial}{\partial x_1}. \tag{2.4}$$

arises as a model kinetic equation with a trapping external force; the heat kernel for (2.4) was computed by Aarão [1]. These two heat kernels have a form similar to (2.2), but with more complicated quadratic forms in the exponential.

Chandrasekhar, although he used probabilistic considerations for the actual calculations in [16], noted in passing that the heat kernels for operators like (2.1), (2.3), and (2.4) should have the general form

$$P(x, y, t) = \varphi(t) \exp\{-Q_t(x, y)\} \tag{2.5}$$

where Q_t is a nonnegative quadratic form. The *Ansatz* (2.5) was rediscovered later, for a larger class of operators, including the Hermite operator. Applying one of the operators above, for example, leads to a Riccati system of equations for the coefficients of the quadratic form and a linear equation for $\log \varphi$ [4]. Combining this approach with the partial Fourier transform gives another way to derive some of the results described in the next two sections. For example, after Fourier transformation with respect to the variable t, in the case of (3.2), or with respect to the x_3 in the case of (4.1) or (4.2b), the resulting operator has heat kernel of the form (2.5) with the Fourier transform variable as a parameter. The same is true of the example

$$\frac{\partial^2}{\partial x_1{}^2} + \frac{\partial^2}{\partial x_2{}^2} + \left(x_1 \frac{\partial}{\partial t_1} + x_2 \frac{\partial}{\partial t_2}\right)^2 \tag{2.6}$$

which is neither group invariant, like (3.3) below, nor degenerate elliptic, like (4.1) and (4.2) below.

The transport operators (2.1), (2.3), and (2.4) are of mixed type, in the sense that they are parabolic with x_2 as "time" variable, but the preferred time direction depends on the sign of x_1. The classical operator of mixed type is the Tricomi operator,

$$\frac{\partial^2}{\partial x_1{}^2} + x_1 \frac{\partial^2}{\partial x_2{}^2} \tag{2.7}$$

in simplest form. Fundamental solutions for this operator, related Euler–Poisson–Darboux operators, and other operators of Tricomi–Clairaut type have been found by Bader and Germain [2], Davis [20], Delache [22], and by Delache and Leray [23]. The latter paper obtains formulas for higher order equations similar in form to Hadamard's formulas for higher order hyperbolic equations [34], in terms of hypergeometric functions of a defining function of the characteristic conoid.

Each of the operators (2.1), (2.3), and (2.4) has the form $X^2 + Y$, where X and Y are vector fields in $\mathbb{R}^2$. A more general form is

$$L = \sum_{k=1}^{m} X_j^2 + Y \tag{2.8}$$

where the X_j and Y are smooth real vector fields on some d–dimensional manifold M. In particular, the Kohn sub-Laplacian that arises naturally in the study of several complex variables has the form (2.8); see the next section. Like the transport operators, the Kohn operator is not elliptic, but is (often) hypoelliptic: Lu smooth implies u smooth, even locally. Transport operators like those above have this same property; it follows from the fact that the Green's kernel is C^∞ off the diagonal. Kolmogorov's result for (2.1) and the hypoellipticity results for the Kohn operator were cited by Hörmander in introducing his famous "sums of squares" theorem [39]: an operator L of the form (2.8) is hypoelliptic if the vector fields X_j and Y, together with their repeated commutators, generate the full tangent space at each point of M. For operators not in sums-of-squares form, see Oleinik and Radkevič [50].

3   Subelliptic operators; sums of squares II

Exact fundamental solutions for elliptic operators have been found even in relatively recent times, for example by Herglotz [38], John [41], Debiard and Gaveau [21]. However in recent years more attention has been given to subelliptic operators. Most of the examples to follow stem (historically, if not logically) from one particular problem coming from several complex variables. The simplest "strictly pseudoconvex" domain in $\mathbb{C}^{n+1}$ is the unit ball, which is biholomorphically equivalent to the Siegel upper half space

$$\Omega = \left\{ z : \operatorname{Im} z_{n+1} > \sum_{j=1}^{n} |z_j|^2 \right\}. \tag{3.1}$$

The holomorphic vector fields

$$Z_j = \frac{\partial}{\partial z_j} + 2i\bar{z}_j \frac{\partial}{\partial z_{n+1}}, \qquad j = 1, 2, \ldots, n$$

are tangent to the boundary $\partial\Omega$. Setting $z_j = x_j + ix_{n+j}$, the x_j for $j \le 2n + 1$ coordinatize the boundary. The restriction to the boundary of $2Z_j$ is

$$X_j - i X_{n+j} = \left(\frac{\partial}{\partial x_j} + 2x_{n+j} \frac{\partial}{\partial t} \right) - i \left(\frac{\partial}{\partial x_{n+j}} - 2x_j \frac{\partial}{\partial t} \right), \qquad t = x_{2n+1}. \tag{3.2}$$

The Kohn sub-Laplacian for $\partial\Omega$, the tangential operator $L_n = \sum(\bar{Z}_j Z_j + Z_j \bar{Z}_j)$, has the form

$$L_n = \frac{1}{2} \sum_{j=1}^{2n} X_j^2. \tag{3.3}$$

This operator is not elliptic, but the Hörmander condition mentioned above is satisfied. In fact, for any j the commutator $[X_{n+j}, X_j] = 4\partial/\partial t$ provides the one missing direction in the tangent space. The boundary $\partial\Omega$ has a natural group structure as a Heisenberg group H_n; the X_j and $T = \partial/\partial t$ are left-invariant for the group structure. Therefore to compute a Green's function for L_n, it suffices to compute a Green's function with pole at the origin. This was done by Folland [26]; the result is

$$G_n(x, t; 0, 0) = \frac{c_n}{(|x|^4 + t^2)^{n/2}}. \tag{3.4}$$

The operator $L_{n,\alpha} = L_n - \alpha \partial/\partial t$, α a complex constant, is also of interest. Folland and Stein [27] computed a Green's function with pole at the origin for $L_{n,\alpha}$,

$$G_{n,\alpha}(x, t; 0, 0) = c_{n,\alpha} \, (|x|^2 - it)^{-(n+\alpha)/2} \, (|x|^2 + it)^{-(n-\alpha)/2} \tag{3.5}$$

and deduced that $L_{n,\alpha}$ is hypoelliptic precisely for α not of the form $\pm 2i(2m+1)$, $m = n, n+1, \ldots$. Analogous exact solutions were obtained for the corresponding operators on the sphere in $\mathbb{C}^{n+1}$ by Dadok and Harvey [19]. These exact results led to many other developments, including detailed analysis of regularity properties of the Kohn sub-Laplacian for general strictly pseudoconvex domains [27] and analysis of more general operators and systems that are transversally elliptic and have multiple characteristics, e.g., [13], [14], [53], [54], [55].

Another consequence was the determination of the heat kernel for the operator L_n. Here we use u for the time variable. The heat kernel was first computed by Gaveau [28] and Hulanicki [40]; with pole at the origin it is

$$P_n(x, t; 0, 0; u) = \frac{c_n}{u^{n+1}} \int_{-\infty}^{\infty} e^{-f(x,t,\tau)/u} \, V(\tau) \, d\tau \, ;$$

$$f(x, t, \tau) = \frac{\tau}{2} \coth(2\tau)|x|^2 - it\tau, \qquad V(\tau) = \frac{(2\tau)^n}{\sinh(2a\tau)^n}. \tag{3.6}$$

(Recomputing P_n in different ways is something of an industry: see [4], [5], [11], [42].) Gaveau's study of this problem led in several directions, including the introduction of what is now called a sub-Riemannian metric or Carnot–Caratheodory metric on the Heisenberg group; see [12] for an indication of further developments.

If we break the symmetry of L_n above by taking

$$X_j = \frac{\partial}{\partial x_j} + 2a_j x_{n+j} \frac{\partial}{\partial t}, \qquad X_{n+j} = \frac{\partial}{\partial x_{n+j}} - 2a_j x_j \frac{\partial}{\partial t},$$

with positive $a_j = a_{n+j}$ not all the same, then (with a few discrete exceptions) no closed-form formulas like (3.4) or (3.5) are known for the Green's functions. One can obtain instead an integral formula

$$G(x, t; 0, 0) = \int_{-\infty}^{\infty} \frac{V(\tau)\,d\tau}{f(x, t, \tau)^n} \tag{3.7}$$

where

$$f(x, t, \tau) = \frac{\tau}{2} \sum_{j=1}^{2n} \cosh(2a_j\tau)x_j^2 - i\tau t, \qquad V(\tau) = \prod_{j=1}^{n} \frac{2a_j\tau}{\sinh(2a_j\tau)}. \tag{3.8}$$

The simplest example of a domain in $\mathbb{C}^2$ that is pseudoconvex but not strictly pseudoconvex is analogous to (3.1):

$$\Omega = \Omega_{2,4} = \{z \in \mathbb{C}^2 : \operatorname{Im} z_2 > |z_1|^2\}. \tag{3.9}$$

A holomorphic vector field tangent to the boundary is

$$Z = \frac{\partial}{\partial z} + 2i|z|^2 \, \bar{x} \frac{\partial}{\partial t},$$

so the Kohn sub-Laplacian $\bar{Z}Z + Z\bar{Z}$ is

$$\frac{1}{2}\frac{\partial^2}{\partial x^2} + \frac{1}{2}\frac{\partial^2}{\partial y^2} + 4(x^2 + y^2)\left[y\frac{\partial}{\partial x} - x\frac{\partial}{\partial y}\right]\frac{\partial}{\partial t} + 8(x^2 + y^2)^3\frac{\partial^2}{\partial t^2}.$$

The Green's function was computed by Greiner [31]. Identifying the boundary with $\mathbb{C} \times \mathbb{R} = \{(z, t)\}$, the Green's function with pole at (w, s) is

$$G(z, t; w, s) = \frac{c}{\sigma} \log\left(\frac{p + \bar{p} + i|1 - p^2|}{1 + |p|^2}\right), \tag{3.10}$$

where

$$\sigma = \left[|z|^4 + |w|^4 - 2\operatorname{Im} z\bar{w} - i(t - s)\right]^{1/2},$$

$$p = \frac{2z\bar{w}}{\left[|z|^4 + |w|^4 + i(t - s)\right]^{1/2}}.$$

This allows for a very detailed study of regularity properties of the operator near the line $z = 0$ where strict pseudoconvexity fails.

It is not clear, on inspection, what formulas like (3.4), (3.7) and (3.9) have in common. None of them can be explained by the most obvious generalization from the classical case, the associated subelliptic or Carnot–Caratheodory geometry. Comparison with classical formulas led Gaveau and Greiner to observe that f in (3.6), (3.7) satisfies an equation of Hamilton–Jacobi type,

$$\frac{\partial f}{\partial \tau} + H(x, t, \nabla f) = f, \tag{3.11}$$

where the Hamiltonian H is the principal symbol of the operator L, while the "volume element" V satisfies a transport equation

$$\tau \frac{\partial V}{\partial \tau} - n V = Lf .\tag{3.12}$$

In particular, f is a sort of complex action, which suggests an approach through "complex Hamiltonian mechanics." This idea led to generalizations in three directions.

In one direction, the approach through complex Hamiltonian mechanics shows how to construct a parametrix of Hadamard type for the analogous operator on a general contact manifold [7]. In a second direction, the approach leads to heat kernel and Green's function formulas for the analogous sums of squares operators $\sum X_j^2$ on an arbitrary two-step nilpotent group, i.e., one for which the left-invariant vector fields and their first commutators $[X_j, X_k]$ generate the tangent space at each point [5]. In the third direction, geometric understanding of the variables in (3.8) led to the calculation of exact formulas for the Green's functions of the Kohn sub-Laplacians for the weakly pseudoconvex boundaries

$$\{(z, w) \in \mathbb{C}^n \times \mathbb{C} : \text{Im } w = |z|^{2k}\}, \qquad n = 1, 2, \ldots, \quad k = 2, 3, \ldots;\tag{3.13}$$

see [6], [8]. These Green's functions are integrals of somewhat complicated algebraic functions, similar to (4.4) below.

4 Degenerate elliptic operators; sums of squares III

A striking example from the study of sums-of-squares operators is due to Baouendi and Goulaouic [3]. It is the degenerate elliptic operator

$$L = \frac{\partial^2}{\partial x_1^2} + \frac{\partial^2}{\partial x_2^2} + x_1^2 \frac{\partial^2}{\partial x_3^2},\tag{4.1}$$

which is hypoelliptic but not analytic-hypoelliptic: Lu may be analytic even if u is not. This example led to considerable effort to isolate those features that are associated with analytic-hypoellipticity and with intermediate regularity such as Gevrey-hypoellipticity with various exponents; see, for example, [15], [17], [18], [24], [25], [33], [35], [37], [54], [55], [56]. (Gevrey classes were originally introduced in order to study PDEs such as parabolic equations, where the classical Cauchy–Kovalevskaya theorem fails [29]. Their resurgence seems to be due to Ohya's work [49] on weakly hyperbolic equations, so here in degenerate elliptic equations one completes the triple play.)

In contrast to (4.1), the operators

$$\frac{\partial^2}{\partial x_1{}^2} + x_1^2\left(\frac{\partial^2}{\partial x_2{}^2} + \frac{\partial^2}{\partial x_3{}^2}\right), \tag{4.2a}$$

$$\frac{\partial^2}{\partial x_1{}^2} + \frac{\partial^2}{\partial x_2{}^2} + (x_1^2 + x_2^2)\frac{\partial^2}{\partial x_3{}^2}, \tag{4.2b}$$

$$\frac{\partial^2}{\partial x_1{}^2} + \frac{\partial^2}{\partial x_2{}^2} + x_1^2\frac{\partial^2}{\partial x_3{}^2} + x_2^2\frac{\partial^2}{\partial x_4{}^2} \tag{4.2c}$$

are analytic-hypoelliptic. (We remark that (4.2a) and (4.2b) have smooth characteristic varieties, but (4.2c) does not.) The approach via complex Hamiltonian mechanics as in §3 produces Green's functions with the same general form as (3.7) for a large class of degenerate elliptic operators with second order degeneracy, including the Baouendi–Goulaouic operator (4.1) and the three operators of (4.2); see [10]. Careful analysis of the Green's functions gives a necessary and sufficient condition for analytic hypoellipticity, and gives the (micro)–location of the failure of analytic hypoellipticity in cases like (4.1). As noted in [10], the specific analytic results were not new or surprising, but the exact formulas may allow the testing of more refined conjectures.

Degenerate elliptic operators with a higher order of degeneracy have also been studied extensively, e.g., in some of the references above. Examples of operators of this type are

$$L_{nmk} = \sum_{j=1}^{n} \frac{\partial^2}{\partial t_j{}^2} + |t|^{2k-2}\sum_{j=1}^{m}\frac{\partial^2}{\partial x_j{}^2}, \quad n, m, k = 1, 2, 3, \ldots. \tag{4.3}$$

Following ideas and methods in the calculation of the Green's kernels for the operators associated to boundaries (3.10) leads to explicit Green's kernels for the operators (4.3); [9]. (We modify the notation in [9], for use in §5.) These kernels are algebraic functions when m is even, and integrals of algebraic functions when m is odd.

For even m the Green's function for L_{nmk}, which is homogeneous with respect to the natural dilation structure and has pole at (s, y), is

$$G_{nmk}(t, x; s, y) =$$

$$\frac{c_{nmk}}{\sigma^{(n-2)/2k}}\left(\frac{1}{R_e}\frac{\partial}{\partial\sigma}\right)^{(m-2)/2}\left\{\frac{\sigma^{(m-2)/2+(n-2)/2k}\psi_-(R_e, \sigma)}{[\psi_+(R_e, \sigma, v) - 2q(\sigma, v)]^{n/2}}\right\}, \tag{4.4}$$

where the variables are

$$R_e = R_e(t, x; s, y) = \frac{1}{2}\left(|t|^{2k} + |s|^{2k} + k^2|x - y|^2\right),$$

$$\sigma = \sigma(t, s) = |t|^k|s|^k,$$

$$v = v(t, s) = \frac{t \cdot s}{|s||t|}, \tag{4.5}$$

and the associated functions are

$$\psi_+(R_e, \sigma) = \left(R_e + \sqrt{R_e^2 - \sigma^2}\right)^{1/k} + \left(R_e - \sqrt{R_e^2 - \sigma^2}\right)^{1/k},$$

$$\psi_-(R_e, \sigma) = \frac{\left(R_e + \sqrt{R_e^2 - \sigma^2}\right)^{1/k} - \left(R_e - \sqrt{R_e^2 - \sigma^2}\right)^{1/k}}{\sqrt{R_e^2 - \sigma^2}},$$

$$q(\sigma, v) = \sigma^{1/k} v = t \cdot s. \tag{4.6}$$

For odd m the Green's function is given by

$$G_{nmk}(t, x; s, y) = \frac{c_{nmk}}{\sigma^{(n-2)/2k}} \left(\frac{1}{R_e} \frac{\partial}{\partial \sigma}\right)^{(m-1)/2} \left\{\sigma^{(m-1)/2} H_{n1k}\right\}, \tag{4.7}$$

where H_{n1k} is a fractional integral of total degree $1/2$ of the kernel H_{n2k}:

$$H_{n1k}(v, \rho) = R^{1/2} \int_0^1 \frac{(u\sigma)^{(n-2)/2k} u^{1/2} \psi_-(R_e, u\sigma)}{[\psi_+(R_e, u\sigma, v) - 2q(u\sigma, v)]^{n/2}} \frac{du}{u(1-u)}. \tag{4.8}$$

(This replaces a somewhat more complicated expression in [9] that is not free of typographical errors.)

5   Hyperbolic operators

Hyperbolic equations were of particular interest to Leray. He devoted much effort to making as explicit as possible the fundamental solutions, first of strictly hyperbolic equations, and then of weakly hyperbolic equations, in order to obtain precise information on well-posedness and propagation of singularities, building on classical work of Hadamard [34], Herglotz [38], and Riesz [51], and taking an active interest in the work of others; [44], [45], [46], [35], [57], [30], [36], [52]. As noted above, the work of Ohya [49] and of Leray and Ohya [47] seems to have brought Gevrey spaces back into general awareness, with consequences for other types of equations. I will mention here only some new results that have grown out of some of the constructions described above.

The wave equation for the Kohn sub-Laplacian on the Heisenberg group, a weakly hyperbolic operator, was first investigated by Nachman [48]. Recently Greiner, Holcman, and Kannai [31] have exploited some of the developments described above to derive explicit wave kernels for various operators with degeneracies of order 2, including the Grushin operator

$$\frac{\partial^2}{\partial x_1^2} + x_1^2 \frac{\partial^2}{\partial x_2^2},$$

the Heisenberg sub-Laplacian (3.3), and degenerate elliptic operators like (4.1), (4.2), using three different methods: separation of variables, a transmutation formula from the heat kernel, and analytic continuation of $\partial^2/\partial s^2 + L$ with respect to s. For example, they show that at time u the wave kernel with pole at the origin is

$$W_u(x, t; 0, 0) = c_n \int_{\Gamma_{x,t}} \left(\frac{1}{u}\frac{\partial}{\partial u}\right)^n \left\{\frac{H(u - \sqrt{2f(x, t, \tau)})}{\sqrt{u^2 - 2f(x, t, \tau)}}\right\} V(\tau)\, d\tau, \quad (5.1)$$

where f and V are given by (3.8) and $\Gamma_{x,t}$ is a certain contour in the complex plane.

In this section we give a still simpler derivation of the fundamental solution for a class of degenerate hyperbolic operators

$$L_{mk} = \frac{\partial^2}{\partial t^2} - t^{2k-2} \sum_{j=1}^{m} \frac{\partial^2}{\partial x_j^2} \quad k = 2, 3, 4, \ldots, \quad m = 2, 4, \ldots, \quad (5.2)$$

by analytic continuation from the Green's function for the degenerate elliptic operator

$$L_{1mk} = \frac{\partial^2}{\partial t^2} + t^{2k-2} \sum_{j=1}^{m} \frac{\partial^2}{\partial x_j^2}, \quad m = 2, 4, \ldots \quad (5.3)$$

as given in §4. Here we modify the notation of the previous section and set

$$R = R(t, x; s, y) = \frac{1}{2}\left(t^{2k} + s^{2k} - k^2|x - y|^2\right),$$
$$\sigma = \sigma(t, s) = (st)^k, \quad (5.4)$$

and

$$\psi_+(R, \sigma) = \left(R + \sqrt{R^2 - \sigma^2}\right)^{1/k} + \left(R - \sqrt{R^2 - \sigma^2}\right)^{1/k},$$
$$\psi_-(R, s) = \frac{\left(R + \sqrt{R^2 - \sigma^2}\right)^{1/k} - \left(R - \sqrt{R^2 - \sigma^2}\right)^{1/k}}{\sqrt{R^2 - \sigma^2}}. \quad (5.5)$$

Theorem 5.1. *Suppose m is even. The wave kernel for the degenerate hyperbolic operator (5.2) in the region $st \geq 0$ is*

$$W(x, y, t, s) =$$
$$(-1)^{m/2}\mathrm{sgn}\,|s - t|\frac{2\,k^{m-2}}{(2\pi)^{m/2}}\sigma^{1/2}\frac{1}{R^{\frac{mk-2}{2k}}}\left(\frac{\partial}{\partial\sigma}\right)^{(m-2)/2}\left\{\sigma^{(m-2)/2}\sigma^{-1/2}\,H_{2k}\right\},$$
$$(5.6)$$

where

$$H_{2k} = H_{2k}(R, s) = \frac{\psi_-(R, \sigma)}{[\psi_+(R, \sigma) - 2\sigma^{1/k}]_+^{1/2}}. \quad (5.7)$$

Here $[X]_+^{-1/2}$ denotes the function of one variable

$$[X]_+^{-1/2} = 0, \quad x \leq 0, \qquad [X]_+^{-1/2} = X^{-1/2}, \quad X > 0,$$

and the derivatives in (5.6) are taken in the sense of distributions.
  This means that if if $f : \mathbb{R}^m \to \mathbb{C}$ is smooth enough, then the function

$$u(x, t) = \int_{\mathbb{R}^m} W(x, y, t, s) f(y) \, dy \qquad (5.8)$$

is the solution of the Cauchy problem

$$\frac{\partial^2 u}{\partial t^2} = t^{2k-2} \sum_{j=1}^{m} \frac{\partial^2 u}{\partial x_j^2}, \quad t \geq 0, \qquad (5.9)$$

$$u(x, s) = 0, \qquad (5.10)$$

$$\frac{\partial u}{\partial t}(x, s) = f(x). \qquad (5.11)$$

The proof of Theorem 5.1 has two parts: verification of the differential equation (5.9), and verification of the boundary conditions (5.10), (5.11).

Proof of (5.9). We need to show that

$$L_{mk} W_{mk} = 0 \qquad (5.12)$$

in the sense of distributions. We assume that $st > 0$, $s \neq t$, and continue the functions (4.5), (4.6) of §4 analytically in the variables $x, y \in \mathbb{C}^m$ subject to the condition

$$k^2 |x - y|^2 < (t^k - s^k)^2. \qquad (5.13)$$

Specifically, let

$$x^\theta = e^{i\theta} x, \quad y^\theta = e^{i\theta} y \qquad 0 \leq \theta \leq \pi. \qquad (5.14)$$

Then in the notation of §4,

$$R_\theta \equiv R_e(s, t, x^\theta, y^\theta) = \frac{1}{2} \left(t^{2k} + s^{2k} + e^{2i\theta} k^2 |x - y|^2 \right),$$

$$L_{mk}^\theta \equiv \frac{\partial^2}{\partial t^2} + |t|^{2k-2} \sum_{j=1}^{m} \frac{\partial^2}{\partial x_j^{\theta 2}} = \frac{\partial^2}{\partial t^2} + e^{-2i\theta} |t|^{2k-2} \sum_{j=1}^{m} \frac{\partial^2}{\partial x_j^{\theta 2}}. \qquad (5.15)$$

The various functions in §4 extend analytically, under the assumption (5.13). In fact

$$2(R_\theta - \sigma) = (t^k - s^k)^2 + e^{2i\theta} k^2 |x - y|^2,$$

from which it follows that R_θ and $R_\theta - \sigma$ have positive real part. It follows in turn that $R_\theta^2 - \sigma^2 = (R_\theta - \sigma)(R + \sigma)$ is never negative, so $\sqrt{R_\theta^2 - \sigma^2}$ has positive real part. Therefore both

$$\left(R_\theta + \sqrt{R_\theta^2 - \sigma^2} \right)^{1/k}$$

and

$$\left(R_\theta - \sqrt{R_\theta^2 - \sigma^2}\right)^{1/k} = \frac{st}{(R_\theta + \sqrt{R_\theta^2 - \sigma^2})^{1/k}}$$

continue analytically. Moreover

$$\psi_+(R_\theta) - 2q = \left\{(R_\theta + \sqrt{R_\theta^2 - \sigma^2})^{1/2k} - (R_\theta - \sqrt{R_\theta^2 - \sigma^2})^{1/2k}\right\}^2. \quad (5.16)$$

It follows that this function also continues analytically and does not vanish. Consequently the kernels G_{1mk} of (4.4) have continuations G_{mk}^θ, and

$$L_{mk}^\theta G_{mk}^\theta = 0$$

in the region (5.13). Now $R_\pi = R$, so G_{mk}^θ is a multiple of W_{mk}. Also, $L_{mk}^\theta = L_{mk}$, so $L_{mk} W_{mk} = 0$ in the region (5.13). Moreover, by definition the support of W_{mk} is the closure of the region (5.13). Therefore the distribution $L_{mk} W_{mk}$ is supported on the hypersurface

$$(t^k - s^k)^2 - k^2|x - y|^2 = 0 \quad (5.17)$$

when $t \neq s$, $st > 0$. However the distribution derivatives of the function (5.12) have no part supported at the origin (by consideration of homogeneity), so $L_{mk} W_{mk}$ has no part supported on the hypersurface (5.17). This proves (5.12) and (5.9).

Proof of (5.10) and (5.11). By symmetry we may assume $s, t \geq 0$. The result when $s = 0$ follows by continuity, so we assume $s > 0$. In the region (5.13), as $t \to 0$ it follows that $|x - y| \to 0$ and therefore $R \sim s^{2k}$. Therefore (5.16) implies

$$\sigma \sim R; \quad \psi_- \sim 2R^{1/k-1} \sim 2s^{2-2k},$$

while (5.13) and (5.15) imply

$$\psi_+ - 2st = R^{1/k}\left\{(1 + \sqrt{1 - \sigma^2/R^2})^{1/k} - (1 - \sqrt{1 - \sigma^2/R^2})^{1/k}\right\}^2$$

$$\sim \frac{R^{1/k}}{k^2}\left(1 - \frac{\sigma^2}{R^2}\right)$$

$$= \frac{R^{1/k-2}}{k^2}(R + \sigma)(R - \sigma)$$

$$\sim \frac{2R^{1/k-1}}{k^2}(R - \sigma). \quad (5.18)$$

We shall compute (5.8) up to terms of higher order in $t - s$. Up to the constant factor from (5.7), we have shown that the (formal) integral in (5.8) may be replaced by

$$2R^{1/k-1} \cdot \frac{k\,R^{1/2-1/2k}}{\sqrt{2}} \cdot \int_{\mathbb{R}^m}\left(\frac{\partial}{\partial\sigma}\right)^{(m-2)/2}\left\{(R - \sigma)_+^{-1/2}\right\}f(y)\,dy. \quad (5.19)$$

Now

$$R - \sigma = \frac{1}{2}\{(t^k - s^k)^2 - k^2 |x - y|^2\} \sim \tau - \frac{k^2 r^2}{2},$$

where

$$\tau = \frac{1}{2} k^2 (t - s)^2 s^{2k-2}, \qquad r = |x - y|.$$

By the preceding observations, up to terms of higher order in $t - s$ the integral in (5.19) is

$$(-1)^{(m-2)/2} f(x) \sigma_{m-1} \frac{\partial^{(m-2)/2}}{\partial \tau^{(m-2)/2}} \int_{r^2 < 2\tau/k^2} \frac{r^{m-1} \, dr}{(\tau - k^2 r^2/2)^{1/2}}, \qquad (5.20)$$

where $\sigma_{m-1} = 2\pi^{m/2}/\Gamma(\frac{m}{2})$ is the volume of the $m - 1$ sphere. A change of variables to $w = k^2 r^2 / 2\tau$ shows that the integral in (5.20) is

$$\frac{2^{m/2} \tau^{(m-1)/2}}{2 k^m} \int_0^1 \frac{w^{m/2-1}}{\sqrt{1-w}} \, dw = \frac{2^{m/2} \tau^{(m-1)/2}}{2 k^m} \cdot B\left(\frac{m}{2}, \frac{1}{2}\right).$$

Therefore the derivative in (5.20) is

$$\frac{2^{m/2}}{2 k^m} \cdot B\left(\frac{m}{2}, \frac{1}{2}\right) \cdot \frac{m-1}{2} \cdot \frac{m-3}{2} \cdots \frac{1}{2} \cdot \tau^{1/2}$$
$$= \frac{2^{m/2}}{2 k^m} \cdot \frac{\Gamma\left(\frac{m}{2}\right)\Gamma\left(\frac{1}{2}\right)}{\Gamma\left(\frac{m+1}{2}\right)} \cdot \frac{\Gamma\left(\frac{m+1}{2}\right)}{\Gamma\left(\frac{1}{2}\right)} \cdot \tau^{1/2} = \frac{2^{m/2}}{2 k^m} \Gamma\left(\frac{m}{2}\right) \cdot \tau^{1/2}. \qquad (5.21)$$

Combining (5.19), (5.20), and (5.21), and recalling that $R \sim s^{2k}$, we find that up to terms of higher order in $t - s$, the quantity (5.19) is

$$(-1)^{(m-2)/2} \frac{(2\pi)^{m/2}}{2 k^{m-2}} |t - s| f(x). \qquad (5.22)$$

Multiplying by the constant from (5.6), we obtain

$$u(x, t) = (t - s) f(x) + O\left((t - s)^2\right) \qquad \text{as} \quad t \to s. \qquad (5.23)$$

Both (5.10) and (5.11) follow from (5.23). This completes the proof of Theorem 5.1.

The case of odd m is more complicated. This case, and other examples obtained from the degenerate elliptic operators (4.3), will be taken up elsewhere.

References

[1] J. Aarão, *A transport equation of mixed type*, J. Diff. Equations **150** (1998), 188–202.

[2] R. Bader and P. Germain, *Solutions élémentaires de certaines équations aux dérivées partielles du type mixte*, Bull. Soc. Math. France **81** (1953), 145–174.

[3] M. S. Baouendi and C. Goulaouic, *Non-analytic hypoellipticity for some degenerate elliptic operators*, Bull. Amer. Math. Soc. **78** (1972), 483–486.

[4] R. Beals, *A note on fundamental solutions*, Comm. P. D. E. **24** (1999), 369–376.

[5] R. Beals, B. Gaveau, and P. C. Greiner, *The Green function of model step two hypoelliptic operators and the analysis of certain tangential Cauchy Riemann complexes*, Advances in Math. **121** (1996), 288–345.

[6] R. Beals, B. Gaveau, and P. C. Greiner. *On a geometric formula for the fundamental solutions of subelliptic laplacians*, Math. Nach. **181** (1996), 81–163.

[7] R. Beals, B. Gaveau, and P. C. Greiner, *Complex Hamiltonian mechanics and parametrices for subelliptic Laplacians I, II, III*, Bull. Sci. Math. **121** (1997), 1-36; 97–149; 195–259.

[8] R. Beals, B. Gaveau, and P. C. Greiner, *Uniform hypoelliptic Green's functions*, J. Math. Pures Appl. **77** (1998), 209–248.

[9] R. Beals, B. Gaveau, and P. C. Greiner, *Green's functions for some highly degenerate elliptic operators*, J. Funct. Anal. **165** (1999), 407–429.

[10] R. Beals, B. Gaveau, P. C. Greiner, and Y. Kannai, *Exact fundamental solutions for a class of degenerate elliptic operators*, Comm. P. D. E. **24** (1999), 719–742.

[11] R. Beals and P. C. Greiner, "Calculus on Heisenberg Manifolds," Annals of Math. Studies no. 119, Princeton Univ. Press, Princeton, NJ, 1988.

[12] A. Bellaïche and J.-J. Risler, "Subriemannian Geometry," Progress in Mathematics 144, Birkhäuser, Basel, 1996.

[13] L. Boutet de Monvel, A. Grigis, and B. Helffer, *Paramétrixes d'opérateurs pseudo-differentiels à caractéristiques multiples*, Astérisque **34–35** (1976), 93–121.

[14] L. Boutet de Monvel and F. Treves, *On a class pseudodifferential operators with double characteristics*, Inventiones Math. **24** (1974), 1–34.

[15] A. Bove and D. S. Tartakoff, *Optimal non-isotropic Gevrey exponents for sums of squares of vector fields*, Comm. P. D. E. **22** (1997), 1263–1282.

[16] S. Chandrasekhar, *Stochastic problems in physics and astronomy*, Rev. Mod. Phys. **15** (1943), 1–89.

[17] M. Christ, *Certain sums of squares of vector fields fail to be analytic hypoelliptic*, Comm. P. D. E. **16** (1991), 1695–1707.

[18] M. Christ, *Intermediate Gevrey exponents occur*, Comm. P. D. E. **22** (1997), 225–235.

[19] J. Dadok and R. Harvey, *The fundamental solution for the Kohn-Laplacian $\Box_b$ on the sphere in $\mathbb{C}^n$*, Math. Annalen **244** (1979), 89–104.

[20] R. M. Davis, *On a regular Cauchy problem for the Euler-Poisson-Darboux equation*, Ann. Mat. Pura Appl. **42** (1956), 205–226.

[21] A. Debiard and B. Gaveau, *Analysis on root systems*, Can. J. Math. **39** (1987), 1281–1404.

[22] S. Delache, *Calcul des solutions élémentaires des opérateurs de Tricomi-Clairaut auto-adjoints, strictement hyperboliques*, Bull. Soc. Math. France **97** (1969), 5–79.

[23] S. Delache and J. Leray, *Calcul de la solution élémentaire de l'opérateur d'Euler-Poisson-Darboux et de l'opérateur de Tricomi-Clairaut hyperbolique d'ordre 2*, Bull. Soc. Math. France **99** (1971), 313–336.

[24] M. Derridj and D. S. Tartakoff, *Local analyticity for $\Box_b$ and the $\bar{\partial}$-Neumann problem at certain weakly pseudo-convex boundary points*, Comm. P. D. E. **13** (1988), 1847–1868.

[25] M. Derridj and C. Zuily, *Régularité analytique et Gevrey pour des classes d'opérateurs élliptiques paraboliques dégénérés du second ordre*, Astérisque **2–3** (1973), 309–336.

[26] G. Folland, *A fundamental solution for a subelliptic operator*, Bull. Amer. Math. Soc. **79** (1973), 373–376.

[27] G. Folland and E. M. Stein, *Estimates for the $\bar{\partial}_b$-complex and analysis on the Heisenberg group*, Comm. Pure Appl. Math. **27** (1974), 429–522.

[28] B. Gaveau, *Principe de moindre action, propagation de la chaleur et estimées sous-elliptiques sur certains groupes nilpotents*, Acta Math. **139** (1977), 95–153.

[29] M. Gevrey, *Sur la nature analytique des solutions des équations aux dérivées partielles*, Ann. Sci. École Norm. Sup. **35** (1918), 129–189.

[30] D. Gourdin, *Systèmes faiblement hyperboliques à caracteristiques multiples*, CRAS **278** (1974), 269–272.

[31] P. C. Greiner, *A fundamental solution for a nonelliptic partial differential operator*, Can. J. Math. **31** (1979), 1107–1120.

[32] P. C. Greiner, D. Holcman, and Y. Kannai, *Wave kernels related to second order operators*, preprint.

[33] A. Grigis and J. Sjöstrand, *Front d'ondes analytique et sommes de carrés de champs de vecteurs*, Duke Math. J. **52** (1985), 35–51.

[34] J. Hadamard, "Le Problème de Cauchy et les Équations aux Dérivées Partielles Linéaires Hyperboliques," Hermann, Paris 1932.

[35] Y. Hamada, *On the propagation of singularities of the solution of the Cauchy problem*, Publ. RIMS Kyoto Univ. **6**, no. 2 (1970), 357–384.

[36] Y. Hamada, J. Leray, and C. Wagschal, *Systèmes d'équations aux dérivées partielles à caractéristiques multiples: problème de Cauchy ramifié: hyerbolicité partielle*, J. Math. Pures Appl. **55** (1976), 297–352.

[37] N. Hanges and A. Himonas, *Singular solutions for some sums of squares of vector fields*, Comm. P. D. E. **16** (1991), 1503–1511.

[38] G. Herglotz, *Über die Integration linearer partieller Differentielgleichungen mit konstanten Koeffizienten, I, II, III*, Bericht. Sächs. Akad. Wiss. zu Leipzig, Math. Phys. Kl. **78** (1926), 93–126, 2870-318; **80** (1928), 69–116.

[39] L. Hörmander, *Hypoelliptic second order differential equations*, Acta Math. **119** (1967), 147–171.

[40] A. Hulanicki, *The distribution of energy in the Brownian motion in the Gaussian field and analytic-hypoellipticity of certain subelliptic operators on the Heisenberg group*, Studia Math. **56** (1976), 165–173.

[41] F. John, "Plane Waves and Spherical Means Applied to Partial Differential Equations." Interscience, New York, 1955.

[42] A. Klingler, *New derivation of the Heisenberg kernel*, Comm. PDE **22** (1997), 2051–2060.

[43] A. N. Kolmogorov, *Zufällige Bewegungen*, Acta Math. **35** (1934), 116–117.

[44] J. Leray, *Les solutions élémentaires d'une équation aux dérivées partielles à coefficients constants*, C. R. Acad. Sci. Paris, Sér. I, **234** (1952), 1112–1114.

[45] J. Leray, *Intégrales abéliennes et solutions élémentaires des équations hyperboliques*, Colloque CBRM de Bruxelles 'Equations aux Dérivées Partielles, Thorne and Gauthier-Villars, 1954, pp. 37–43.

[46] J. Leray, *Le problème de Cauchy pour une équation linéaire à coefficients polynomiaux*, C. R. Acad. Sci. Paris, Sér. I, **242** (1956), 953–957.

[47] J. Leray and Y. Ohya, *Systémes linéaires hyperboliques non stricts*, Colloque CBRM de Liège d'Analyse fonctionelle, Thorne and Gauthier-Villars, 1965, pp. 105–144.

[48] A. I. Nachman, *The wave equation on the Heisenberg group*, Comm. P. D. E. **6** (1982), 675–714.

[49] Y. Ohya, *Le problème de Cauchy pour les équations hyperboliques à caractéristiques multiples*, J, Math. Soc. Japan **16** (1964), 268–286.

[50] O. A. Oleinik and E. V. Radkevič, "Second Order Equations with Non-Negative Characteristic Form," Moscow, 1971; English translation, Plenum, New York, London, 1973.

[51] M. Riesz, *L'Intégrale de Riemann-Liouville et le problème de Cauchy*, Acta Math. **81**, (1949), 1–223.

[52] D. Schiltz, J. Vaillant, and C. Wagschal, *Problème de Cauchy ramifié: racine caractéristique double ou triple en involution*, J. Math. Pure Appl. **61** (1982), 423–443.

[53] J. Sjöstrand, *Parametrices for pseudodifferential operators with multiple characteristics*, Ark. för Mat. **12** (1974), 85–130.

[54] D. S. Tartakoff, *On the local Gevrey and quasi-analytic hypoellipticity for $\Box_b$*, Comm. Pure Appl. Math. **26** (1973), 699–712.

[55] D. S. Tartakoff, *Local analytic hypoellipticity for $\Box_b$ on npn-degenerate Cauchy-Riemann manifolds*, Proc. Nat. Acad. Sci. U. S. A. **75** (1978), 3027–3028.

[56] F. Treves, *Analytic hypoellipticity of a class of pseudodifferential operators with double characteristics and applications to the $\bar{\partial}$-Neumann problem*, Comm. P. D. E. **3** (1978), 475–642.

[57] C. Wagschal, *Problème de Cauchy analytique à données méromorphes*, J. Math. Pure Appl. **51** (1972), 373–397.

Richard Beals
Yale University
Department of Mathematics
Box 208283
New Haven, CT 06520-8283
beals@math.yale.edu

Necessary Conditions for Hyperbolic Systems

Antonio Bove and Tatsuo Nishitani

1 Introduction

In this article we study the Cauchy problem for a first order system

$$L(x, D) = D_0 + \sum_{j=1}^{n} A_j(x)D_j + B(x) = L_1(x, D) + L_0(x)$$

where $A_j(x)$ and $B(x)$ are $r \times r$ smooth matrices and

$$L_1(x, D) = D_0 + \sum_{j=1}^{n} A_j(x)D_j, \quad L_0(x) = B(x).$$

Our aim is to obtain general necessary conditions at multiple characteristics in order that the Cauchy problem for $L(x, D)$ is C^∞ well posed.

Let ρ be a characteristic of order r and assume that the rank of $L_1(\rho)$ is $r - 1$. This case is so close to the scalar case and detailed studies are done in [2]. In this note we study the simplest case among truly vectorial cases, that is assuming

$$\operatorname{rank} L_1(\rho) = r - 2$$

we look for general necessary conditions for the C^∞ well posedness. Let us denote

$$h(x, \xi) = \det L_1(x, \xi)$$

and we always assume that $h(x, \xi)$ has only real roots with respect to ξ_0 when x is near the origin and $\xi' = (\xi_1, \ldots, \xi_n) \in \mathbf{R}^n$.

Without restrictions we may assume that $\rho = (0, e_n)$. To get necessary conditions we make a dilation around the reference characteristic point ρ. This procedure localizes the operator around that point. The symbol of the dilated (localized) operator can be thought of as a formal expansion in the dilation parameter and winds up to be in some noncommutative field. We then introduce a determinant "Det" on this noncommutative field (the precise definition will be given in Section 2). The general picture can be sketched as follows: the leading part of the noncommutative determinant "Det" of the dilation (localization) of the complete symbol should be the dilation (localization) of the usual determinant of the principal symbol.

In Section 2 we define $\mathrm{Det}_{(s)}$, depending on s, for matrix valued symbols

$$A(x, \xi; \lambda) = \sum_{j=n(A)} \lambda^{-\theta j} A_j(x, \xi),$$

where $A_j(x, \xi)$ are polynomials in (x, ξ). Then $\mathrm{Det}_{(s)} A$ has the form

$$\sum_{j=p(A)}^{p(A)+s-1} \lambda^{-\theta j} f_j(x, \xi),$$

where $f_j(x, \xi)$ are meromorphic at $(0, 0)$ and verifies, for instance, if $A(x, \xi; \lambda)$ and $B(x, \xi; \lambda)$ are polynomials in (x, ξ) and

$$\mathrm{Det}_{(s)} C = \mathrm{Det}_{(s)} A \cdot \mathrm{Det}_{(s)} B$$

then

$$C(x, \lambda^{-s\theta} D; \lambda) = A(x, \lambda^{-s\theta} D; \lambda) B(x, \lambda^{-s\theta} D; \lambda).$$

We note that

$$\det L_1(\lambda^{-\theta} x, e_n + \lambda^{-\theta} \xi) = \lambda^{-r\theta} [h_\rho(x, \xi) + O(\lambda^{-\theta})] \qquad (1.1)$$

because $(0, e_n)$ is a characteristic of order r.

Theorem 1.1. *Let $2s + 2 \geq r$ and put*

$$G = L_1(\lambda^{-\theta} x, e_n + \lambda^{-\theta} \xi) + \lambda^{-(s+2)\theta} L_0(\lambda^{-\theta} x).$$

Then in order that the Cauchy problem for $L_1(x, D) + L_0(x)$ is C^∞ well posed it is necessary that

$$\mathrm{Det}_{(s)} G = O(\lambda^{-r\theta}), \quad \sigma(\mathrm{Det}_{(s)} G) = h_\rho \qquad (1.2)$$

where $\sigma(A)$ denotes the leading part of the λ-expansion of A.

The method of proof of the above theorem consists in constructing an asymptotic solution u_λ of the equation

$$L(\lambda^{-\theta+\delta\theta} x, \lambda e_n + \lambda^{\theta-\delta\theta} D) u_\lambda$$
$$= \lambda\{L_1(\lambda^{-\theta}(\lambda^{\delta\theta} x), e_n + \lambda^{\theta-1-\delta\theta} D) + \lambda^{-1} L_0(\lambda^{-\theta}(\lambda^{\delta\theta} x))\} u_\lambda$$
$$= \lambda\{L_1(\lambda^{-\theta}(\lambda^{\delta\theta} x), e_n + \lambda^{-\theta}(\lambda^{-s\theta-\delta\theta} D)) + \lambda^{-1} L_0(\lambda^{-\theta}(\lambda^{\delta\theta} x))\} u_\lambda$$
$$= \lambda G(\lambda^{\delta\theta} x, \lambda^{-s\theta-\delta\theta} D; \lambda) u_\lambda \sim 0,$$

where s verifies $\theta - 1 - \delta\theta = -\theta - s\theta - \delta\theta$, that is $1 = (s+2)\theta$, which contradicts an a priori estimate (depending on λ) resulting from the C^∞ well posedness of the Cauchy problem for $L_1(x, D) + L_0(x)$.

Here is an outline of the paper: in Section 2 we give the precise definition of $\text{Det}_{(s)}$ and state several properties proved in [4] which will be used in later sections. In Section 3 we reduce the construction of a null asymptotic solution to the construction of a null asymptotic solution for a 2×2 system of higher order (Proposition 3.1). In Section 4, we prove that, under some additional conditions, one can construct an asymptotic solution for a 2×2 system $F(x, \lambda^{-s\theta} D; \lambda)$ provided $\det F = \lambda^{-k\theta}[g(x, \xi) + O(\lambda^{-\theta})]$ with $g(x, \xi) \neq 0$ and $k < s$ (Proposition 4.1). We prove Theorem 1.1 in Section 5 applying Proposition 4.1.

In what follows we put $q = s + 2$.

2   Definition of Det

We say that

$$K(\theta) = \{f(x, \xi; \lambda) = \sum_{j=n_f}^{\infty} \lambda^{-\theta j} f_j(x, \xi) \mid f_j(x, \xi)$$

is meromorphic in a neighborhood of $(0, 0)\}$

where the sum is a formal sum and $n_f \in \mathbf{Z}$. We define $f \# g$ for $f, g \in K(\theta)$ by

$$f \# g = \sum \frac{1}{\alpha!} \lambda^{-\theta(i+j+s|\alpha|)} f_i^{(\alpha)}(x, \xi) g_{j(\alpha)}(x, \xi)$$

$$= \sum_k \left[\sum_{i+j+s|\alpha|=k} \frac{1}{\alpha!} f_i^{(\alpha)}(x, \xi) g_{j(\alpha)}(x, \xi) \right] \lambda^{-\theta k}$$

where

$$f_{(\beta)}^{(\alpha)}(x, \xi) = \partial_\xi^\alpha D_x^\beta f(x, \xi), \qquad D_{x_j} = \frac{1}{i} \frac{\partial}{\partial x_j}.$$

In particular, if $f(x, \xi; \lambda)$, $g(x, \xi; \lambda) \in K(\theta)$ are given by a finite sum of $f_j(x, \xi)$ and $g_j(x, \xi)$ respectively and both $f_j(x, \xi)$ and $g_j(x, \xi)$ are polynomials in ξ, then it is clear that

$$f(x, \lambda^{-s\theta} D; \lambda) g(x, \lambda^{-s\theta} D; \lambda) = (f \# g)(x, \lambda^{-s\theta} D; \lambda).$$

It is easy to see that $K(\theta)$ is a noncommutative field when equipped with the product $\#$. Let us set

$$\overline{K} = (K(\theta)^\times / [K(\theta)^\times, K(\theta)^\times]) \cup \{0\}$$

where $[K(\theta)^\times, K(\theta)^\times]$ denotes the commutator subgroup of the multiplicative group $K(\theta)^\times = K(\theta) \setminus \{0\}$. Recall that every $f \neq 0$ has a canonical image $\bar{f}$ in $\overline{K}$.

Let us denote by $M(m; K(\theta))$ the set of all $m \times m$ matrices with entries in $K(\theta)$. Dieudonné ([6]) (see [1] for an exposition of the theory) proved that there exists a unique multiplicative morphism

$$\mathrm{Det} : M(m; K(\theta)) \to \overline{K}$$

verifying the following properties:

1. Let $A \in M(m; K(\theta))$ and let A' be obtained from A by multiplying one row by $f \in K(\theta)$; then

$$\mathrm{Det}\, A' = \bar{f} \cdot \mathrm{Det}\, A.$$

2. Let A' be obtained from A by adding one row to another; then

$$\mathrm{Det}\, A' = \mathrm{Det}\, A.$$

3.
$$\mathrm{Det}\, I = \bar{1}.$$

Since Det depends on s, we denote it by $\mathrm{Det}_{(s)}$ whenever it is necessary to make reference to its dependence on the parameter s. We also recall that

$$\mathrm{Det}(A \# B) = \mathrm{Det}\, A \cdot \mathrm{Det}\, B \tag{2.1}$$

and

$$\mathrm{Det}\, A = \mathrm{Det}\, A_{11} \cdot \mathrm{Det}\, A_{22} \tag{2.2}$$

$$\text{if} \quad A = \begin{pmatrix} A_{11} & A_{12} \\ 0 & A_{22} \end{pmatrix} \quad \text{or} \quad A = \begin{pmatrix} A_{11} & 0 \\ A_{21} & A_{22} \end{pmatrix}.$$

Lemma 2.1. *Let f, $g \in K(\theta)$. Assume $f \# g = 1 + O(\lambda^{-s\theta})$; then we have $fg = 1 + O(\lambda^{-s\theta})$ and vice versa.*

Lemma 2.2. *Let f, $g \in K(\theta)$ and assume that*

$$f \# g^{-1} = 1 + O(\lambda^{-s\theta}).$$

Then we have

$$n_f = n_g, \quad f_i = g_i, \quad n_f \leq i < n_f + s. \tag{2.3}$$

Let us now take a look at the commutator subgroup. It is clear that

$$f \# g = fg + O(\lambda^{(n_f+n_g-s)\theta}),$$
$$f^{-1} \# g^{-1} = f^{-1}g^{-1} + O(\lambda^{-(n_f+n_g+s)\theta}).$$

This shows that

$$f \# g \# f^{-1} \# g^{-1} = (fg + O(\lambda^{(n_f+n_g-s)\theta})) \# (f^{-1}g^{-1} + O(\lambda^{-(n_f+n_g+s)\theta}))$$
$$= (fg) \# (f^{-1}g^{-1}) + O(\lambda^{-s\theta}).$$

From Lemma 2.1 the right-hand side is $1 + O(\lambda^{-s\theta})$. This proves that the commutator subgroup is generated by $f \in K(\theta)$ of the form

$$f = 1 + O(\lambda^{-s\theta}).$$

Then Lemma 2.2 shows that $f \# g^{-1} \in [K(\theta)^\times, K(\theta)^\times]$ implies (2.3). Thus we may regard $\overline{K}$ as the set

$$\left\{ \sum_{j=n_f}^{n_f+s-1} \lambda^{-\theta j} f_j(x, \xi) \mid f_j(x, \xi) \text{ is meromorphic in a neighborhood of } (0, 0) \right\}.$$

Of course for $\bar{f} = \sum_{j=n_f}^{n_f+s-1} \lambda^{-\theta j} f_j(x, \xi)$, $\bar{g} = \sum_{j=n_g}^{n_g+s-1} \lambda^{-\theta j} g_j(x, \xi) \in \overline{K}$ we have

$$\bar{f} \cdot \bar{g} = \sum_{k=n_f+n_g}^{n_f+n_g+s-1} \lambda^{-\theta k} \left(\sum_{i+j=k} f_i g_j \right).$$

Definition 2.3. We say that $A \in M(m; K(\theta))$ belongs to $\mathcal{H}(k)$ if $A = \sum \lambda^{-j\theta} A_j$ and A_j is a sum of terms $A_{j\ell}$ which are homogeneous of degree $j + k - \ell q \geq 0$ for some $\ell \geq 0$. We define $\delta(A)$ for $A \in \mathcal{H}(k)$ by

$$\delta(A) = \sum \lambda^{-j\theta} A_{j0},$$

that is, $\delta(A)$ is the sum of the terms with the highest degree of homogeneity.

It is easy to check that if $A_i \in \mathcal{H}(k_i)$, then $A_1 \# A_2 \in \mathcal{H}(k_1 + k_2)$.

Lemma 2.3. *Let $A_i \in \mathcal{H}(k_i)$, $i = 1, 2$. Then we have*

$$\delta(A_1 \# A_2) = \delta(A_1)\delta(A_2).$$

Lemma 2.4. *Assume that $G = \sum_{j=0} \lambda^{-j\theta} G_j$. Then we have*

$$\text{Det} G = \det G + O(\lambda^{-s\theta})$$

where $s = q - 2$. In particular $\text{Det} G = O(1)$.

Let f be holomorphic at $(0, 0)$. We say that $F \in K(\theta)$, $F = \sum_j \lambda^{-\theta j} F_j$ is holomorphic outside $\{f = 0\}$ if for every j, with $F_j = f_j / g_j$ where f_j and g_j are relatively prime, the irreducible factors of g_j coincide with those of f. We say that F is holomorphic on $\{f = 0\}$ if for every j, g_j and f have no common irreducible factor.

Let $F = (F_{ij}) \in M(m; K(\theta))$. Then we say that F is holomorphic outside $\{f = 0\}$ if every F_{ij} is holomorphic outside $\{f = 0\}$ and we say that F is holomorphic on $\{f = 0\}$ if every F_{ij} is holomorphic on $\{f = 0\}$.

Proposition 2.5. *If F is holomorphic outside $\{f = 0\}$, then $\mathrm{Det}\, F$ is holomorphic outside $\{f = 0\}$.*

Corollary 2.6. *If F is holomorphic at $(0, 0)$, then $\mathrm{Det}\, F$ is also holomorphic at $(0, 0)$.*

3   First reduction

Let us write

$$G(x, \tilde{\xi}; \lambda) = \sum_{j=0} \lambda^{-\theta j} G_j(x, \tilde{\xi})$$

where

$$G_j(x, \tilde{\xi}) = \sum_{|\alpha+\beta|=j} \frac{1}{\alpha! \beta!} L^{(\alpha)}_{1(\beta)}(0, e_n) x^\beta \tilde{\xi}^\alpha, \quad j < s + 2$$

$$G_j(x, \tilde{\xi}) = \sum_{|\alpha+\beta|=j} \frac{1}{\alpha! \beta!} L^{(\alpha)}_{1(\beta)}(0, e_n) x^\beta \tilde{\xi}^\alpha$$

$$+ \sum_{|\beta|=j-s-2} \frac{1}{\beta!} L_{0(\beta)}(0) x^\beta, \quad j \geq s + 2$$

so that $G \in \mathcal{H}(0)$. Recall that

$$h_\rho(x, \tilde{\xi}) = \sum_{|\alpha+\beta|=r} \frac{1}{\alpha! \beta!} h^{(\alpha)}_{(\beta)}(0, e_n) x^\beta \tilde{\xi}^\alpha.$$

Our purpose is to construct an asymptotic solution for $G(x, \lambda^{-s\theta} D; \lambda)$. To do so we reduce the problem to the same problem for a 2×2 system. Choose M, N to be nonsingular constant matrices such that

$$M G_0(0, e_n) N = \begin{pmatrix} I_{r-2} & 0 \\ 0 & 0 \end{pmatrix}$$

where I_{r-2} is the identity matrix of order $r - 2$. We denote $M G(x, \lambda^{-s\theta} D; \lambda) N$ by $G(x, \lambda^{-s\theta} D; \lambda)$ again so that

$$G(x, \tilde{\xi}; \lambda) = \sum_{j=0} \lambda^{-\theta j} G_j(x, \tilde{\xi}), \quad G_0(x, \tilde{\xi}) = \begin{pmatrix} I_{r-2} & 0 \\ 0 & 0 \end{pmatrix}. \tag{3.1}$$

Let us write G in block matrix notation

$$G = \begin{pmatrix} G_{11} & G_{12} \\ G_{21} & G_{22} \end{pmatrix}$$

where the blocks correspond to those in (3.1). Note that with

$$G_{ij} = \sum_{p=0} \lambda^{-\theta p} G_{ij,p}(x, \tilde{\xi})$$

one has

$$G_{11} = I - \sum_{k=1} \lambda^{-k\theta} H_k(x, \tilde{\xi}) = I - H(x, \tilde{\xi}; \lambda).$$

Define

$$R_N(x, \tilde{\xi}; \lambda) = \sum_{n=0}^{N} \overbrace{H \# \cdots \# H}^{n} = \sum_{n=0}^{N} H^{\#n}$$

so that

$$G_{11} \# R_N = I - H^{\#(N+1)}$$

and introduce a differential operator (symbol) Λ_N:

$$\Lambda_N(x, \tilde{\xi}; \lambda) = \begin{pmatrix} I_{r-2} & -(R_N \# G_{12})(x, \tilde{\xi}; \lambda) \\ 0 & I_2 \end{pmatrix}. \tag{3.2}$$

We remark that

$$G\# \begin{pmatrix} I_{r-2} & -G_{11}^{-1} \# G_{12} \\ 0 & I_2 \end{pmatrix} = \begin{pmatrix} G_{11} & 0 \\ G_{21} & G_{22} - G_{21} \# G_{11}^{-1} \# G_{12} \end{pmatrix}$$

and hence, from (2.2), it follows that

$$\mathrm{Det}\, G = \mathrm{Det}\, G_{11} \cdot \mathrm{Det}(G_{22} - G_{21} \# G_{11}^{-1} \# G_{12}). \tag{3.3}$$

Multiply G with Λ_N on the right; we get

$$G\#\Lambda_N = \begin{pmatrix} G_{11} & G_{12} - G_{11} \# R_N \# G_{12} \\ G_{21} & G_{22} - G_{21} \# R_N \# G_{12} \end{pmatrix} = \begin{pmatrix} G_{11} & S_N \\ G_{21} & F_N \end{pmatrix}$$

where

$$F_N = G_{22} - G_{21} \# R_N \# G_{12}, \quad S_N = G_{12} - G_{11} \# R_N \# G_{12}. \tag{3.4}$$

Note that

$$S_N = G_{12} - G_{11} \# R_N \# G_{12} = G_{12} - (I - H^{\#(N+1)}) \# G_{12}$$
$$= H^{\#(N+1)} \# G_{12} = \lambda^{-(N+1)\theta} K(x, \tilde{\xi}; \lambda)$$

with some $K(x, \tilde{\xi}; \lambda) = \sum_{j=0} \lambda^{-\theta j} K_j(x, \tilde{\xi})$ where $K_j(x, \tilde{\xi})$ are polynomials in $(x, \tilde{\xi})$.

If

$$u = \exp\left\{i \sum_{k=0}^{p} \lambda^{\sigma_k} \phi_k(x)\right\} \sum_{j=0} \lambda^{-\theta j} u_j(x), \quad s\theta \geq \sigma_0 > \cdots > \sigma_p > 0$$

is an asymptotic solution for $F_N(x, \lambda^{-s\theta} D; \lambda)$ then it is clear that

$$\begin{pmatrix} 0 \\ u \end{pmatrix}$$

is also an asymptotic solution for $G(x, \lambda^{-s\theta} D; \lambda)$, (taking N large) because we have

$$e^{-i \sum_{k=0}^{p} \lambda^{\sigma_k} \phi_k(x)} S_N(x, \lambda^{-s\theta} D; \lambda) e^{i \sum_{k=0}^{p} \lambda^{\sigma_k} \phi_k(x)} = O(\lambda^{-(N+1)\theta}).$$

We summarize what has been proved up to now:

Proposition 3.1. *The construction of a null asymptotic solution for* $G(x, \lambda^{-s\theta} D; \lambda)$ *with* $\sigma_0 \leq s\theta$ *is reduced to the same problem for*

$$F_N(x, \lambda^{-s\theta} D; \lambda).$$

Let us fix $0 < \delta < 1$ and consider

$$G(\lambda^{\delta\theta} x, \lambda^{-\delta\theta - s\theta} D; \lambda). \tag{3.5}$$

To construct an asymptotic solution for (3.5) we note that

$$G(\lambda^{\delta\theta} x, \lambda^{-\delta\theta - s\theta} D; \lambda) \Lambda_N(\lambda^{\delta\theta} x, \lambda^{-\delta\theta - s\theta} D; \lambda)$$
$$= \begin{pmatrix} G_{11}(\lambda^{\delta\theta} x, \lambda^{-\delta\theta - s\theta} D; \lambda) & S_N(\lambda^{\delta\theta} x, \lambda^{-\delta\theta - s\theta} D; \lambda) \\ G_{21}(\lambda^{\delta\theta} x, \lambda^{-\delta\theta - s\theta} D; \lambda) & F_N(\lambda^{\delta\theta} x, \lambda^{-\delta\theta - s\theta} D; \lambda) \end{pmatrix}.$$

Since $S_N(x, \tilde{\xi}; \lambda) \in \mathcal{H}(0)$ and $S_N(x, \tilde{\xi}; \lambda) = O(\lambda^{-(N+1)\theta})$ it is clear that

$$S_N(\lambda^{\delta\theta} x, \lambda^{\delta\theta} \tilde{\xi}; \lambda) = O(\lambda^{-(N+1)(1-\delta)\theta}). \tag{3.6}$$

Let

$$u = e^{i \sum_{k=0}^{p} \lambda^{\sigma_k} \phi_k(x)} \sum_{j=0} \lambda^{-\theta j} u_j(x), \quad \sigma_0 = s\theta + 2\delta\theta > \sigma_1 > \cdots > \sigma_p > 0$$

be an asymptotic solution for $F_N(\lambda^{\delta\theta} x, \lambda^{-\delta\theta - s\theta} D; \lambda)$. Then it is clear that $^t(0, u)$ is also an asymptotic solution to $G(\lambda^{\delta\theta} x, \lambda^{-\delta\theta - s\theta} D; \lambda)$ since we have

$$e^{-i \sum_{k=0}^{p} \lambda^{\sigma_k} \phi_k(x)} S_N(\lambda^{\delta\theta} x, \lambda^{-\delta\theta - s\theta} D; \lambda) e^{i \sum_{k=0}^{p} \lambda^{\sigma_k} \phi_k(x)} = O(\lambda^{-(N+1)(\theta - \delta)}).$$

Proposition 3.2. *The construction of a null asymptotic solution with* $\sigma_0 = s\theta + 2\delta\theta$ *for the operator* $G(\lambda^{\delta\theta} x, \lambda^{-\delta\theta - s\theta} D; \lambda)$ *is reduced to the same problem for*

$$F_N(\lambda^{\delta\theta} x, \lambda^{-\delta\theta - s\theta} D; \lambda).$$

We now take a closer look at $F_N(x, \lambda^{-s\theta} D; \lambda)$:

$$F_N(x, \tilde{\xi}; \lambda) = \sum_{j=0} \lambda^{-\theta j} F_j(x, \tilde{\xi}).$$

Note that $F_0 = 0$ because $G_{ij,0} = 0$ if $(i, j) \neq (1, 1)$.

Lemma 3.3. *Taking N large we have*

$$\mathrm{Det}\, F_N = [1 + O(\lambda^{-\theta})] \cdot \mathrm{Det}\, G.$$

Let us write

$$G^0(x, \xi; \lambda) = L_1(\lambda^{-\theta} x, e_n + \lambda^{-\theta} \xi).$$

Lemma 3.4. *We have*

$$\delta(\det G_{11}^0) \delta(\det F_N) = \delta(\det G^0) + O(\lambda^{-N\theta}).$$

Corollary 3.5. *We have*

$$\delta(\det F_N) = \lambda^{-r\theta}[h_\rho + O(\lambda^{-\theta})].$$

4 A lemma

In this section we are interested in constructing an asymptotic null solution for an operator whose symbol has the form

$$F(x, \tilde{\xi}; \lambda) = \sum_{j=0} \lambda^{-\theta j} F_j(x, \tilde{\xi}), \quad F_0(x, \tilde{\xi}) \neq 0,$$

where $F_j(x, \tilde{\xi})$ is a 2×2 matrix which is a polynomial in $\tilde{\xi}$. Our purpose is to prove the following.

Proposition 4.1. *Assume that*

$$\det F(x, \tilde{\xi}) = \lambda^{-k\theta}[g(x, \tilde{\xi}) + O(\lambda^{-\theta})]$$

with some $0 \leq k < s$ where $g(x, 0) = 0$ and $\partial_{\xi_0} g(x, 0) \neq 0$. Then one can construct an asymptotic null solution for

$$F(x, \lambda^{-s\theta} D; \lambda).$$

Proof. We first treat the case that $F_0(x, 0) \neq 0$. It is easy to see that

$$\det F_0(x, 0) = 0$$

for if $k > 0$ then it is obvious and if $k = 0$ it follows from $g(x, 0) = 0$. Hence there are nonsingular smooth matrices $M(x)$, $N(x)$ defined in some open set such that

$$M(x) F_0(x, 0) N(x) = \begin{pmatrix} 1 & 0 \\ 0 & 0 \end{pmatrix}.$$

Writing

$$\tilde{F}(x, \lambda^{-s\theta} D; \lambda) = M(x) F(x, \lambda^{-s\theta} D; \lambda) N(x),$$

one gets $\tilde{F}(x, \tilde{\xi}; \lambda) = M(x) F(x, \tilde{\xi}; \lambda) N(x) + O(\lambda^{-s\theta})$ and this shows that

$$\det \tilde{F}(x, \tilde{\xi}; \lambda) = \lambda^{-k\theta}[c(x)g(x, \tilde{\xi}) + O(\lambda^{-\theta})],$$

with $c(x) = \det M(x) \det N(x) \neq 0$. Obviously it is enough to construct an asymptotic null solution for $\tilde{F}(x, \lambda^{-s\theta} D; \lambda)$. Denote $\tilde{F}(x, \tilde{\xi}; \lambda)$, $c(x)g(x, \tilde{\xi})$ by $F(x, \tilde{\xi}; \lambda)$ and $g(x, \tilde{\xi})$ again. Let us set

$$F = \begin{pmatrix} F_{11} & F_{12} \\ F_{21} & F_{22} \end{pmatrix}, \quad F_0(x, 0) = \begin{pmatrix} 1 & 0 \\ 0 & 0 \end{pmatrix}$$

so that

$$F_{11}(x, \tilde{\xi}; \lambda) = 1 - B(x, \tilde{\xi}; \lambda), \quad B(x, \tilde{\xi}; \lambda) = \sum_{j=0} \lambda^{-\theta j} B_j(x, \tilde{\xi})$$

where $B_0(x, 0) = 0$.

Define a differential operator (or rather symbol) $R_N(x, \tilde{\xi}; \lambda)$ by

$$R_N(x, \tilde{\xi}; \lambda) = \sum_{n=0}^{N} B(x, \tilde{\xi}; \lambda) \overbrace{\# \cdots \#}^{n} B(x, \tilde{\xi}; \lambda) = \sum_{n=0}^{N} B(x, \tilde{\xi}; \lambda)^{\#n}$$

and $\Lambda(x, \tilde{\xi}; \lambda)$ by

$$\Lambda(x, \tilde{\xi}; \lambda) = \begin{pmatrix} 1 & -R_N \# F_{12} \\ 0 & 1 \end{pmatrix}.$$

Then it follows that

$$(F \# \Lambda)(x, \tilde{\xi}; \lambda) = \begin{pmatrix} F_{11} & S \\ F_{21} & K \end{pmatrix}$$

where

$$K = F_{22} - F_{21} \# R_N \# F_{12}, \quad S = F_{12} - F_{11} \# R_N \# F_{12}. \tag{4.1}$$

Lemma 4.2. *Let $0 \leq k < s$ and let $\phi(x)$ be a smooth function. Then we have*

$$e^{-i\lambda^{k\theta}\phi(x)} S(x, \lambda^{-s\theta} D; \lambda) e^{i\lambda^{k\theta}\phi(x)} = O(\lambda^{-N(s-k)\theta}).$$

Due to Lemma 4.2 our problem is reduced to the problem of constructing an asymptotic null solution for $K(x, \lambda^{-s\theta} D; \lambda)$ with $\sigma_0 < s\theta$.

Lemma 4.3. *We have*

$$F_{11}(x, \tilde{\xi}; \lambda) K(x, \tilde{\xi}; \lambda) = \det F(x, \tilde{\xi}; \lambda) + O(\lambda^{-s\theta}) + O(|\tilde{\xi}|^{N+1}).$$

From Lemma 4.3, due to the fact that $k < s$, we obtain that

$$F_{11}K = \lambda^{-k\theta}[g(x, \tilde{\xi}) + \lambda^{-\theta}r(x, \tilde{\xi}; \lambda)] + O(|\tilde{\xi}|^{N+1}) \tag{4.2}$$

where $r(x, \tilde{\xi}; \lambda)$ is a suitable symbol. Let us define $\tilde{g}^{(1)}(x, \tilde{\xi})$ by

$$g(x, \lambda^{-\theta}\tilde{\xi}) + \lambda^{-\theta}r(x, \lambda^{-\theta}\tilde{\xi}; \lambda) = \lambda^{-\theta}[\tilde{g}^{(1)}(x, \tilde{\xi}) + O(\lambda^{-\theta})]$$

so that

$$\tilde{g}^{(1)}(x, \tilde{\xi}) = \sum_{|\alpha|=1} \partial_\xi^\alpha g(x, 0)\tilde{\xi}^\alpha + r_0(x, 0). \tag{4.3}$$

Then from (4.2) it follows that

$$[1 + O(\lambda^{-\theta})]K(x, \lambda^{-\theta}\tilde{\xi}; \lambda)$$
$$= \lambda^{-k\theta}[g(x, \lambda^{-\theta}\tilde{\xi}) + \lambda^{-\theta}r(x, \lambda^{-\theta}\tilde{\xi}; \lambda)] + O(\lambda^{-(N+1)\theta})$$

and hence

$$K(x, \lambda^{-\theta}\tilde{\xi}; \lambda) = \lambda^{-(k+1)\theta}[\tilde{g}^{(1)}(x, \tilde{\xi}) + O(\lambda^{-\theta})]. \tag{4.4}$$

Take $\psi(x)$ so that

$$\tilde{g}^{(1)}(x, \psi_x(x)) = 0$$

and consider

$$e^{-i\lambda^{(s-1)\theta}\psi(x)}K(x, \lambda^{-s\theta}D; \lambda)e^{i\lambda^{(s-1)\theta}\psi(x)}$$
$$= K(x, \lambda^{-\theta}(\psi_x(x) + \lambda^{-(s-1)\theta}D); \lambda) + \lambda^{-(s+1)\theta}R(x, \lambda^{-(s-1)\theta}D; \lambda)$$

which follows from Lemma 4.4 below. Let us denote the right-hand side of the above equality by $F^{(1)}(x, \lambda^{-(s-1)\theta}D; \lambda)$ and set $g^{(1)}(x, \tilde{\xi}) = \tilde{g}^{(1)}(x, \psi_x(x) + \tilde{\xi})$ which is linear in $\tilde{\xi}$ so that we have

$$F^{(1)}(x, \tilde{\xi}; \lambda) = K(x, \lambda^{-\theta}(\psi_x(x) + \tilde{\xi}); \lambda) + \lambda^{-(s+1)\theta}R(x, \tilde{\xi}; \lambda)$$
$$= \lambda^{-(k+1)\theta}[\tilde{g}^{(1)}(x, \psi_x(x) + \tilde{\xi}) + O(\lambda^{-\theta})] + \lambda^{-(s+1)\theta}R(x, \tilde{\xi}; \lambda)$$
$$= \lambda^{-(k+1)\theta}[g^{(1)}(x, \tilde{\xi}) + O(\lambda^{-\theta})]$$

because $k + 1 < s + 1$.

Now our construction of an asymptotic solution for $K(x, \lambda^{-s\theta}D; \lambda)$ is reduced to the same problem for $F^{(1)}(x, \lambda^{-(s-1)\theta}D; \lambda)$ or, equivalently, to the same problem for the operator $\lambda^{(k+1)\theta}F^{(1)}(x, \lambda^{-(s-1)\theta}D; \lambda)$ which is

$$g^{(1)}(x, \lambda^{-(s-1)\theta}D) + \lambda^{-\theta}r(x, \lambda^{-(s-1)\theta}D; \lambda).$$

Since this is a scalar operator the construction of an asymptotic null solution for the latter is standard (see [7] or, in the actual framework, [3]).

We now turn to the other case, that is $F_0(x, 0) = 0$, where $k > 0$, and we show that the problem is reduced to the case $k = 0$. We first need to improve slightly Lemma 3.1 in [3].

Lemma 4.4. *Let $G(x, D)$ be a differential operator with smooth coefficients, defined in a neighborhood of the origin U, and let $0 < k < s$, s, $k \in \mathbb{N}$. Let $\phi(x) \in C^\infty(U)$. Then we have*

$$e^{-i\lambda^{k\theta}\phi(x)} G(x, \lambda^{-s\theta} D) e^{i\lambda^{k\theta}\phi(x)}$$
$$= G(x, \lambda^{-(s-k)\theta}(\phi_x(x) + \lambda^{-k\theta} D)) + \lambda^{-s\theta-(s-k)\theta} R(x, \lambda^{-(s-k)\theta}\lambda^{-k\theta} D; \lambda)$$

with $R(x, \tilde{\xi}; \lambda) = \sum_{j=0} \lambda^{-\theta j} R_j(x, \tilde{\xi})$ where $R_j(x, \tilde{\xi})$ are polynomials in $\tilde{\xi}$ with coefficients in $C^\infty(U)$, G is given by a finite expansion with respect to the parameter λ and $G(x, \lambda^{-(s-k)\theta}(\phi_x(x) + D))$ denotes the differential operator with symbol

$$G(x, \lambda^{-(s-k)\theta}(\phi_x(x) + \xi)).$$

For later use, we consider also the case when $k > s$.

Lemma 4.5. *Let $G(x, \xi)$ be a polynomial in (x, ξ) of degree m. Then we have*

$$e^{-i\lambda^{s\theta+\delta\theta}\phi(x)} G(\lambda^{\delta\theta} x, \lambda^{-s\theta} D; \lambda) e^{i\lambda^{s\theta+\delta\theta}\phi(x)}$$
$$= G(\lambda^{\delta\theta} x, \lambda^{\delta\theta}\phi_x(x) + \lambda^{-s\theta} D; \lambda) + \lambda^{(m-1)\delta\theta-s\theta} R(x, \lambda^{-s\theta} D; \lambda).$$

We return to the construction of an asymptotic solution for F. Note that

$$e^{-i\lambda^{(s-1)\theta}\psi(x)} F(x, \lambda^{-s\theta} D; \lambda) e^{i\lambda^{(s-1)\theta}\psi(x)} \tag{4.5}$$
$$= F(x, \lambda^{-\theta}(\psi_x(x) + \lambda^{-(s-1)\theta} D); \lambda) + \lambda^{-(s+1)\theta} R(x, \lambda^{-(s-1)\theta} D; \lambda)$$

by Lemma 4.4. Taking into account the fact that $F_0(x, \lambda^{-\theta}\tilde{\xi}) = O(\lambda^{-\theta})$, let us denote the right-hand side of (4.5) by

$$\lambda^{-\theta} F^{(1)}(x, \lambda^{-(s-1)\theta} D; \lambda), \quad F^{(1)}(x, \tilde{\xi}; \lambda) = \sum_{j=0} \lambda^{-\theta j} F_j^{(1)}(x, \tilde{\xi}),$$

that is

$$\lambda^{-\theta} F^{(1)}(x, \tilde{\xi}; \lambda) = F(x, \lambda^{-\theta}(\psi_x(x) + \tilde{\xi}); \lambda) + \lambda^{-(s+1)\theta} R(x, \tilde{\xi}; \lambda). \tag{4.6}$$

From (4.6) it follows that

$$\lambda^{-2\theta} \det F^{(1)}(x, \tilde{\xi}; \lambda) = \det F(x, \lambda^{-\theta}(\psi_x(x) + \tilde{\xi}); \lambda) + O(\lambda^{-(s+2)\theta}) \tag{4.7}$$

because $F(x, \lambda^{-\theta}(\psi_x(x) + \tilde{\xi}); \lambda) = O(\lambda^{-\theta})$. On the other hand, repeating the same arguments as before one can write

$$\det F(x, \lambda^{-\theta}(\psi_x + \xi); \lambda) = \lambda^{-(k+1)\theta}[g^{(1)}(x, \xi) + O(\lambda^{-\theta})] \tag{4.8}$$

where we can choose $\psi(x)$ so that $g^{(1)}(x, 0) = 0$ and $\partial_{\xi_0} g^{(1)}(x, 0) \neq 0$. Then we have from (4.7) and (4.8)

$$\det F^{(1)}(x, \xi; \lambda) = \lambda^{-(k-1)\theta}[g^{(1)}(x, \xi) + O(\lambda^{-\theta})] + O(\lambda^{-s\theta})$$
$$= \lambda^{-(k-1)\theta}[g^{(1)}(x, \xi) + O(\lambda^{-\theta})] \qquad (4.9)$$

because $k < s$.

Summing up, the construction of an asymptotic solution for $F(x, \lambda^{-s\theta} D; \lambda)$ is reduced to the same problem for the operator

$$F^{(1)}(x, \lambda^{-(s-1)\theta} D; \lambda),$$

verifying (4.9), where $g^{(1)}(x, 0) = 0$ and $\partial_{\xi_0} g^{(1)}(x, 0) \neq 0$.

Assume that

$$F_j^{(1)}(x, \xi) = 0, \quad j < \ell, \quad F_\ell^{(1)}(x, \xi) \neq 0.$$

Then one can write

$$F^{(1)}(x, \xi; \lambda) = \lambda^{-\ell\theta} \tilde{F}^{(1)}(x, \xi; \lambda) = \lambda^{-\ell\theta} \sum_{j=0} \lambda^{-\theta j} \tilde{F}_j^{(1)}(x, \xi)$$

with $\tilde{F}_0^{(1)}(x, \xi) \neq 0$. Since $k - \ell - 1 < s - 1$ one can reduce the problem to the same one with smaller indices $(k - \ell - 1, s - 1)$. We apply the same arguments to $\tilde{F}^{(1)}(x, \lambda^{-(s-1)\theta} D; \lambda)$. Then either we arrive at the first case—where our problem becomes a scalar problem—or we reach the point when

$$F^{(p)}(x, \xi; \lambda) = \sum_{j=0} \lambda^{-\theta j} F_j^{(p)}(x, \xi), \quad F_0^{(p)}(x, \xi) \neq 0,$$

$$\det F^{(p)}(x, \xi; \lambda) = g^{(p)}(x, \xi) + O(\lambda^{-\theta})$$

where $g^{(p)}(x, 0) = 0$ and $\partial_{\xi_0} g^{(p)}(x, 0) \neq 0$ and we have to construct an asymptotic solution for

$$F^{(p)}(x, \lambda^{-s_p\theta} D; \lambda), \quad s_p > 0.$$

Thus our problem is reduced to the case $k = 0$. Since

$$\det F_0^{(p)}(x, \xi) = g^{(p)}(x, \xi)$$

it is clear that $F_0^{(p)}(x, 0) \neq 0$ because if $F_0^{(p)}(x, 0) = 0$ we would have $\partial_{\xi_0} \det F_0^{(p)}(x, 0) = 0$ which contradicts the assumption. Hence we are again in the first case and the assertion can be easily proved. □

5 Proof of Theorem

First of all we recall that

$$F = G_{22} - G_{21} \# R \# G_{12}$$

and we assume that

$$\mathrm{Det}\, F = \lambda^{-t\theta}[g_0 + O(\lambda^{-\theta})], \quad g_0 \neq 0 \tag{5.1}$$

for some $t \geq 0$. Let us also assume that, for some $\ell \geq 0$,

$$F_j = 0, \quad j < \ell, \quad F_\ell \neq 0.$$

Without any loss of generality we may assume that the $(1, 1)$-th entry of F_ℓ is different from zero. Let us set

$$F^{(0)} = \lambda^{\ell\theta} F, \quad F^{(0)} = \sum_{j=0} \lambda^{-\theta j} F_j^{(0)}$$

so that

$$\mathrm{Det}\, F^{(0)} = \lambda^{2\ell\theta} \mathrm{Det}\, F, \quad \det F^{(0)} = \lambda^{2\ell\theta} \det F. \tag{5.2}$$

Note that $2\ell \leq t$ for, if $2\ell > t$, then

$$\mathrm{Det}\, F = \lambda^{-2\ell\theta} \mathrm{Det}\, F^{(0)} = O(\lambda^{-2\ell\theta})$$

which contradicts (5.1).

Lemma 5.1. *There exist $w^{(p)}$, $0 \leq p \leq t - 2\ell - 1$ such that with*

$$\Lambda^{(p)} = \begin{pmatrix} 1 & w^{(p)} \\ 0 & w \end{pmatrix}, \quad \Gamma = \begin{pmatrix} \lambda^{-\theta/2} & 0 \\ 0 & \lambda^{\theta/2} \end{pmatrix}$$

we have

$$F \# \Lambda^{(0)} \# \Gamma \# \cdots \# \Lambda^{(t-2\ell-1)} \# \Gamma = \lambda^{-t\theta/2} F^*$$

where

$$F^* = \sum_{j=0} \lambda^{-j\theta} F_j^*, \quad w^{t-2\ell} \delta(\det F) = \lambda^{-t\theta} \delta(\det F^*)$$

and w is the $(1, 1)$–th entry of F_ℓ.

Lemma 5.2. *We have*

$$\det F^* = g_0^* + \lambda^{-\theta} g_1^* + \cdots + \lambda^{-(s-1)\theta} g_{s-1}^* + \cdots, \quad g_0^* \neq 0$$

where g_i^* is a sum of terms which are homogeneous polynomials of degree $\ell(t - 2\ell) + t + i - jq \geq 0$ for some suitable j's. Moreover we have

$$w^{t-2\ell} \mathrm{Det}\, F = \lambda^{-t\theta} \mathrm{Det}\, F^*, \tag{5.3}$$

$$\delta(\det F^*) = \lambda^{-(r-t)\theta} w^{t-2\ell}[h_\rho + O(\lambda^{-\theta})]. \tag{5.4}$$

Corollary 5.3. *We have $t \leq r$.*

Proof of Theorem 1.1. We first prove that

$$\mathrm{Det}\, F = O(\lambda^{-r\theta}) \tag{5.5}$$

is necessary for the C^∞ well posedness of the Cauchy problem for $L_1(x, D) + L_0(x)$. Let us study

$$e^{-i\lambda x_n}[L_1(x, D) + L_0(x)]e^{i\lambda x_n}$$

which yields

$$\lambda\{L_1(x, e_n + \lambda^{-1}D) + \lambda^{-1}L_0(x)\}.$$

We perform the following dilation: $x \to \lambda^{-(1-\delta)\theta}x$, where $0 < \delta < 1$ will be determined later. Defining s by $-1+(1-\delta)\theta = -\theta-\delta\theta-s\theta$, that is $1 = (s+2)\theta$, we have

$$L_1(\lambda^{-(1-\delta)\theta}x, e_n + \lambda^{(1-\delta)\theta-1}D) + \lambda^{-1}L_0(\lambda^{-(1-\delta)\theta}x)$$

$$= L_1(\lambda^{-\theta}(\lambda^{\delta\theta}x), e_n + \lambda^{-\theta}(\lambda^{-\delta\theta-s\theta}D)) + \lambda^{-(s+2)\theta}L_0(\lambda^{-\theta}(\lambda^{\delta\theta}x))$$

$$= \sum_{j=0}\lambda^{-\theta j}G_j(\lambda^{\delta\theta}x, \lambda^{-\delta\theta-s\theta}D) = G(\lambda^{\delta\theta}x, \lambda^{-\delta\theta-s\theta}D; \lambda).$$

Let us recall here the a priori estimate

$$|u|_{C^0(W^t)} \le C\lambda^M |G(\lambda^{\delta\theta}x, \lambda^{-\delta\theta-s\theta}D; \lambda)u|_{C^P(W^t)}, \quad \forall u \in C_0^\infty(W), \ \lambda \ge \bar{\lambda} \tag{5.6}$$

resulting from the C^∞ well posedness of the Cauchy problem for $L_1(x, D)+L_0(x)$, where $W^t = \{x \in W \mid x_0 \le t\}$ and W is any given compact set in $\mathbf{R}^{n+1}$. To prove (5.5), supposing (5.1) with $t < r$, we construct an asymptotic solution for $G(\lambda^{\delta\theta}x, \lambda^{-\delta\theta-s\theta}D; \lambda)$ contradicting (5.6). By Proposition 3.2 it is enough to construct an asymptotic null solution for $F(\lambda^{\delta\theta}x, \lambda^{-\delta\theta-s\theta}D; \lambda)$ with $\sigma_0 = s\theta + 2\delta\theta$. Setting

$$\Lambda = \Lambda^{(0)}\#\Gamma\#\cdots\#\Lambda^{(t-2\ell-1)}\#\Gamma$$

from Lemma 5.1 it follows that

$$F(\lambda^{\delta\theta}x, \lambda^{-\delta\theta-s\theta}D; \lambda)\Lambda(\lambda^{\delta\theta}x, \lambda^{-\delta\theta-s\theta}D; \lambda)$$
$$= \lambda^{-t\theta/2}F^*(\lambda^{\delta\theta}x, \lambda^{-\delta\theta-s\theta}D; \lambda).$$

We construct an asymptotic null solution u with $\sigma_0 = s\theta + 2\delta\theta$ for $F^*(\lambda^{\delta\theta}x, \lambda^{-\delta\theta-s\theta}D; \lambda)$ and then verify that

$$\Lambda(\lambda^{\delta\theta}x, \lambda^{-\delta\theta-s\theta}D; \lambda)u$$

is nontrivial in such a way that $\Lambda(\lambda^{\delta\theta}x, \lambda^{-\delta\theta-s\theta}D; \lambda)u$ itself is actually an asymptotic null solution for $F(\lambda^{\delta\theta}x, \lambda^{-\delta\theta-s\theta}D; \lambda)$. Recall that

$$\det F^* = g_0^* + \cdots + \lambda^{-(r-t)\theta}g_{r-t}^* + \cdots$$

where $g_{r-t}^* = w(x, \xi)^{t-2\ell}h_\rho(x, \xi) + \cdots$ by Lemma 5.2.

Lemma 5.4. *We have* $s + 2 \leq t$ *and for* $0 \leq i < r - t$

$$g_i^*(x, \xi) = w(x, \xi)^{t-2\ell} h_i(x, \xi)$$

where $h_i(x, \xi)$ *are homogeneous polynomials of degree* $t + i - q$.

Thanks to Lemma 5.4 one can write

$$\det F^* = w^{t-2\ell}[h_0 + \lambda^{-\theta} h_1 + \cdots + \lambda^{-(r-t)\theta} h_\rho] + O(\lambda^{-(r-t+1)\theta})$$

where $h_j(x, \xi)$ are homogeneous polynomials of degree $t + j - q$. Let us define δ by

$$\delta = \max_{h_i \neq 0, 0 \leq i < r-t} \frac{r - t - i}{r - t - i + q} \tag{5.7}$$

so that $\delta \leq 1/2$ and

$$-i\theta + (t + i - q)\delta\theta \leq -(r - t)\theta + r\delta\theta$$

for $0 \leq i < r - t$ because $s + 2 \leq t$ and $q = s + 2$. Assume that the maximum in (5.7) is attained when $i = i^*$.

Lemma 5.5. *There is* $(\bar{x}, \bar{\xi}')$ *such that* $h_\rho(\bar{x}, \xi_0, \bar{\xi}') + h_{i^*}(\bar{x}, \xi_0, \bar{\xi}') = 0$ *has a simple non real root* τ *verifying* $w(\bar{x}, \tau, \bar{\xi}') \neq 0$.

Let ϕ be a solution to

$$\phi_{x_0} = \tau(x, \phi_{x'}(x)), \quad \phi(\bar{x}_0', x') = \langle \bar{\xi}', x' \rangle + i|x' - \bar{x}'|^2$$

where τ is given in Lemma 5.5. Let us consider

$$e^{-i\lambda^{s\theta+2\delta\theta}\phi(x)} F^*(\lambda^{\delta\theta} x, \lambda^{-\delta\theta-s\theta} D; \lambda) e^{i\lambda^{s\theta+2\delta\theta}\phi(x)}$$
$$= H^*(x, \lambda^{-\delta\theta-s\theta} D; \lambda). \tag{5.8}$$

Write

$$F^*(x, \xi; \lambda) = \sum_{k=0}^{\infty} \lambda^{-k\theta} F_k^*(x, \xi), \quad F_k^* = (F_{k,ij}^*)_{1 \leq i, j \leq 2}$$

and apply Lemma 4.4. Then we get

$$e^{-i\lambda^{s\theta+2\delta\theta}\phi(x)} F_{k,ij}^*(\lambda^{\delta\theta} x, \lambda^{-\delta\theta-s\theta} D) e^{i\lambda^{s\theta+2\delta\theta}\phi(x)}$$
$$= F_{k,ij}^*(\lambda^{\delta\theta} x, \lambda^{\delta\theta} \phi_x(x) + \lambda^{-\delta\theta-s\theta} D)$$
$$+ \lambda^{(m_{ij}+k-1)\delta\theta-s\theta-\delta\theta} R_{k,ij}(x, \lambda^{-\delta\theta-s\theta} D; \lambda)$$
$$= H_{k,ij}^*(x, \lambda^{-\delta\theta-s\theta} D; \lambda)$$

where $m_{i1} = \ell$ and $m_{i2} = t - 2\ell + \ell(t - 2\ell + 1)$ which follows from (5.5). We then see that the first column of $H_k^*(x, \xi; \lambda)$ is

$$\lambda^{-k\theta+(k+\ell)\delta\theta}[c_{i1}^k(x) + o(1)] + O(\lambda^{-k\theta+(k+\ell-2)\delta\theta-s\theta})$$

while the second column of $H_k^*(x, \xi; \lambda)$ is

$$\lambda^{-k\theta + [k + \ell(t - 2\ell) + t - \ell]\delta\theta}[c_{i2}^k(x) + o(1)] + O(\lambda^{-k\theta + [k + \ell(t - 2\ell) + t - \ell - 2]\delta\theta - s\theta}).$$

Taking these estimates into account we put

$$\Gamma^* = \begin{pmatrix} \lambda^{-\ell\delta\theta} & 0 \\ 0 & \lambda^{-[\ell(t - 2\ell) + t - \ell]\delta\theta} \end{pmatrix}$$

and set

$$H^*(x, \lambda^{-\delta\theta - s\theta} D; \lambda)\#\Gamma^* = G^*(x, \lambda^{-\delta\theta - s\theta} D; \lambda).$$

Then it is clear that

$$G_{k,j1}^*(x, \xi; \lambda) = \lambda^{-\ell\delta\theta} F_{k,j1}^*(\lambda^{\delta\theta} x, \lambda^{\delta\theta} \phi_x + \xi)$$
$$+ \lambda^{(k-2)\delta\theta - s\theta} R_{k,j1}(x, \xi; \lambda),$$

$$G_{k,j2}^*(x, \xi; \lambda) = \lambda^{-[\ell(t - 2\ell) + t - \ell]\delta\theta} F_{k,j2}^*(\lambda^{\delta\theta} x, \lambda^{\delta\theta} \phi_x + \xi)$$
$$+ \lambda^{(k-2)\delta\theta - s\theta} R_{k,j2}(x, \xi; \lambda).$$

Let us put

$$\tilde{s}\kappa\theta = \delta\theta + s\theta, \quad \kappa = \frac{1}{r - t - i^* + q}.$$

Lemma 5.6. *We have*

$$\det G^*(x, \xi; \lambda) = \lambda^{-\tilde{i}\kappa\theta}[g^*(x, \xi) + O(\lambda^{-\kappa\theta})]$$

with some $\tilde{i} < \tilde{s}$ where $g^*(x, 0) = 0$ and $\partial_{\xi_0} g^*(x, 0) \neq 0$.

We are now ready to prove (5.5). Let us take ψ so that

$$g^*(x, \psi_x) = 0$$

and consider

$$e^{-i\lambda^{\tilde{s}\kappa\theta} \psi(x)} G^*(x, \lambda^{-\tilde{s}\kappa\theta} D; \lambda) e^{i\lambda^{\tilde{s}\kappa\theta} \psi(x)}$$
$$= G^*(x, \psi_x + \lambda^{-\tilde{s}\kappa\theta} D; \lambda) + \lambda^{-\tilde{s}\kappa\theta} R^*(x, \lambda^{-\tilde{s}\kappa\theta} D; \lambda)$$
$$= G^{**}(x, \lambda^{-\tilde{s}\kappa\theta} D; \lambda).$$

Then it is clear that

$$\det G^{**}(x, \xi; \lambda) = \lambda^{-\tilde{i}\kappa\theta}[g^{**}(x, \xi) + O(\lambda^{-\kappa\theta})]$$

where $g^{**}(x, 0) = 0$ and $\partial_{\xi_0} g^{**}(x, 0) \neq 0$. Therefore in order to construct an asymptotic null solution for $G^{**}(x, \lambda^{-\tilde{s}\kappa\theta} D; \lambda)$ it suffices to apply Proposition 4.1.

Furthermore let

$$U(x, \lambda) = e^{i\lambda^{s\theta + 2\delta\theta} \Phi(x,\lambda)} \sum_{j=0} \lambda^{-\kappa\theta j} u_j(x)$$

be an asymptotic null solution of

$$F(\lambda^{\delta\theta} x, \lambda^{-\delta\theta - s\theta} D; \lambda) \Lambda(\lambda^{\delta\theta} x, \lambda^{-\delta\theta - s\theta} D; \lambda),$$

constructed as above, where

$$\Phi(x, \lambda) = \phi(x) + \lambda^{-\delta\theta} \psi(x) + \cdots.$$

In order to exhibit an asymptotic null solution for $F(\lambda^{\delta\theta} x, \lambda^{-\delta\theta - s\theta} D; \lambda)$ we must still show that $\Lambda(\lambda^{\delta\theta} x, \lambda^{-\delta\theta - s\theta} D; \lambda)U$ is nontrivial. Let us argue by contradiction: suppose that

$$e^{-i\lambda^{s\theta + 2\delta\theta} \Phi(x,\lambda)} \Lambda(\lambda^{\delta\theta} x, \lambda^{-\delta\theta - s\theta} D; \lambda) e^{i\lambda^{s\theta + 2\delta\theta} \Phi(x,\lambda)} u \sim 0.$$

This would imply that

$$w(\lambda^{\delta\theta} x, \lambda^{\delta\theta} \phi_x) = \lambda^{\ell\delta\theta} w(x, \phi_x) = 0,$$

which contradicts Lemma 5.4.

Assuming that $\mathrm{Det}\, F = O(\lambda^{-r\theta})$ we next show that

$$\sigma(\mathrm{Det}\, F) = h_\rho,$$

which clearly proves that $\sigma(\mathrm{Det}\, G) = h_\rho$. We argue in a way analogous to, but impler than, the above. Let us take $\delta = 0$ and construct an asymptotic solution for

$$G(x, \lambda^{-s\theta} D; \lambda).$$

The same arguments as in the proof of Lemma 5.2 and Lemma 5.3 show that

$$\det F^* = g_0^* + \lambda^{-\theta} g_1^* + \cdots, \quad g_0^* = w^{r-2\ell}(h_\rho + f)$$

where f is a homogeneous polynomial of degree $r - q$. Assuming $\mathrm{Det}\, G \neq h_\rho$ we would have $f \neq 0$. We repeat the same arguments as in the proof of Lemma 5.4 to conclude that there is a nonreal simple root τ of $h_\rho(\bar{x}, \tau, \bar{\xi}') + f(\bar{x}, \tau, \bar{\xi}') = 0$ verifying $w(\bar{x}, \tau, \bar{\xi}') \neq 0$. The rest of the proof is a repetition of the preceding arguments. $\qquad\square$

References

[1] E. Artin, *Geometric Algebra*, Interscience Publisher Inc., New York, 1957.

[2] S. Benvenuti, E. Bernardi and A. Bove, *The Cauchy problem for hyperbolic systems with multiple characteristics*, Bull. Sci. Math. **122** (1998), 603–634.

[3] A. Bove and T. Nishitani, *Necessary conditions for the well-posedness of the Cauchy problem for hyperbolic systems*, Osaka J. Math. **39** (2002), 149–179.

[4] A. Bove and T. Nishitani, *Necessary conditions for hyperbolic systems*, to appear in Bull. Sci. Math.

[5] A. D'Agnolo and G. Taglialatela, *Sato-Kashiwara determinant and Levi conditions for systems*, J. Math. Sci. Univ. Tokyo **7** (2000), 401–422.

[6] J. Dieudonné, *Les déterminants sur un corps non commutatif*, Bull. Soc. Math. France **71** (1943), 27–45.

[7] V. Ivrii and V.M. Petkov, *Necessary conditions for the correctness of the Cauchy problem for non-strictly hyperbolic equations*, Uspehi Mat. Nauk. **29** (1974), 3–70.

[8] W. Matsumoto, *The Cauchy problem for systems–through the normal form for systems and theory of weighted determinant*, Séminaire Equations aux derivées partielles, 1998–1999, Exp. no. XVIII, 30 pp, Ecole Polytechnique, Palaiseau.

[9] T. Nishitani, *Hyperbolicity of localizations*, Ark. Mat. **31** (1993), 377–393.

[10] T. Nishitani, *Strongly hyperbolic systems of maximal rank*, Publ. RIMS. **33** (1997), 765–773.

[11] T. Nishitani, *Necessary conditions for strong hyperbolicity of first order systems*, Jour. d'Analyse Math. **61** (1993), 181–229.

[12] M. Sato and M. Kashiwara, *The Determinant of matrices of pseudodifferential operators*, Proc. Japan Acad. **51** (1975), 17–19.

Antonio Bove
Università degli Studi di Bologna
Dipartimento di Matematica
Piazza di Porta S.Donato 5
40127, Bologna, Italy
and
Istituto Nazionale di Fisica Nucleare
Sezione di Bologna, Italy

Tatsuo Nishitani
Department of Mathematics
Osaka University
Machikaneyama 1-16, Toyonaka
Osaka 560-0043, Japan
tatsuo@math.wani.osaka-u.ac.jp

Monodromy of the Ramified Cauchy Problem

Renaud Camales and Claude Wagschal

ABSTRACT In this paper, we try to understand the action of the homotopy group on the solution of the ramified Cauchy problem. R. Camales has obtained very simple results about the spectrum of the monodromy. All the proofs will be given in [C].

1 Notation

Let X be a connected holomorphic manifold and let E be a complex Banach space, we denote by:

$\mathcal{R}(X)$: the universal covering space of X,

$\mathcal{H}(X; E)$: the space of holomorphic functions $f : X \to E$,

$\Gamma_a(X)$ where $a \in X$: the space of loops $\gamma : I \to X, \gamma(0) = \gamma(1) = a$,

$\mathcal{O}_a(X; E)$: the vector space of germs of holomorphic functions at a point $a \in X$.

If $u \in \mathcal{O}_a$ admits analytic continuation along all loops $\gamma \in \Gamma_a$, u_γ is the germ at the point a obtained by analytic continuation of u along γ and F_a^u is the vector subspace of $\mathcal{O}_a$ spanned by all the determinations $(u_\gamma)_{\gamma \in \Gamma_a}$. We define an automorphism $A_\gamma^u \in GL(F_a^u)$ by

$$A_\gamma^u : \theta \in F_a^u \mapsto \theta_\gamma \in F_a^u$$

which depends only on the homotopy class of the loop γ and so we obtain a linear representation of the homotopy group

$$A^u : [\gamma] \in \pi_1(X) \mapsto A_\gamma^u \in GL(F_a^u) \text{ where } \gamma \in [\gamma],$$

called the monodromy of the germ u.

If F_a^u is a finite-dimensional vector space, we say that u is of finite determination and we denote $\mathcal{O}_a^f$ the vector space of all germs of finite determination. Let $\sigma(A_\gamma^u)$ be the spectrum of A_γ^u, we put

$$\sigma_\gamma(u) = \sigma(A_\gamma^u) \text{ for } u \in \mathcal{O}_a^f.$$

Note. For example, let $u \in \mathcal{H}(X; E)$, then $\sigma_\gamma(u) = \{1\}$ if $u \neq 0$ and $\sigma_\gamma(u) = \emptyset$ if $u = 0$.

Remark 1.1. The monodromy of $u \in \mathcal{O}_a^f$ is said to be resoluble if there exists a basis of F_a^u such that, for all $\gamma \in \Gamma_a$, the matrix of A_γ^u in this basis is an upper triangular matrix. Let $\mathcal{O}_a^r$ be the set of all these germs.

2   Monodromy of an integro-differential Cauchy problem

We consider the following integro-differential problem :

$$\begin{cases} \left(D_0^m - A_m(x, D)\right)u(t, x) = \sum_{l \in \mathcal{L}} A_l^m(x, D) D_t^{-l} u(t, x) + w_m(t, x), \\[2mm] \left(D_0^h - A_h(x, D)\right)u(t, x) \\[2mm] = \sum_{l \in \mathcal{L}} A_l^h(x, D) D_t^{-l} u(t, x) + w_h(t, x) \text{ pour } x_0 = 0, 0 \le h < m \end{cases} \tag{2.1}$$

with the assumptions of [PW] :

$t \in \mathbf{C}, x = (x_0, \dots, x_n) \in \mathbf{C}^{n+1}$,

$\mathcal{L}$ is a finite subset of $\mathbf{Z}$,

E and F are two complex Banach spaces and $(a, u) \in E \times F \mapsto au \in F$ is a continuous bilinear map,

$A_h(x, D)$ and $A_l^h(x, D)$ are linear differential operators with holomorphic coefficients in an open neighbourhood of the origin of $\mathbf{C}^{n+1}$ with values in E and, for $0 \le h \le m$,

$$\text{order } A_h \le h, \quad \text{order}_{x_0} A_h < h, \tag{2.2}$$

$$\text{order } A_l^h \le \begin{cases} h + l - 1 & \text{si } l < 0, \\[2mm] h + l & \text{si } l \ge 0. \end{cases} \tag{2.3}$$

Let O be a connected open set of $\mathbf{C}$ and $b \in O$, for $u \in \mathcal{O}_{(b,0)}(O \times \mathbf{C}^{n+1})$, we define the primitive of u with respect to t by

$$D_t^{-1} u(t, x) = \int_b^t u(\tau, x) \, d\tau.$$

Then, we have the following theorem [PW].

Theorem 2.1. *Let Ω be a simply connected open neighbourhood of the origin of $\mathbf{C}^{n+1}$. Then there exists a simply connected open neighbourhood $\Omega' \subset \Omega$ of the origin of $\mathbf{C}^{n+1}$ and $\delta > 0$ as follows : let O be a connected open subset of $\mathbf{C}$ whose diameter is $\le \delta$, $w_h : \mathcal{R}(O) \times \Omega \to F$, $0 \le h \le m$, holomorphic functions; then the problem (2.1) has a unique holomorphic solution $u : \mathcal{R}(O) \times \Omega' \to F$.*

Put $w = (w_h)_{0 \le h \le m}$ and denote by $u = Sw$ the solution of (2.1). Consider w as a germ at the point $(b, 0) \in O \times \mathbf{C}^{n+1}$.

Suppose the germs w_h for $0 \le h < m$ are independent of x_0.

Theorem 2.2. *If the group $\pi_1(O)$ is of finite type, we have*

$$w \in \mathcal{O}_{(b,0)}^f(O \times \Omega; F^{m+1}) \implies u \in \mathcal{O}_{(b,0)}^f(O \times \Omega'; F),$$

$$w \in \mathcal{O}_{(b,0)}^r(O \times \Omega; F^{m+1}) \implies u \in \mathcal{O}_{(b,0)}^r(O \times \Omega'; F),$$

and

$$\sigma_\gamma(u) \cup \{1\} = \sigma_\gamma(w) \cup \{1\} \text{ for all } \gamma \in \Gamma_{(b,0)}(O \times \Omega').$$

3 Monodromy of the ramified Cauchy problem

Let $a(x, D)$ be a linear differential operator with holomorphic coefficients in a neighbourhood of the origin of $\mathbf{C}^{n+1}$; we suppose that the hyperplane $S : x_0 = 0$ is not characteristic and that the operator a has multiple characteristics of constant multiplicity in the sense of $[HLW]$ or $[PW]$. We write $K_i : k_i(x) = 0, 1 \le i \le d$, the characteristic hypersurfaces issued from $T : x_0 = x_1 = 0$ and

$$K = \bigcup_{i=1}^{d} K_i.$$

We consider the Cauchy problem

$$\begin{cases} a(x, D)u(x) = v(x), \\ D_0^h u(x) \quad = w_h(x') \text{ pour } x_0 = 0, \ 0 \le h < m. \end{cases} \tag{3.1}$$

Theorem 3.1. ($[L]$, $[PW]$) *Let Ω be a simply connected open neighbourhood of the origin of $\mathbf{C}^{n+1}$ such that $\Omega \cap S$ is connected. Then there exists a simply connected open neighbourhood $\Omega' \subset \Omega$ of the origin of $\mathbf{C}^{n+1}$ as follows: let $a \in \Omega' \cap S - T$ and $v, w_h \in \mathcal{O}_a$ germs which admit analytic continuation to holomorphic functions $v : \mathcal{R}(\Omega - K) \to \mathbf{C}$ and $w_h : \mathcal{R}(\Omega \cap S - T) \to \mathbf{C}$; then the problem (3.1) has a unique solution $u \in \mathcal{O}_a$ which admits an extension in a holomorphic function on $\mathcal{R}(\Omega' - K)$.*

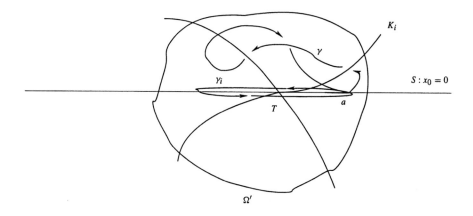

With the same assumptions, we can choose the neighbourhood Ω' as follows:

Theorem 3.2. *If the germs v and w_h are of finite determination, the germ u is also of finite determination and*

$$\sigma_\gamma(v) \subset \sigma_\gamma(u) \subset \sigma_\gamma(v) \cup \bigcup_{i=1}^{d} \sigma_{\gamma_i}(v) \cup \bigcup_{i=1}^{d} \sigma_{\gamma_i}(w) \text{ for all } \gamma \in \Gamma_a(\Omega' - K) \quad (3.2)$$

where $w = (w_h)_{0 \le h < m}$ and $\gamma_i \in \Gamma_a(\Omega' \cap S - T)$ is the loop $s \mapsto (0, k_i(\gamma(s)),$ $a_2, \ldots, a_n)$,

$$\sigma_\delta(w) \subset \sigma_\delta(u) \text{ for all } \delta \in \Gamma_a(\Omega' \cap S - T).$$

If the monodromy of v and w is resoluble, the monodromy of u is resoluble.

4   Methods

The arguments used by R. Camales are very ingenious. From an algebraic viewpoint, we need the following elementary lemma.

Lemma 4.1. *Let F be a finite-dimensional vector space, $A \in \mathcal{L}(E)$ an endomorphism which admits in a system of generators of F a representation by a matrix M; then $\sigma(A) \subset \sigma(M)$.*

For example, let $u, v \in \mathcal{O}_a^f(X)$, (φ_i) a basis of F_a^u and (ψ_j) a basis of F_a^v. We have

$$F_a^{u+v} \subset F_a^u + F_a^v,$$

so $u + v$ is of finite determination ; the vector space $F_a^u + F_a^v$ is invariant by analytic continuation and in the system of generators (φ_i, ψ_j) the automorphism $\theta \mapsto \theta_\gamma$ admits a representation of the form

$$\begin{pmatrix} M_\gamma^u & 0 \\ 0 & M_\gamma^v \end{pmatrix}$$

where M_γ^u and M_γ^v are the matrices of A_γ^u and A_γ^v in the basis (φ_i) and (ψ_j). This proves that

$$\sigma_\gamma(u + v) \subset \sigma_\gamma(u) \cup \sigma_\gamma(v).$$

Sketch of the proof of Theorem 3.1. We can write

$$(D_0^m - A_m)u_\gamma = \sum_{l \in \mathcal{L}_-} A_l^m D_t^{-l} u_\gamma + \sum_{l \in \mathcal{L}_+} A_l^m (D_t^{-l} u)_\gamma + (w_m)_\gamma$$

where

$$\mathcal{L}_+ = \{l \in \mathcal{L} ; l > 0\}, \quad \mathcal{L}_- = \{l \in \mathcal{L} ; l \le 0\}.$$

If we put

$$P_\gamma^l u = (D_t^{-l} u)_\gamma - D_t^{-l}(u_\gamma), l > 0,$$

we obtain

$$(D_0^m - A_m)u_\gamma = \sum_{l \in \mathcal{L}} A_l^m D_t^{-l} u_\gamma + \sum_{l \in \mathcal{L}_+} A_l^m P_\gamma^l u + (w_m)_\gamma$$

and we can make the same thing for the Cauchy data. This prove that

$$u_\gamma = u_1^\gamma + u_2^\gamma$$

where

$$u_1^\gamma = S(w_\gamma), \; u_2^\gamma = S(Q_\gamma u)$$

with

$$(Q_\gamma u)_h = \sum_{l \in \mathcal{L}_+} A_l^h(x, D) P_\gamma^l u.$$

1. Let $(\theta_j)_{1 \le j \le q}$ be a basis of $F_{(b,0)}^w$, $u_j = S\theta_j$ and let H be the finite dimensional vector space spanned by (u_j); u_1^γ belongs to H.

2. Let C be the unbounded component of $\mathbf{C} - O$; the open $O' = \mathbf{C} - C$ is simply connected and diam $O' =$ diam $O \le \delta$. Futhermore, the germ $Q_\gamma u$ is a polynomial in t with holomorphic coefficients in Ω'. By the existence theorem, with a smaller Ω', we deduce that u_2^γ is holomorphic in $O' \times \Omega'$ and a fortiori in $O \times \Omega'$.

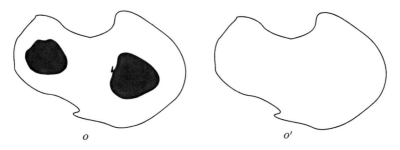

$$O \qquad\qquad\qquad O'$$

Next, there exist loops $\gamma_i \in \Gamma_{(b,0)}(O \times \Omega')$, $1 \le i \le p$, such that all loops $\gamma \in \Gamma_{(b,0)}(O \times \Omega')$ are homotopic to a loop of the form

$$\gamma_{i_1} \cdots \gamma_{i_k}, \; 1 \le i_j \le p.$$

We put

$$u_{ij} = S(Q_{\gamma_i} u_j) \in \mathcal{H}(O \times \Omega'), 1 \le i \le p, 1 \le j \le q.$$

We can prove that u_2^γ belongs to the vector space G spanned by (u_{ij}). This proves that

$$F_{(b,0)}^u \subset G + H \, ;$$

thus, u is of finite determination.

The vector space $G + H$ is invariant by analytic continuation and the automorphism $\theta \mapsto \theta_\gamma$ has in the system of generators (u_{ij}, u_j) a representation by a matrix of the form

$$\begin{pmatrix} Id & B \\ 0 & M_\gamma^w \end{pmatrix}$$

where M_γ^w is the matrix of A_γ^w in the basis (θ_j). So, we obtain that

$$\sigma_\gamma(u) \subset \sigma_\gamma(G + H) \subset \sigma_\gamma(w) \cup \{1\},$$

which proves the theorem.

For Theorem 3.2, one of the difficulties is to prove that 1 does not appear in the formula (3.2). For this, we use the curious lemma which follows.

Lemma 4.2. *Let X be a connected holomorphic manifold such that the group $\pi_1(X)$ is isomorphic to $\mathbf{Z}$ and generated by the homotopy class of a loop γ and let $u \in \mathcal{O}_a^f(X)$ such that $1 \notin \sigma_\gamma(u)$ and $v \in \mathcal{H}(X)$ is a holomorphic function, then*

$$1 \notin \sigma_\gamma(u + v) \Longleftrightarrow v = 0.$$

References

[C]   Camales, R., Monodromie du problème de Cauchy ramifié, in preparation.

[HLW] Hamada, Y., Leray, J. et Wagschal, C., Systèmes d'équations aux dérivées partielles à caractéristiques multiples : problème de Cauchy ramifié ; hyperbolicité partielle, *J. Math. Pures Appl.* **55** (1976), 297–352.

[L]   Leichtnam, E., Le problème de Cauchy ramifié, *Ann. scient. Éc. Norm. Sup.* **23** (1990), 369–443.

[PW]   Pongérard, P. and Wagschal, C., Ramification non abélienne, *J. Math. Pures Appl.* **77** (1998), 51–88.

Need affiliation for Renaud Camales

Claude Wagschal
24, rue Luchet
31200 Toulouse
wagschal@mip.ups-tlse.fr

Nonlinear Stability of an Expanding Universe with the S^1 Isometry Group

Yvonne Choquet-Bruhat and Vincent Moncrief

ABSTRACT We prove the existence for an infinite proper time in the expanding direction of spacetimes satisfying the vacuum Einstein equations on a manifold of the form $\Sigma \times S^1 \times R$ where Σ is a compact surface of genus $G > 1$. The Cauchy data are supposed to be invariant with respect to the group S^1 and sufficiently small, but we do not impose a restrictive hypothesis made in the previous work [1].

1 Introduction

An Einsteinian universe is a pair $(V, {}^{(4)}g)$, with V a smooth 4-dimensional manifold and ${}^{(4)}g$ a Lorentzian metric on V which satisfies the Einstein equations. Such a universe satisfies the classical causality requirements if it is globally hyperbolic, equivalently if V is a product $V = M \times R$ with each $M_t \equiv M \times \{t\}$ space like and a Cauchy surface, i.e., intersected once by each causal curve (timelike or null). It is well known that given a 3-dimensional manifold endowed with a properly Riemannian metric $\bar{g}$ and a symmetric 2-tensor K satisfying the Einstein constraint equations, there exists (modulo appropriate functional hypotheses on the data) a globally hyperbolic vacuum (the theorem extends to classical sources which admit of a well-posed Cauchy problem) Einsteinian universe such that M_{t_0} is a Cauchy surface, and the spacetime metric ${}^{(4)}g$ induces on M_{t_0} the metric $\bar{g}$ while K is the extrinsic curvature of M_{t_0}. This solution is unique, up to isometry, in the class of maximal spacetimes (i.e., which cannot be embedded in a larger one). In spite of its formulation using R this solution is a local one: t is just a coordinate, it has no intrinsic meaning: the physically meaningful quantity is the proper time, determined by the metric ${}^{(4)}g$. The main problems which remain open in this field are the infinite proper time existence, or the formation of singularities and their nature.

In this article we will prove the existence for an infinite proper time, in the expanding direction, of vacuum Einsteinian universes with Cauchy data which are in a neighbourhood of a vacuum Einsteinian universe defined as follows.

- $V = M \times R$ is such that M is a compact manifold of the form $M = S^1 \times \Sigma$,

Supported in part by the NSF contract no. PHY-9732629 and PHY-0098084 of Yale University.

with Σ a smooth, orientable surface.

- The spacetime metric is invariant under the group S^1. It is given by

$$^{(4)}g = -4dt^2 + 2t^2\sigma + \theta^2 \tag{1.1}$$

with σ a metric on Σ independent of t and of scalar curvature -1, and θ a 1-form on S^1. In local coordinates x^a on Σ and dx^3 on S^1 we have

$$\sigma = \sigma_{ab}dx^a dx^b, \quad a,b = 1,2 \quad \theta = dx^3.$$

The property

$$R(\sigma) = -1 \tag{1.2}$$

implies, by the Gauss–Bonnet theorem, that the surface Σ has genus $G > 1$.

Remark. (Important fact for the Thurston classification.) The Lorentzian 3-metric $-4dt^2 + 2t^2\sigma$ is homogeneous in t, but not the 4-metric.

The above universe is a particular case of the ones described in the next section.

2 S^1 invariant Einsteinian universes

The spacetime manifold is a product $S^1 \times \Sigma \times R$, where Σ is a smooth, compact, orientable 2-manifold of genus $G > 1$. The spacetime 4-metric is invariant under the action of the group S^1, with spacelike orbits $S^1 \times \{x\} \times \{t\}$. We restrict our study here to the so-called *polarized case* where the orbits are orthogonal to 3-dimensional Lorentzian sections. The metric can be written, without loss of generality and for later convenience, in the form

$$^{(4)}g = e^{-2\gamma}\,^{(3)}g + e^{2\gamma}(\theta)^2, \tag{2.1}$$

with γ a scalar function and $^{(3)}g$ an arbitrary Lorentzian 3-metric on $\Sigma \times R$ which we write under the usual form adapted to the Cauchy $2 + 1$ splitting

$$^{(3)}g = -N^2 dt^2 + g_{ab}(dx^a + v^a dt)(dx^b + v^b dt); \tag{2.2}$$

$N, v, g = g_{ab}dx^a dx^b$ are t dependent scalar (lapse), vector (shift), metric on Σ and $\theta = dx^3$ a 1-form on S^1, x^3 a periodic coordinate. We denote by ∂_α the Pfaff derivatives in the coframe $\theta^0 = dt$, $\theta^a = dx^a + v^a dt$, Greek indices taking the values 0, 1 or 2. It holds that

$$\partial_0 = \frac{\partial}{\partial t} - v^a \frac{\partial}{\partial x^a}. \tag{2.3}$$

We introduce the extrinsic curvature k_t of Σ_t i.e., we set

$$k_{ab} = -\frac{1}{2N}\hat{\partial}_0 g_{ab}, \tag{2.4}$$

with $\hat{\partial}_0$ the operator on time dependent tensors on Σ given by

$$\hat{\partial}_0 = \frac{\partial}{\partial t} - L_{v_t} \tag{2.5}$$

with L_{v_t} the Lie derivative with respect to v_t. The mean curvature τ of Σ_t, which will play a fundamental role in our later estimates, is

$$\tau \equiv g^{ab} k_{ab}. \tag{2.6}$$

3 Equations

The metric $^{(4)}g$ is supposed to satisfy the vacuum Einstein equations,

$$Ricci(^{(4)}g) = 0.$$

The equations $^{(4)}R_{\alpha 3} = 0$ are identically satisfied by the metric (2.1).

3.1 Wave equation for γ

The equation $^{(4)}R_{33} = 0$ implies that the function γ satisfies the wave equation on $(\Sigma \times R, ^{(3)}g)$. This equation reads

$$-N^{-1}\partial_0(N^{-1}\partial_0\gamma) + N^{-1}g^{ab}\nabla_a(N\partial_b\gamma) + N^{-1}\tau\partial_0\gamma = 0. \tag{3.1}$$

3.2 3-dimensional Einstein equations

When $^{(4)}R_{3\alpha} = 0$ and $^{(4)}R_{33} = 0$ the equations $^{(4)}R_{\alpha\beta} = 0$ are equivalent to Einstein's equations on $\Sigma \times R$ for the metric $^{(3)}g$ with source the stress energy tensor of the scalar field γ, namely

$$^{(3)}R_{\alpha\beta} = 2\partial_\alpha\gamma\partial_\beta\gamma . \tag{3.2}$$

In dimension 3, the Einstein equations are nondynamical, except for the conformal class of g. We set on $\Sigma \times R$,

$$g_{ab} = e^{2\lambda}\sigma_{ab} .$$

We impose that, on each Σ_t, σ_t has scalar curvature $R(\sigma_t) = -1$, i.e., $\sigma_t \in M_{-1}$, which is no restriction since any metric σ on Σ, which is of genus greater than 1, is conformal to such a metric, and $e^{2\lambda}$ is to be determined.

As a gauge condition we suppose that the mean extrinsic curvature τ is constant on each Σ_t, i.e., depends only on t. We will construct an expanding space time, i.e., we take τ to be negative and increasing from a value $\tau_0 < 0$. The moment of maximum expansion will be attained if τ reaches the value $\tau = 0$. For convenience we define the parameter t by

$$t = -\tau^{-1}, \quad t \geq t_0. \tag{3.3}$$

The following equations (momentum constraint) hold on Σ_t:

$$-Ne^{-2\lambda(3)}R_{0a} \equiv D_b h_a^b = L_a \equiv -2D_a \gamma \dot{\gamma} \tag{3.4}$$

with $h_{ab} = k_{ab} - \frac{1}{2}g_{ab}\tau$ the traceless part of k_t, D_a the covariant derivative in the metric σ_t and indices raised with σ^{ab}, and we have set

$$\dot{\gamma} = e^{2\lambda}\gamma' \text{ with } \gamma' = N^{-1}\partial_0\gamma.$$

Given σ, γ and $\dot{\gamma}$, this is a linear equation for h on Σ_t, independent of λ. The general solution is the sum of a solution q of the homogeneous equation, a trace and divergence free tensor called a TT tensor, and a conformal Lie derivative r. Such tensors are $L^2(\sigma)$ orthogonal.

The so-called Hamiltonian constraint on Σ_t, on the other hand, is given by

$$2N^{-2(3)}S_{00} \equiv R(^{(3)}g) - g^{ac}g^{bd}h_{bc}h_{ad} + \frac{1}{2}\tau^2$$
$$= 2(N^{-2}\partial_0\gamma\partial_0\gamma + g^{ab}\partial_a\gamma\partial_b\gamma). \tag{3.5}$$

When $g_{ab} = e^{2\lambda}\sigma_{ab}$ it becomes a semilinear elliptic equation in λ:

$$\Delta\lambda = f(x, \lambda) \equiv p_1 e^{2\lambda} - p_2 e^{-2\lambda} + p_3, \tag{3.6}$$

with

$$p_1 \equiv \frac{1}{4}\tau^2, \, p_2 \equiv |\dot{u}|^2 + \frac{1}{2}|h|^2), \, p_3 \equiv -(\frac{1}{2} + |Du|^2).$$

The equation $^{(3)}R_{00} = \rho_{00}$ gives for N the linear elliptic equation

$$\Delta N - \alpha N = -e^{2\lambda}\partial_t\tau \text{ with } \alpha \equiv e^{-2\lambda}(|h|^2 + |\dot{u}|^2) + \frac{1}{2}e^{2\lambda}\tau^2. \tag{3.7}$$

The shift ν satisfies the equation (resulting from the expression for h)

$$(L_\sigma n)_{ab} \equiv D_a n_b + D_b n_a - \sigma_{ab}D_c v^c = f_{ab} \text{ with } n_a \equiv v_a e^{-2\lambda}, \tag{3.8}$$

$$f_{ab} \equiv 2Ne^{-2\lambda}h_{ab} + \partial_t\sigma_{ab} - \frac{1}{2}\sigma_{ab}\sigma^{cd}\partial_t\sigma_{cd}.$$

We require the metric σ_t to lie in some chosen cross section $Q \to \psi(Q)$ of the fiber bundle $M_{-1} \to T_{eich}$ with T_{eich} (Teichmüller) the space of classes of conformally inequivalent Riemannian metrics, identified with R^{6G-6}. The solvability condition for the shift equation determines dQ/dt in terms of h. One obtains an ordinary differential system for the evolution of Q by requiring that the t-dependent spatial tensor $^{(3)}R_{ab} - 2\partial_a u\partial_b u$, which is TT by the previously solved equations, be L^2-orthogonal to TT tensors.

4   Local existence theorem

The Cauchy data on Σ_{t_0} are:

1. A C^∞ Riemannian metric σ_0 which projects onto a point $Q(t_0)$ of T_{eich} and a C^∞ tensor q_0 which is TT in the metric σ_0. The spaces W_s^p and $H_s \equiv W_s^2$ are the usual Sobolev spaces of tensor fields on the Riemannian manifold (Σ, σ_0).

2. Cauchy data for γ and γ on Σ_0, i.e.,

$$\gamma(t_0, .) = \gamma_0 \in H_2, \, \dot\gamma(t_0, .) = \dot\gamma_0 \in H_1$$

Theorem 4.1. *The Cauchy problem with the above data for the Einstein equations with S^1 isometry group (polarized case) has, if $T - t_0$ is small enough, a solution with $\gamma \in C^0([t_0, T], H_2) \cap C^1([t_0, T], H_1)$; $\lambda, N, \nu \in C^0([t_0, T], W_3^p)$, $1 < p < 2$ and $N > 0$ while $\sigma \in C^1([t_0, T], C^\infty)$ with σ_t uniformly equivalent to σ_0. This solution is unique up to t parametrization of τ and choice of a cross section of M_{-1} over T_{eich}.*

The proof is by solving alternately elliptic systems for h, λ, N, ν on each Σ_t, the wave equation for u, and the differential system satisfied by Q. The iteration converges if $T - t_0$ is small enough.

We will prove that the solution exists for all $t \geq t_0 > 0$, and for an infinite proper time, by obtaining a priori estimates of the various norms, and of a strictly positive lower bound independent of T for N.

5   Energy estimates

We omit the index t when the context is clear. We denote by $\| \cdot \|_\sigma$ and $\| \cdots \|_g$ the L^2 norms in the σ and g metric. We denote by C_σ numbers depending only on (Σ, σ). A lower case index m or M denotes respectively the lower or upper bound of a scalar function on Σ_t. It follows from the equations satisfied by N and λ, by the maximum principle, that

$$0 < N_m \leq N_M \leq 2\frac{\partial_t \tau}{\tau^2} \quad \text{and} \quad e^{-2\lambda_m} \leq \frac{1}{2}\tau^2. \tag{5.1}$$

5.1   First energy estimate

Inspired by the Hamiltonian constraint we define the *energy* by

$$E(t) \equiv \|\gamma'\|_g^2 + \|D\gamma\|_g^2 + \frac{1}{2}\|h\|_g^2. \tag{5.2}$$

By integrating the Hamiltonian constraint over (Σ_t, g) and using the constancy of τ and the Gauss–Bonnet theorem we find, with χ the Euler characteristic of Σ, that

$$E(t) = \frac{\tau^2}{4} Vol_g(\Sigma_t) + 4\pi\chi \tag{5.3}$$

and after some manipulations, that

$$\frac{dE(t)}{dt} = \tau \int_t (\frac{1}{2}|h|_g^2 + |\gamma'|^2) N \mu_g. \tag{5.4}$$

We see that $E(t)$ is a nonincreasing function of t if τ is negative.

5.2   Second energy estimate

We define the *energy* of *gradient* γ by

$$E^{(1)}(t) \equiv \int_{\Sigma_t} (J_0 + J_1)\mu_g, \quad J_1 = |\Delta_g \gamma|^2, \quad J_0 = |D\gamma'|^2. \tag{5.5}$$

After a lengthy but straightforward calculation we have found (see [1], indices raised with g)

$$\frac{dE^{(1)}}{dt} - 2\tau E^{(1)} = \tau \int_{\Sigma_t} \{N J_0 + (N-2)(J_0 + J_1)\}\mu_g + Z, \tag{5.6}$$

$$Z \equiv 2 \int_{\Sigma_t} \{N h^{ab} \partial_a u' \partial_b \gamma' + 2N h^{ab} \nabla_a \partial_b \gamma \Delta_g \gamma + (\nabla_b(\partial^a N \partial_a \gamma)$$
$$+ \tau \partial_b N \gamma')(\partial^b \gamma') + [(2\partial_a N h^{ac} - 4N\partial^c \gamma \gamma')\partial_c \gamma$$
$$+ 2\partial^a N \partial_a \gamma' + \gamma' \Delta_g N]\Delta_g \gamma\}\mu_g. \tag{5.7}$$

We set

$$E(t) = \varepsilon^2, \quad E^{(1)}(t) = \tau^{-2}\varepsilon_1^2. \tag{5.8}$$

6   Elliptic estimates

We have shown in [1] that if the sum $\varepsilon^2 + \varepsilon_1^2$ is bounded by a number c — *hypothesis* H_c — the quantities N, h, λ, v can be bounded on Σ_t by using the elliptic equations they satisfy. In particular, denoting by λ_M the supremum of λ on Σ_t,

$$\frac{1}{\sqrt{2}}|\tau|e^{\lambda_M} \leq 1 + CC_{\sigma_t}(\varepsilon^2 + \varepsilon_1^2) \tag{6.1}$$

where C denotes a number depending on c, and C_{σ_t} a Sobolev constant of (Σ_t, σ_t). We have also obtained estimates for h, and for N, namely

$$\|h\|_{L^\infty(g_t)} \leq CC_{\sigma_t}|\tau|\{\varepsilon + (\varepsilon + \varepsilon_1)^2\}, \tag{6.2}$$

$$0 \leq 2 - N_m \leq CC_{\sigma_t}(\varepsilon^2 + \varepsilon\varepsilon_1), \quad \|DN\|_{L^\infty(g_t)} \leq CC_{\sigma_t}|\tau|(\varepsilon^2 + \varepsilon\varepsilon_1). \tag{6.3}$$

All these estimates would be sufficient to prove that the energies remain uniformly bounded for all $t \geq t_0$, if small enough initially, if we knew a uniform (in t) bound of the Sobolev constants C_{σ_t}. We can obtain such a bound only if the total energy, $\varepsilon^2 + \varepsilon_1^2$ decays when t increases.

7   Corrected energy estimates

The decay we are looking for will be obtained through the introduction of "corrected energies" whose t derivatives take advantage of the negative part of the derivative of the corresponding original energy to give a negative definite contribution.

7.1   First corrected energy

One defines a *corrected first energy* by the formula, where α is a positive number,

$$E_\alpha(t) = E(t) - 2\alpha\tau \int_{\Sigma_t} (\gamma - \bar{\gamma})\gamma'\mu_g, \quad \text{with} \quad \bar{\gamma} = \frac{1}{Vol_\sigma\Sigma_t} \int_{\Sigma_t} \gamma\mu_\sigma. \quad (7.1)$$

We estimate the complementary term using the Poincaré inequality which gives

$$\|\gamma - \bar{\gamma}\|_{g_t} \le e^{\lambda_M}\|\gamma - \bar{\gamma}\|_{\sigma_t} \le e^{\lambda_M}\Lambda_{\sigma_t}^{-1/2}\|D\gamma\|_{\sigma_t}. \quad (7.2)$$

Therefore, by the Cauchy–Schwarz inequality, since $\|D\gamma\|_{\sigma_t} = \|D\gamma\|_{g_t}$,

$$|\tau \int_{\Sigma_t} \gamma'(\gamma - \bar{\gamma})\mu_g| \le |\tau|e^{\lambda_M}\Lambda_{\sigma_t}^{-1/2}\|\gamma'\|_g\|D\gamma\|_g. \quad (7.3)$$

We deduce from this inequality that

$$E(t) \le \frac{1}{1-a_t}E_\alpha(t) \quad \text{with} \quad a_t \equiv \frac{\alpha|\tau|e^{\lambda_M}}{\Lambda_{\sigma_t}^{\frac{1}{2}}}.$$

We will have $a_t < 1$ if

$$\alpha < \frac{\Lambda_{\sigma_t}^{\frac{1}{2}}}{|\tau|e^{\lambda_M}}. \quad (7.4)$$

We have seen that there exist numbers C and C_σ such that

$$|\tau|e^{\lambda_M} \le \sqrt{2}(1 + CC_\sigma(\varepsilon^2 + \varepsilon_1^2)). \quad (7.5)$$

We suppose that there exist numbers $\Lambda > 0$ and $\delta > 0$, independent of t, such that for all t it holds that

$$\Lambda_{\sigma_t}^{\frac{1}{2}} \ge \Lambda^{\frac{1}{2}}(1+\delta); \quad (7.6)$$

then it holds that

$$\frac{\Lambda_{\sigma_t}^{\frac{1}{2}}}{|\tau|e^{\lambda_M}} \ge \frac{\Lambda^{\frac{1}{2}}(1+\delta)}{\sqrt{2}(1 + CC_\sigma(\varepsilon^2 + \varepsilon_1^2))} > \frac{1}{\sqrt{2}}\Lambda^{\frac{1}{2}} \quad (7.7)$$

as soon as

$$CC_\sigma(\varepsilon^2 + \varepsilon_1^2) < \delta. \quad (7.8)$$

When this inequality is satisfied we can choose any number α such that

$$\alpha \leq \frac{1}{\sqrt{2}} \Lambda^{\frac{1}{2}} \tag{7.9}$$

and so insure that $a_t < 1$. For instance if

$$CC_\sigma(\varepsilon^2 + \varepsilon_1^2) < \frac{\delta}{2}, \tag{7.10}$$

then

$$1 - a_t \equiv 1 - \frac{\alpha|\tau|e^{\lambda M}}{\Lambda_{\sigma_t}^{\frac{1}{2}}} \geq \frac{\delta}{2(1+\delta)}. \tag{7.11}$$

7.2 Decay of the corrected energy

We set

$$\frac{dE_\alpha}{dt} = \frac{dE}{dt} - R_\alpha$$

with (the terms explicitly containing the shift ν give an exact divergence which integrates to zero)

$$R_\alpha = 2\alpha\tau \int_{\Sigma_t} \{\partial_0\gamma'(\gamma - \gamma) + \gamma'\partial_0(\gamma - \bar{\gamma}) - N\tau\gamma'(\gamma - \bar{\gamma})\}\mu_g$$

$$+ 2\alpha\frac{d\tau}{dt} \int_{\Sigma_t} \gamma'(\gamma - \bar{\gamma})\mu_g. \tag{7.12}$$

To simplify the writing we suppose that $\int_{\Sigma_t} \gamma'\mu_g = 0$; this quantity is conserved in time if γ satisfies the wave equation (3.1). Some elementary computations using (3.1) and integration by parts show that, using also $\frac{d\tau}{dt} = \tau^2$,

$$R_\alpha = 2\alpha\tau \int_{\Sigma_t} \{[|\gamma'|^2 - |D\gamma|_g^2]N + \tau(\gamma - \bar{\gamma})\gamma'\}\mu_g. \tag{7.13}$$

We write $\frac{dE_\alpha}{dt}$ in the form

$$\frac{dE_\alpha}{dt} = \tau \int_{\Sigma_t} \{2[\frac{1}{2}|h|^2 + (1 - 2\alpha)|\gamma'|^2 + 2\alpha|D\gamma|_g^2]$$

$$- 2\alpha\tau\gamma'(\gamma - \bar{\gamma})\}\mu_g + \tau A \tag{7.14}$$

where A can be estimated with higher order terms in the energies, using the inequality (6.3) satisfied by $N - 2$, since A reads

$$A \equiv \int_{\Sigma_t} (N - 2)[\frac{1}{2}|h|^2 + (1 - 2\alpha)|\gamma'|^2 + 2\alpha|D\gamma|_g^2]\mu_g. \tag{7.15}$$

We look for a positive number k such that the difference $\frac{dE_\alpha}{dt} - k\tau E_\alpha$ can be estimated with higher order terms in the energies. We have

$$\frac{dE_\alpha}{dt} - k\tau E_\alpha = 2\tau \int_{\Sigma_t} \{[\frac{1}{2}|h|^2 + (1 - 2\alpha - \frac{k}{2})|\gamma'|^2 + (2\alpha - \frac{k}{2})|D\gamma|_g^2] \quad (7.16)$$

$$-\alpha(1 - k)\tau\gamma'(\gamma - \bar\gamma)\}\mu_g + \tau A. \quad (7.17)$$

We have treated in [1] the case where $\Lambda \geq \frac{1}{8}$, $\alpha = \frac{1}{4}$. In this case it is possible to take $k = 1$ and obtain immediately

$$\frac{dE_{\frac{1}{4}}}{dt} - \tau E_{\frac{1}{4}} \leq |\tau A|. \quad (7.18)$$

In the general case we have

$$\frac{dE_\alpha}{dt} - k\tau E_\alpha \leq 2\tau \int_{\Sigma_t} \{[(1 - 2\alpha - \frac{k}{2})|\gamma'|^2 + (2\alpha - \frac{k}{2})|D\gamma|_g^2] \quad (7.19)$$

$$-\alpha(1 - k)\tau\gamma'(\gamma - \bar\gamma)\}\mu_g + |\tau A|. \quad (7.20)$$

The estimate (7.3) together with the inequality (7.7) gives

$$|\tau \int_{\Sigma_t} \gamma'(\gamma - \bar\gamma)\mu_g| \leq \sqrt{2}\Lambda^{-1/2}\|\gamma'\|_g\|D\gamma\|_g. \quad (7.21)$$

Therefore it holds that

$$\int_{\Sigma_t} \{[(1 - 2\alpha - \frac{k}{2})|\gamma'|^2 + (2\alpha - \frac{k}{2})|D\gamma|_g^2] - \alpha(1 - k)\tau\gamma'(\gamma - \bar\gamma)\}\mu_g \geq$$

$$(1 - 2\alpha - \frac{k}{2})\|\gamma'\|_g^2 + (2\alpha - \frac{k}{2})\|D\gamma\|_g^2 - \alpha(1 - k)\sqrt{2}\Lambda^{-1/2}\|\gamma'\|_g\|D\gamma\|_g. \quad (7.22)$$

The above quadratic form in $\|\gamma'\|_g$ and $\|D\gamma\|_g$ is nonnegative if

$$k \leq 4\alpha, \quad k \leq 2(1 - 2\alpha) \quad (7.23)$$

and k is such that

$$2\alpha^2\Lambda^{-1}(1 - k)^2 - 4(2\alpha - \frac{k}{2})(1 - 2\alpha - \frac{k}{2}) \leq 0; \quad (7.24)$$

the inequalities (7.23) imply that

$$k \leq 1. \quad (7.25)$$

The inequality (7.24) reads

$$(1 - 2\Lambda^{-1}\alpha^2)k^2 - (1 - 2\Lambda^{-1}\alpha^2)2k - 2\Lambda^{-1}\alpha^2 + 8\alpha(1 - 2\alpha) > 0. \quad (7.26)$$

That is, since $1 - 2\Lambda^{-1}\alpha^2 > 0$,

$$k^2 - 2k + 1 - \frac{(1 - 4\alpha)^2}{(1 - 2\Lambda^{-1}\alpha^2)} > 0; \tag{7.27}$$

equivalently

$$k < 1 - \frac{1 - 4\alpha}{(1 - 2\Lambda^{-1}\alpha^2)^{\frac{1}{2}}}. \tag{7.28}$$

There will exist such a $k > 0$ if

$$\frac{1 - 4\alpha}{(1 - 2\Lambda^{-1}\alpha^2)^{\frac{1}{2}}} < 1, \tag{7.29}$$

that is

$$-2\Lambda^{-1}\alpha - 16\alpha + 8 > 0, \tag{7.30}$$

i.e.,

$$\alpha < \frac{4}{8 + \Lambda^{-1}}. \tag{7.31}$$

We have

$$\frac{4}{8 + \Lambda^{-1}} \leq \frac{1}{4}, \quad \text{if } \Lambda \leq \frac{1}{8}. \tag{7.32}$$

An elementary computation shows that

$$\frac{4}{8 + \Lambda^{-1}} \leq \frac{\Lambda^{\frac{1}{2}}}{\sqrt{2}} \tag{7.33}$$

with the equality satisfied only when $\Lambda = \frac{1}{8}$.

We choose α such that it satisfies the inequality (7.31), and then $k > 0$ such that it satisfies (7.28).

7.3   Corrected second energy

We define a *corrected second energy* $E_\alpha^{(1)}$ by the formula

$$E_\alpha^{(1)}(t) = E^{(1)}(t) + 2\alpha^{(1)}\tau \int_{\Sigma_t} \Delta_g \gamma \gamma' \mu_g.$$

Using again the Cauchy–Schwarz inequality, and the Poincaré inequality to estimate $\|\gamma'\|_{g_t}$ in terms of $\|D\gamma'\|_{g_t}$ (the hypothesis $\bar{\gamma}' = 0$ is not necessary here because on a compact manifold $\int_{\Sigma_t} \Delta_g \gamma \mu_g = 0$) we find, (with the same a_t as in the previous subsection)

$$E^{(1)}(t) \leq \frac{1}{1 - a_t} E_\alpha^{(1)}(t). \tag{7.34}$$

7.4 Decay of the second corrected energy

We have found in [1], by straightforward but lengthy computations with the use of the wave equation for γ and $^{(3)}R_0^c \equiv -N\nabla_a h^{ac} = 2\partial_0\gamma\partial^c\gamma$ together with Cauchy–Schwarz and Sobolev theorems, an equality of the form

$$\frac{dE_\alpha^{(1)}}{dt} \equiv \frac{dE^{(1)}}{dt} + R_\alpha^{(1)},$$

where, with the choice $\tau = \frac{-1}{t}$,

$$R_\alpha^{(1)} \equiv 2\alpha^{(1)}\tau \int_{\Sigma_t} \{-N|D\gamma'|^2 + N|\Delta_g\gamma|^2 + (N+1)\tau\Delta_g\gamma\gamma'\}\mu_g + \tau Z_\alpha$$

where Z_α, given by

$$Z_\alpha \equiv 2\alpha \int_{\Sigma_t} \partial^a N(\partial_a\gamma\Delta_g\gamma + \gamma'\partial_a\gamma') + 2Nh^{ab}\nabla_a\partial_b\gamma\gamma'$$
$$+ 2\gamma'\partial_c\gamma(\partial_a Nh^{ac} - \gamma'\partial^c\gamma)\}\mu_g, \tag{7.35}$$

can be estimated with higher order terms in the energies. Using (5.6) we obtain that

$$\frac{dE_\alpha^{(1)}}{dt} - (2+k)\tau E_\alpha^{(1)}$$

$$= \tau \int_{\Sigma_t} \{(2N - 2 - k - 2\alpha N)J_0 + (2\alpha N + N - 2 - k)J_1\}\mu_g$$

$$+ 2\alpha\tau^2 \int_{\Sigma_t} \{(N + 1 - 2 - k)\Delta_g\gamma\gamma'\}\mu_g + Z + \tau Z_\alpha \tag{7.36}$$

which we write as

$$\frac{dE_\alpha^{(1)}}{dt} - (2+k)\tau E_\alpha^{(1)} = \tau \int_{\Sigma_t} \{(2 - k - 4\alpha)J_0 + (4\alpha - k)J_1\}\mu_g$$

$$+ 2\alpha\tau^2 \int_{\Sigma_t} \{(1 - k)\Delta_g uu'\}\mu_g + Z + \tau Z_\alpha + Z_N \tag{7.37}$$

where Z_N can be estimated with higher order terms in the energies through the estimate of $N - 2$.

The same estimates as those done for the first corrected energy show that, under the hypothesis made previously, it holds that the term linear in the energies on the right-hand side of the above equality is always nonnegative for the following choices of α and k:

- 1. $\Lambda \geq \frac{1}{8}$. We can choose $\alpha = \frac{1}{4}$ and $k = 1$.

- 2. $\Lambda < \frac{1}{8}$. We must then choose α and k satisfying the inequalities (7.31) and (7.28).

In all cases the estimate of the higher order terms in the energies are the same ones as obtained in [1], and the following equality holds:

$$\frac{dE_\alpha^{(1)}}{dt} = (2+k)\tau E_\alpha^{(1)} + |\tau|^3 B$$

where B is a polynomial in first and second derivatives of γ, h, Dh, DN, $D^2 N$ whose many terms can all be bounded using previous estimates by a polynomial in ε and ε_1 whose terms are at least of degree 3 and the coefficients bounded by CC_{σ_t}, under the H_c hypothesis, with $c > 0$, a given appropriate number.

8   Decay of the total energy

We define $y(t)$ to be the *total corrected energy* namely,

$$y(t) \equiv E_\alpha(t) + \tau^{-2} E_\alpha^{(1)}.$$

It bounds the total energy $x(t) \equiv E_{tot}(t) \equiv \varepsilon^2 + \varepsilon_1^2$ by

$$x(t) \equiv \varepsilon^2 + \varepsilon_1^2 \leq \frac{1}{1 - a_t} y(t) \quad \text{with} \quad a_t \equiv \frac{\alpha |\tau| e^{\lambda M}}{\Lambda_{\sigma_t}^{\frac{1}{2}}}.$$

We make the following a priori hypothesis, for all $t \geq t_0$ for which the considered quantities exist:

- **Hypothesis H_σ 1.** The numbers C_{σ_t} are uniformly bounded by a constant M_σ.

2. There exist $\Lambda > 0$ and $\delta > 0$ such that the inequality (7.6) is satisfied.

- **Hypothesis H_E** The energies ε_t^2 and $\varepsilon_{1,t}^2$ satisfy the inequality (7.8).

We choose α such that (the case $\Lambda \geq \frac{1}{8}$, $\alpha = \frac{1}{4}$ was considered in [1]).

$$\alpha < \frac{4}{8 + \Lambda^{-1}} < \frac{1}{4} \quad \text{with} \quad \Lambda < \frac{1}{8}. \tag{8.1}$$

Under the hypotheses H_c, H_E and H_σ there exists a number $M > 0$ such that, for all t

$$1 - a_t \geq M > 0. \tag{8.2}$$

Under the hypothesis H_E, all powers of E_{tot} greater than 3/2 are bounded by the product of $E_{tot}^{3/2}$ by a constant.

We denote by M_i any given number dependent on the bounds of these H's hypothesis but independent of t.

Under the hypotheses H_c, H_σ and H_E the function y satisfies a differential inequality of the form

$$\frac{dy}{dt} \leq -\frac{k}{t}(y - M_1 y^{3/2}).$$

(8.3)

We suppose that $y_0 \equiv y(t_0)$ satisfies

$$y_0^{1/2} < \frac{1}{2M_1}.$$

(8.4)

Then y starts decreasing, continues to decrease as long as it exists and satisfies an inequality which gives by integration, after setting $y = z^2$,

$$\log\left\{\frac{z(1 - M_1 z_0)}{(1 - M_1 z)z_0}\right\} + \frac{1}{2}k \log \frac{t}{t_0} \leq 0 \quad \text{a fortiori} \quad t^k y \leq \frac{t_0^k y_0}{(1 - M_1 z_0)^2};$$

hence, using the hypotheses and previous bounds, the *decay estimate*

$$t^k x(t) \leq M_2 x_0 \quad \text{with} \quad M_2 \leq \frac{4t_0^k}{(1 - a_t)(1 - a_0)} \leq \frac{4t_0^k}{M^2}.$$

9   Teichmuller parameters

We require the metric σ_t to remain, when t varies, in some cross section of M_{-1} over the Teichmuller space $\mathcal{T}$.

Given a metric $s \in M_{-1}$ the *Dirichlet energy* $D_s(\sigma)$ of the metric $\sigma \in M_{-1}$ is the energy of the (unique) harmonic diffeomorphism homotopic to the identity $\phi : (\Sigma, \sigma) \to (\Sigma, s)$. It can be written by conformal invariance as

$$D_s(\sigma) \equiv \int_\Sigma g^{ab} \partial_a \Phi^A \partial_b \Phi^B s_{AB}(\Phi)\mu_g.$$

Let σ_0 satisfy the hypothesis H_σ; then there exists a number D such that if $|D(\sigma) - D(\sigma_0)| \leq D$, called *Hypothesis* H_D, then σ satisfies also the hypothesis H_σ. We now estimate $D(\sigma)$.

We have if Φ is a harmonic map that

$$\frac{d}{dt}D_s(\sigma) = \int_{\Sigma_t} \{\bar{\partial}_0 g^{ab} \partial_a \Phi^A \partial_b \Phi^B - N\tau g^{ab}\partial_a \Phi^A \partial_b \Phi^B\}s_{AB}(\Phi)\mu_g$$

with

$$\bar{\partial}_0 g^{ab} = 2Ne^{-4\lambda}h^{ab} + Ne^{-2\lambda}h^{ab}\tau,$$

hence

$$\frac{d}{dt}D_s(\sigma) = \int_{\Sigma_t} 2Ne^{-2\lambda}h^{ab}\partial_a \Phi^A_\sigma \partial_b \Phi^B_\sigma s_{AB}(\Phi_\sigma)\mu_\sigma.$$

Using $0 < N \leq 2$ and $e^{-2\lambda} \leq \frac{\tau^2}{2}$ and the bound of $\|h\|_\infty$ we find

$$\frac{d}{dt} D_s(\sigma) \leq |\tau| C C_\sigma [\varepsilon + (\varepsilon + \varepsilon_1)^2] D_s(\sigma).$$

Under the hypotheses that we have made the Dirichlet energy satisfies the differential inequality

$$\frac{1}{D_s(\sigma)} \frac{d}{dt} D_s(\sigma) \leq \frac{C M_\sigma M_2^{\frac{1}{2}} x_0^{\frac{1}{2}}}{t^{1+\frac{k}{2}}}.$$

By integration and elementary calculus we obtain the inequality, valid for all $t \geq t_0$ since k is a strictly positive number,

$$|D(\sigma_t) - D(\sigma_0)| \leq M_3 x_0^{\frac{1}{2}}$$

10   Global existence

Theorem 10.1. *Let* $(\sigma_0, q_0) \in C^\infty(\Sigma_0)$ *and* $(u_0, \dot{u}_0) \in H_2(\Sigma_0, \sigma_0) \times H_1(\Sigma_0, \sigma_0)$ *be initial data for the polarized Einstein equations with* $U(1)$ *isometry group on the initial manifold* $\Sigma_0 \times U(1)$; *suppose that* σ_0 *is such that* $R(\sigma_0) = -1$. *Then there exists a number* $\eta > 0$ *such that if*

$$x_0 \equiv E_{tot}(t_0) < \eta,$$

these Einstein equations have a solution on $\Sigma \times S^1 \times [t_0, \infty)$, *with initial values determined by* $\sigma_0, q_0, u_0, \dot{u}_0$. *The solution has an infinite proper time extension since* $N_m > 0$. *It is unique with* $\tau = -t^{-1}$ *and* σ_t *in a chosen cross section of* M_{-1} *over* T_{eich}.

Proof. The same continuity argument as in [1].

It can be proved that when t tends to infinity the obtained solution tends to a metric of the type (1.1).

Acknowledgements. We thank L. Andersson for suggesting the use of corrected energies. We thank the University Paris VI, the ITP in Santa Barbara, the University of the Aegean in Samos, the IHES in Bures and the Schroedinger Institute in Vienna for their hospitality during our collaboration.

References

[1] Y. Choquet-Bruhat and V. Moncrief, *Future global einsteinian spacetimes with U(1) isometry group*, C.R. Acad. Sci. Paris t. 332 serie I (2001), 137–144; Ann. H. Poincaré **2** (2001), 1007–1064.

[2] V. Moncrief, *Reduction of Einstein equations for vacuum spacetimes with U(1) spacelike isometry group*, Annals of Physics **167** (1986), 118–142.

[3] Y. Choquet-Bruhat and V. Moncrief, *Existence theorem for solutions of Einstein equations with 1 parameter spacelike isometry group*, (H. Brezis and I.E. Segal, eds.), Proc. Symposia in Pure Math. **59** (1996), 67–80.

[4] L. Andersson, V. Moncrief and A. Tromba, *On the global evolution problem in 2+1 gravity*, J. Geom. Phys. **23**, no. 3–4 (1997), (1991), 205.

Yvonne Choquet-Bruhat
YCB Université Paris 6
4 Place Jussieu 75232 Paris France
ycb@ccr.jussieu.fr

Vincent Moncrief
VM Department of Physics
Yale University
New Haven CT, USA
vincent.moncrief@yale.edu

On the Cauchy Problem for a Weakly Hyperbolic Operator: An Intermediate Case Between Effective Hyperbolicity and Levi Condition

Ferruccio Colombini, Mariagrazia Di Flaviano and Tatsuo Nishitani

1 Introduction

We are interested in the Cauchy problem for second order hyperbolic operators of the form

$$\begin{cases} P(t, x, \partial_t, \partial_x)u(t, x) = f(t, x), \\ u(0, x) = u_0(x), \quad \partial_t u(0, x) = u_1(x) \end{cases} \tag{1.1}$$

on $[0, T] \times \mathbf{R}^n$ where

$$\begin{aligned} P(t, x, \partial_t, \partial_x) &= P_2(t, \partial_t, \partial_x) + P_1(t, \partial_t, \partial_x) + c(t, x), \\ P_2(t, \partial_t, \partial_x) &= \partial_t^2 - \sum_{i,j=1}^n a_{ij}(t)\partial_{x_i}\partial_{x_j}, \\ P_1(t, \partial_t, \partial_x) &= \sum_{j=1}^n b_j(t)\partial_{x_j} + b_0(t)\partial_t. \end{aligned} \tag{1.2}$$

We assume that $a_{ij} \in C^\infty([0, T])$ $b_j \in C([0, T])$ $c \in C([0, T]; C^\infty(\mathbf{R}^n))$; moreover

$$a(t, \xi) = \sum_{i,j=1}^n a_{ij}(t)\xi_i\xi_j \geq 0, \quad \forall \xi \in \mathbf{R}^n, \ t \in [0, T]. \tag{1.3}$$

This problem has been extensively studied by many authors, starting from the work of E.E. Levi [6], who, for the first time, pointed out that the Cauchy problem for the operator

$$\partial_t^2 - \partial_x$$

is solvable neither in C^∞ nor in the Gevrey classes γ^s for $s > 2$. Successively many authors studied the Cauchy problem for hyperbolic operators with roots of variable multiplicity, starting from the well-known treatise of Hadamard [5], who gave the decisive orientation and influence to the progress in this field. In any case, from the example of E.E. Levi it is clear that the Cauchy problem (1.1) may fail to be C^∞ well posed, unless one imposes some additional condition on lower order terms (Levi condition). On the other hand, even for a homogeneous operator as (1.2), i.e., the case $P_1(t, \partial_t, \partial_x) = c(t, x) = 0$, it is possible to find (see [4]) examples of Cauchy problems which are not C^∞ well posed: this phenomenon is due to the very rapid oscillations of the coefficients with respect to the time variable. Thus, in order to obtain positive results concerning the C^∞ well-posedness, we are constrained to impose some additional assumptions on the principal part (in order to avoid the phenomenon of oscillations), and, on the other hand, on lower order terms (a kind of Levi condition). To this aim we introduce the following notation:

$$b(t, \xi) = \sum_{j=1}^n b_j(t)\xi_j.$$

It is well known that the Cauchy problem (1.1) is C^∞ well posed without any contraint on lower order terms if and only if the homogeneous part P_2 is effectively hyperbolic (see [10] and its bibliography). We recall that the effective hyperbolicity on our operator is equivalent to

$$\partial_t^2 a(t, \xi) > 0$$

whenever $a(t, \xi) = 0$. This assumption can be expressed by the condition

$$\sum_{j=0}^2 |\partial_t^j a(t, \xi)| \neq 0, \quad \forall |\xi| = 1, \quad t \in [0, T]. \tag{1.4}$$

On the other hand the classical Levi condition for an operator such as (1.2) can be written in the form

$$|b(t, \xi)| \leq C\big(a(t, \xi)\big)^{1/2}, \tag{1.5}$$

for some $C > 0$.

It is then well known that under this condition, the Cauchy problem (1.1) is C^∞ well posed, as soon as we avoid too many oscillations in the principal part, by imposing, for example, a condition of analiticity on the coefficients $a_{ij}(t)$, or, more generally, the so-called "logarithmic condition"

$$\int_0^T \frac{|a_t'(t, \xi)|}{a(t, \xi) + 1} \, dt \leq C_1 \log(1 + |\xi|) + C_2$$

given in [2] and generalized by K. Kajitani [9] to operators with coefficients depending also on x.

In [1] the Cauchy problem (1.1) was studied in the case wherein the coefficients depend only on t, when the condition (1.4) is weakened to an assumption of finite degeneracy, and the Levi condition on the lower order term P_1 is substituted by a more refined condition. More precisely the following theorem was proved in [1]:

Theorem 1.1. *Assume that, for some $k \geq 2$, we have*

$$\sum_{j=0}^{k} |\partial_t^j a(t, \xi)| \neq 0, \quad \forall |\xi| = 1, \quad t \in [0, T]. \tag{1.6}$$

On the other hand let us suppose that for some $C > 0$ and $\gamma \in \left[0, \frac{1}{2}\right]$ we have

$$|b(t, \xi)| \leq C \, a(t, \xi)^\gamma |\xi|^{1-2\gamma}. \tag{1.7}$$

Then, if $\gamma + \frac{1}{k} \geq \frac{1}{2}$, the Cauchy problem (1.1) is C^∞ well-posed.

We remark that, due to the compactness of $[0, T] \times S^{n-1}$, the assumption (1.6) is equivalent to

$$\sum_{j=0}^{\infty} |\partial_t^j a(t, \xi)| \neq 0, \quad \forall |\xi| = 1, \quad t \in [0, T].$$

Moreover we remark that also the case of $\gamma + \frac{1}{k} < \frac{1}{2}$ is treated in [1], obtaining the well-posedness in some Gevrey spaces.

In this paper we study the same type of problem, but allowing the coefficient c of the zero order term to depend on all the variables. More precisely we prove the following

Theorem 1.2. *Assume (1.6) is verified and $\gamma + \frac{1}{k} > \frac{1}{2}$. Then the Cauchy problem (1.1) is C^∞ well-posed.*

We remark that, in comparison with Theorem 1.1, we are unable to treate the case

$$\gamma + \frac{1}{k} = \frac{1}{2}. \tag{1.8}$$

We observe also that for the operator

$$\partial_t^2 - t^{2l} \partial_x^2 + i t^\nu \partial_x$$

the Cauchy problem is C^∞ well-posed if and only if (see [8], [7])

$$\nu \geq l - 1.$$

This shows that the condition (1.8) is optimal.

2 Preliminaries

Let us set

$$\tilde{a}(t, \xi) = a(t, \xi)/|\xi|^2$$

so that $\tilde{a}(t, \xi)$ is homogeneous of degree 0 in ξ. We will use the following.

Lemma 2.1. *There exist $c > 0$, $\delta > 0$ such that for any $|\xi| = 1$ we have*

$$\tilde{a}(t, \xi) \geq c, \quad |t| \leq \delta \tag{2.1}$$

or

$$\partial_t \tilde{a}(t, \xi) = 0, \quad |t| \leq \delta, \ \text{has at most } k \text{ roots with respect to } t. \tag{2.2}$$

Proof. Let $|\bar{\xi}| = 1$ be fixed.

We have 2 cases: 1) $\tilde{a}(0, \bar{\xi}) \neq 0$. In this case we can find $c(\bar{\xi}) > 0$, $\delta(\bar{\xi}) > 0$ and a neighborhood $V(\bar{\xi})$ of $\bar{\xi}$ such that

$$|\tilde{a}(t, \xi)| \geq c(\bar{\xi}), \quad |t| \leq \delta(\bar{\xi}), \ \xi \in V(\bar{\xi}).$$

2) $\tilde{a}(0, \bar{\xi}) = 0$. From (1.6) there is $1 \leq r \leq k$ such that

$$\partial_t^\mu \tilde{a}(0, \bar{\xi}) = 0, \quad 0 \leq \mu < r, \ \partial_t^r \tilde{a}(0, \bar{\xi}) \neq 0.$$

Then we can find $\delta(\bar{\xi}) > 0$ and a neighborhood $V(\bar{\xi})$ of $\bar{\xi}$ such that

$$|\partial_t^r \tilde{a}(t, \xi)| \neq 0, \quad |t| \leq \delta(\bar{\xi}), \ \xi \in V(\bar{\xi}).$$

Thus we conclude that if $\xi \in V(\bar{\xi})$, $|t| \leq \delta(\bar{\xi})$, then $\partial_t \tilde{a}(t, \xi) = 0$ has at most $r - 1$ roots with respect to t.

At this point Lemma 2.1 follows easily from the compactness of $\{|\xi| = 1\}$. $\square$

For $s < t$ we set

$$F_\beta^*(s, t; \xi) = \max \left(\frac{\tilde{a}(t, \xi) + |\xi|^{-2\beta}}{\tilde{a}(s, \xi) + |\xi|^{-2\beta}}, \frac{\tilde{a}(s, \xi) + |\xi|^{-2\beta}}{\tilde{a}(t, \xi) + |\xi|^{-2\beta}} \right), \quad 0 < \beta \leq 1. \tag{2.3}$$

We define $W_\beta^*(t, \xi)$ as follows: let N be fixed (which will be determined later). We set

$$W_\beta^*(t, \xi) = \sup \sum_{i=0}^{N-1} \log F_\beta^*(t_i, t_{i+1}; \xi) \tag{2.4}$$

where supremum is taken over all sequences $\{t_i\}_{i=0}^N$ such that

$$0 \leq t_0 \leq t_1 \leq \cdots \leq t_N \leq t. \tag{2.5}$$

Note that $W_\beta^*(t, \xi)$ is an increasing function in t by definition so that $W_\beta^*(t, \xi)$ is differentiable almost everywhere and

$$\frac{d}{dt} W_\beta^*(t, \xi) \geq 0 \quad \text{a.e.}$$

Lemma 2.2. $e^{W_\beta^*(t,\xi)}$ *is a temperate weight function, that is we have*

$$W_\beta^*(t, \xi) \leq C_\beta \log(1 + |\xi - \eta|) + W_\beta^*(t, \eta)$$

with some $C_\beta > 0$ for $|\xi|, |\eta| \geq 1$.

Proof. We fix a small $0 < \epsilon \ll 1$. When $|\xi - \eta| \geq \epsilon|\xi|$ we proceed as follows. Note that

$$\frac{\widetilde{a}(t, \xi) + |\xi|^{-2\beta}}{\widetilde{a}(s, \xi) + |\xi|^{-2\beta}} \leq C|\xi|^{2\beta} \leq \epsilon^{-2\beta} C|\xi - \eta|^{2\beta}.$$

This shows that

$$F_\beta^*(s, t; \xi) \leq C\epsilon^{-2\beta}|\xi - \eta|^{2\beta} \leq (1 + C'|\xi - \eta|)^{2\beta}.$$

Since $F_\beta^*(s, t; \eta) \geq 1$ one gets

$$F_\beta^*(s, t; \xi) \leq (1 + C'|\xi - \eta|)^{2\beta} F_\beta^*(s, t; \eta). \tag{2.6}$$

Let $\{t_i\}_{i=0}^N$ be any sequence verifying (2.5). Then we have

$$\sum_{i=0}^{N-1} \log F_\beta^*(t_i, t_{i+1}; \xi)$$

$$\leq 2N \log(1 + C'|\xi - \eta|) + \sum_{i=0}^{N-1} \log F_\beta^*(t_i, t_{i+1}; \eta)$$

by (2.6). Since the right-hand side is bounded by

$$2N \log(1 + C'|\xi - \eta|) + W_\beta^*(t, \eta)$$

and $\{t_i\}_{i=0}^N$ is arbitrary we get the desired assertion.

We turn to the case $|\xi - \eta| \leq \epsilon|\xi|$. It is enough to show that

$$\widetilde{a}(t, \xi) + |\xi|^{-2\beta} \leq (1 + C|\xi - \eta|)^3 [\widetilde{a}(t, \eta) + |\eta|^{-2\beta}]. \tag{2.7}$$

Assume that (2.7) is proved. Then exchanging ξ and η and taking $t = s$ one gets

$$[\widetilde{a}(s, \xi) + |\xi|^{-2\beta}]^{-1} \leq (1 + C|\xi - \eta|)^3 [\widetilde{a}(s, \eta) + |\eta|^{-2\beta}]^{-1}. \tag{2.8}$$

Thus from (2.7) and (2.8) we have

$$F_\beta^*(s, t; \xi) \leq (1 + C|\xi - \eta|)^6 F_\beta^*(s, t; \eta). \tag{2.9}$$

The rest of the proof is just a repetition of the case $|\xi - \eta| \geq \epsilon|\xi|$. We now prove (2.7). Let us put $\phi(t, \xi) = \widetilde{a}(t, \xi)|\xi|^{2\beta}$ so that ϕ is homogeneous of degree 2β with respect to ξ. By the Glaeser inequality one has

$$|\partial_{\xi_i} \phi(t, \xi)| \leq C\sqrt{\phi(t, \xi)}|\xi|^{\beta-1}.$$

Hence we have

$$\phi(t, \xi) \leq \phi(t, \eta) + C|\xi - \eta|\sqrt{\phi(t, \eta)}|\eta|^{\beta-1} + C|\xi - \eta|^2|\eta + \theta|\xi - \eta||^{2\beta-2}$$

from the Taylor expansion. Since $(1-\epsilon)|\xi| \leq |\eta| \leq (1+\epsilon)|\xi|$ and hence $C_1^{-1}|\xi| \leq |\eta + \theta(\xi - \eta)| \leq C_1|\xi|$, the right-hand side is bounded by

$$\phi(t, \eta) + C_2|\xi|^{\beta-1}\sqrt{\phi(t, \eta)}|\xi - \eta| + C_2|\xi|^{2\beta-2}|\xi - \eta|^2.$$

Since $2\sqrt{\phi(t, \eta)} \leq \phi(t, \eta) + 1$ it follows that

$$\phi(t, \xi) + 1 \leq [\phi(t, \eta) + 1](1 + C|\xi|^{\beta-1}|\xi - \eta|)^2. \qquad (2.10)$$

Noting $|\xi|^{-2\beta} \leq |\eta|^{-2\beta}(1 + C|\eta|^{-1}|\xi - \eta|)$ and multiplying (2.10) by $|\xi|^{-2\beta}$ we get (2.7). This completes the proof. $\qquad\square$

In what follows we take $N = k + 1$.

Lemma 2.3. *There is $D > 0$ such that we have for any ξ*

$$\frac{d}{dt}\left[W_\beta^*(t, \xi) + Dt\right] \geq \frac{|a'(t, \xi)|}{a(t, \xi) + |\xi|^{2-2\beta}}$$

in $|t| \leq \delta$.

Proof. From Lemma 2.1 for any ξ we have that (2.1) or (2.2) holds. If (2.1) is true, the assertion holds obviously if we take $D > 0$ large because

$$\frac{d}{dt}W_\beta^*(t, \xi) \geq 0.$$

If on the contrary (2.2) holds, choosing $t_0 = 0$, $t_N = t$ and $t_1 \leq t_2 \leq \cdots \leq t_{N-1}$ to be the zeros of $\partial_t\tilde{a}(s, \xi)$ in $(0, t)$ we get

$$\int_0^t \frac{|\partial_t\tilde{a}(s, \xi)|}{\tilde{a}(s, \xi) + |\xi|^{-2\beta}}ds = \sum_{i=0}^{N-1} \log F_\beta^*(t_i, t_{i+1}; \xi) \leq W_\beta^*(t, \xi).$$

On the other hand we have (see [3], proof of Lemma 3.2)

$$W_\beta^*(t, \xi) \leq \int_0^t \frac{|\partial_t\tilde{a}(s, \xi)|}{\tilde{a}(s, \xi) + |\xi|^{-2\beta}}ds.$$

Hence one gets

$$W_\beta^*(t, \xi) = \int_0^t \frac{|\partial_t\tilde{a}(s, \xi)|}{\tilde{a}(s, \xi) + |\xi|^{-2\beta}}ds \qquad (2.11)$$

and this proves the assertion. $\qquad\square$

3   Energy estimate

In this section we prove Theorem 1.2. We apply the Fourier transform with respect to the space variables to the equation, thus we obtain the following ordinary differential equation in t, depending on the parameter ξ,

$$v'' + a(t, \xi)v + ib(t, \xi)v + \widehat{cu} = \widehat{f}, \qquad (3.1)$$

where v denotes the Fourier transform of u with respect to x and the symbol $\frown$ denotes the Fourier transform with respect to x. We consider the energy function

$$\mathcal{E}(t) = \int_{\mathbf{R}^n} \tilde{E}(t,\xi)d\xi = \int_{\mathbf{R}^n} E(t,\xi)K(t,\xi)d\xi \qquad (3.2)$$

where

$$E(t,\xi) = |v'(t,\xi)|^2 + (a(t,\xi) + |\xi|^{2\beta} + 1)|v(t,\xi)|^2$$

and

$$K(t,\xi) = e^{-(W^*_{1-\beta}(t,\xi)+Dt)}$$

with $0 < \beta < 1$ which we will choose later. Differentiating $\mathcal{E}(t)$ with respect to the time, we have

$$\mathcal{E}'(t) = \int_{\mathbf{R}^n} \left(E'(t,\xi) - \frac{d}{dt}(W^*_{1-\beta}(t,\xi) + Dt)E(t,\xi) \right) K(t,\xi)d\xi.$$

Recall that

$$E' = 2\mathrm{Re}(v'',v') + 2\mathrm{Re}(v',v)(a + |\xi|^{2\beta} + 1) + a'|v|^2$$

and using (3.1) we have

$$E' \le 2|b||v||v'| + 2|\widehat{cu}||v'| + 2|\widehat{f}||v'| + 2|\xi|^{2\beta}|v||v'| + 2|v||v'| + |a'||v|^2.$$

Since

$$-\frac{d}{dt}(W^*_{1-\beta}(t,\xi) + Dt) \le -\frac{|a'(t,\xi)|}{a(t,\xi) + |\xi|^{2\beta}},$$

we get

$$E' - \frac{d}{dt}(W^*_{1-\beta}(t,\xi) + Dt)E$$
$$\le 2|b||v||v'| + 2|\widehat{cu}||v'| + 2|\widehat{f}||v'| + 2|\xi|^{2\beta}|v||v'|$$
$$+ 2|v||v'| + |a'||v|^2 - \frac{|a'|}{a + |\xi|^{2\beta}}E$$
$$\le 2|b||v||v'| + 2|\widehat{cu}||v'| + 2|\widehat{f}||v'|$$
$$+ 2|\xi|^{2\beta}|v||v'| + 2|v||v'|.$$

Now we use the estimates

$$2|b||v||v'| \le 2Ca^\gamma |\xi|^{1-2\gamma}|v||v'| \le C|\xi|^{1-2\gamma}(a + |\xi|^{2\beta})^{\gamma-1/2}E,$$
$$2|\widehat{cu}||v'| \le |\widehat{cu}|^2 + E, \quad 2|\widehat{f}||v'| \le |\widehat{f}|^2 + E,$$
$$2|\xi|^{2\beta}|v||v'| \le \frac{|\xi|^{2\beta}}{(a + |\xi|^{2\beta})^{1/2}}E, \quad 2|v||v'| \le E,$$

and we obtain

$$
\begin{aligned}
&\tilde{E}'(t,\xi) \\
&\leq \left(C \frac{|\xi|^{1-2\gamma}}{(a(t,\xi)+|\xi|^{2\beta})^{1/2-\gamma}} + \frac{|\xi|^{2\beta}}{(a(t,\xi)+|\xi|^{2\beta})^{1/2}} + 3 \right) \tilde{E}(t,\xi) \\
&\quad + \left(|\widehat{cu}(t,\xi)|^2 + |\widehat{f}(t,\xi)|^2 \right) K(t,\xi).
\end{aligned}
\tag{3.3}
$$

We put

$$
\alpha(t,\xi) = C \frac{|\xi|^{1-2\gamma}}{(a(t,\xi)+|\xi|^{2\beta})^{1/2-\gamma}} + \frac{|\xi|^{2\beta}}{(a(t,\xi)+|\xi|^{2\beta})^{1/2}} + 3
\tag{3.4}
$$

and we multiply (3.3) by $\exp\left\{ -\int_0^t \alpha(s,\xi)ds \right\}$ obtaining

$$
\begin{aligned}
\partial_t &\left(\tilde{E}(t,\xi) \exp\left\{ -\int_0^t \alpha(s,\xi)ds \right\} \right) \\
&\leq \exp\left\{ -\int_0^t \alpha(s,\xi)ds \right\} \left(|\widehat{cu}(t,\xi)|^2 + |\widehat{f}(t,\xi)|^2 \right) K(t,\xi) \\
&\leq \left(|\widehat{cu}(t,\xi)|^2 + |\widehat{f}(t,\xi)|^2 \right) K(t,\xi).
\end{aligned}
$$

We integrate the inequality with respect to the time and we get

$$
\tilde{E}(t,\xi) \leq \tilde{E}(0,\xi) \exp\left\{ \int_0^t \alpha(s,\xi)ds \right\}
\tag{3.5}
$$

$$
+ \exp\left\{ \int_0^t \alpha(s,\xi)ds \right\} \int_0^t \left(|\widehat{cu}(s,\xi)|^2 + |\widehat{f}(s,\xi)|^2 \right) K(s,\xi)ds.
$$

In order to proceed we need the following lemma.

Lemma 3.1. *Assume (1.6). Let $\tilde{a}(t,\xi) = a(t,\xi)/|\xi|^2$, then for any $\eta \geq 0$ there exist $M_\eta > 0$ and $\varepsilon_0 > 0$ such that for any $\varepsilon \in (0,\varepsilon_0]$ we have*

$$
\int_0^T \frac{1}{(\tilde{a}(s,\xi)+\varepsilon)^\eta}ds \leq
\begin{cases}
M_\eta, & \text{if } \eta < 1/k, \\
M_\eta \log\frac{1}{\varepsilon}, & \text{if } \eta = 1/k, \\
M_\eta \varepsilon^{1/k-\eta}, & \text{if } \eta > 1/k.
\end{cases}
\tag{3.6}
$$

Proof. See [1, Lemma 2]. $\qquad\square$

Applying Lemma 3.1 and recalling that $\gamma + 1/k > \frac{1}{2}$ we have

$$
\int_0^T \frac{|\xi|^{1-2\gamma}ds}{(a(s,\xi)+|\xi|^{2\beta})^{1/2-\gamma}} \leq \int_0^T \frac{ds}{(\tilde{a}(s,\xi)+|\xi|^{2(\beta-1)})^{1/2-\gamma}} \leq M_{1/2-\gamma}
$$

provided $|\xi|^{2(\beta-1)}$ is small enough (this is true since we consider $|\xi|$ large enough and we will choose $\beta < 1$). Moreover Lemma 3.1 shows that

$$\int_0^T \frac{|\xi|^{2\beta} ds}{(a(s,\xi) + |\xi|^{2\beta})^{1/2}} = |\xi|^{2\beta-1} \int_0^T \frac{ds}{(\tilde{a}(s,\xi) + |\xi|^{2(\beta-1)})^{1/2}}$$

$$\leq \begin{cases} M_{1/2} 2(1-\beta)|\xi|^{2\beta-1} \log |\xi| & \text{if } 1/2 = 1/k, \\ M_{1/2} |\xi|^{2\beta-1+2(\beta-1)(1/k-1/2)} & \text{if } 1/2 > 1/k, \end{cases}$$

provided $|\xi|^{2(\beta-1)}$ is small enough. Choosing $\beta = \frac{2}{2+k} < 1$ if $k > 2$ and some $\beta < 1/2$ if $k = 2$, we obtain

$$\int_0^T \frac{|\xi|^{2\beta} ds}{(a(s,\xi) + |\xi|^{2\beta})^{1/2}} \leq M,$$

for some positive constant M depending on k. Coming back to our estimate (3.5), we obtain

$$\int_{\mathbf{R}^n} \tilde{E}(t,\xi) d\xi \leq C_0 \int_{\mathbf{R}^n} \tilde{E}(0,\xi) d\xi \qquad (3.7)$$

$$+ C_0 \int_{\mathbf{R}^n} \int_0^t K(s,\xi) \left(|\widehat{cu}(s,\xi)|^2 + |\widehat{f}(s,\xi)|^2 \right) ds d\xi,$$

for some positive constant C_0.

Now we prove that

$$\int |\widehat{cu}(t,\xi)|^2 K(t,\xi) d\xi \leq C \int |\widehat{u}(t,\xi)|^2 K(t,\xi) d\xi$$

with some $C > 0$. Indeed, since, from Lemma 2.2,

$$K(t,\xi) \leq C_1 (1 + |\xi - \eta|)^M K(t,\eta),$$

we see that

$$\int |\widehat{cu}(s,\xi)|^2 K(s,\xi) d\xi$$

$$\leq \int K(s,\xi) \int |\widehat{c}(s,\xi-\eta)||\widehat{u}(s,\xi)|^2 d\eta d\xi \int |\widehat{c}(s,\eta_1)| d\eta_1$$

$$\leq C_1 \int\int K(s,\eta)|\widehat{u}(s,\eta)|^2 |\widehat{c}(s,\xi-\eta)|(1+|\xi-\eta|)^M d\eta d\xi$$

$$\leq C_2 \int K(s,\eta)|\widehat{u}(s,\eta)|^2 d\eta \int |\widehat{c}(s,\xi)|(1+|\xi|)^M d\xi$$

$$\leq C_3 \int K(s,\eta)|\widehat{u}(s,\eta)|^2 d\eta.$$

We then get

$$\mathcal{E}(t) \leq C_0 \mathcal{E}(0) + C_0 C \int_0^t \mathcal{E}(s) ds + C_0 \int_{\mathbf{R}^n} \int_0^t |\widehat{f}(s,\xi)|^2 ds d\xi$$

and finally we get

$$\mathcal{E}(t) \leq \left(C_0 \mathcal{E}(0) + C_0 \int_{\mathbf{R}^n} \int_0^t |\widehat{f}(s,\xi)|^2 ds d\xi \right) e^{TC_0C}. \qquad \square$$

Acknowledgement. The authors would like express their special thanks to Nicola Orrù for many stimulating discussions during the preparation of this paper.

References

[1] F. Colombini, H. Ishida and N. Orrù, On the Cauchy problem for finitely degenerate hyperbolic equations of second order, *Ark. Mat.* **38** (2000), 223–230.

[2] F. Colombini, E. Jannelli and S. Spagnolo, Well-posedness in the Gevrey classes of the Cauchy problem for a non-strictly hyperbolic equation with coefficients depending on time, *Ann. Scuola Norm. Sup. Pisa* **10** (1983), 291–312.

[3] F. Colombini and T. Nishitani, Two by two Strongly Hyperbolic Systems and Gevrey Classes, *Ann. Univ. Ferrara Sez. VII Sc. Mat. Suppl.* **45** (1999), 79–108.

[4] F. Colombini and S. Spagnolo, An example of a weakly hyperbolic Cauchy problem not well posed in C^∞, *Acta. Math.* **148** (1982), 243–253.

[5] J. Hadamard, *Le problème de Cauchy et les équations aux dérivées partielles linéaires hyperboliques*, Hermann et Cie, Paris, 1932.

[6] E.E. Levi, Sul problema di Cauchy, *Rend. Reale Acad. Lincei* **16** (1907), 105–112.

[7] V. Ya. Ivrii and V. M. Petkov, Necessary conditions for the well posedness of the Cauchy problem for non strictly hyperbolic equations, *Usp. Mat. Nauk* **29**, (1974), 3–70, (Russian). English transl.: *Russ. Math. Surv.* **29**, (1974), 1–70.

[8] V. Ya. Ivrii, Cauchy problem conditions for hyperbolic operators with characteristics of variable multiplicity for Gevrey classes, *Sibirsk. Mat. Zh.* **17**, (1976), 1256–1270, (Russian). English transl.: *Siberian Math. J.* **17**, (1976), 921–931.

[9] K. Kajitani, The well-posed Cauchy problem for hyperbolic operators, Séminaire Jean Vaillant 1989.

[10] T. Nishitani, The effectively hyperbolic Cauchy problem, in *The Hyperbolic Cauchy Problem* (by K. Kajitani and T. Nishitani), Lecture Notes in Math. **1505**, pp. 71–167, Springer-Verlag, Berlin-Heidelberg, 1991.

[11] N. Orrù, On a weakly hyperbolic equation with a term of order zero, *Ann. Fac. Sci. Toul.* **6**, 3 (1997), 525–534.

Ferruccio Colombini
Dipartimento di Matematica
Università di Pisa
Via F. Buonarroti 2, 56127, Italy
colombin@dm.unipi.it

Mariagrazia Di Flaviano
Università dell'Aquila,
Dipartimento di Matematica Pura ed Applicata,
via Vetoio, I-67010 Coppito, Italy

Tatsuo Nishitani
Department of Mathematics, Osaka University,
Machikaneyama 1-16, Toyonaka Osaka, 560-0043, Japan
tatsuo@math.wani.osaka-u.ac.jp

Symplectic Path Intersections and the Leray Index

Maurice de Gosson and Serge de Gosson

ABSTRACT We define a Maslov index for symplectic paths by using the properties of Leray's index for pairs of Lagrangian paths. Our constructions are purely topological, and the index we define satisfies a simple system of five axioms. The fifth axiom establishes a relation between the spectral flow of a family of symmetric matrices and the Maslov index.

1 Introduction

Historically, the embryo of what is called today the "Maslov quantization condition" was already proposed by the physicist Albert Einstein in his 1917 paper [12]. Einstein had remarked that the Bohr quantization rule

$$\oint_\gamma p dr = Nh$$

for periodic electronic orbits (N an integer) should be corrected by including a term $\frac{h}{2}$:

$$\oint_\gamma p dr = \left(N + \tfrac{1}{2}\right) h. \tag{1.1}$$

Einstein's ideas were taken up in 1958 by Keller [13], who constructed approximate solutions to the stationary states of Schrödinger's equation, and later by Maslov and his collaborators [17, 18]. In particular, Maslov generalized Einstein's condition (1.1) by imposing the quantum condition

$$\frac{1}{h} \oint_\gamma p dr - \tfrac{1}{4} m(\gamma) \text{ is an integer} \tag{1.2}$$

for all loops γ in an arbitrary Lagrangian manifold V. Here $m(\gamma)$ is an integer, the *Maslov index*; it is always even when V is oriented (see [21]) and condition (1.2) thus reduces to (1.1) when the Lagrangian manifold is a torus.

Maslov's work was clarified and given a geometrical interpretation by Arnol'd [1], Leray [14] and one of us [5]. The novelty with (1.2), compared to Einstein's condition (1.1), is the appearance of an integer $m(\gamma)$ associated to *any* loop in V (and not just to periodic orbits). The vocation of the Maslov index is to count the number of times the loop γ traverses the caustic of V (i.e., the set of points

$(x, p) \in V$ that do not project diffeomorphically on $\mathbb{R}_x^n$). More generally, one is interested in determining an intersection number for an arbitrary continuous path in a Lagrangian manifold, that is, one wants to associate to every such path Γ and to every Lagrangian plane ℓ an integer $m(\Gamma, \ell)$ satisfying some adequate system of axioms, and coinciding with $m(\gamma)$ when Γ is a loop γ. We have constructed such a topological object in [9] (following previous work in [8, 10]).

A related, but more general problem, is the construction of intersection indices for symplectic paths. This is certainly not only an academic problem: it is, to paraphrase Poincaré, "a problem which poses itself, not a problem one poses". In fact, Maslov's condition (1.2) does not make sense for nonintegrable Hamiltonian systems, because there are no Lagrangian manifolds ("invariant tori") associated to such systems. One tries instead, in this case, to quantize the individual periodic orbits associated to these nonintegrable systems. This is achieved by associating to the periodic orbits a path of symplectic matrices, joining the identity to the monodromy matrix of the orbit. One then tries to define the Maslov index of the orbit in terms of the intersections of that path with a particular locus, the "Maslov cycle" (see [2, 4, 22]). There are also other interesting problems where one wants to define a Maslov index of symplectic paths. For instance, in functional analysis it has become a popular trend to establish formulas relating "spectral flows" of operators to symplectic path intersection indices (see, e.g., [3, 19] and the references therein).

Depending on the envisaged applications, there are several possible requirements for the properties a "Maslov index of symplectic paths" should satisfy. We are going to construct in this article a Maslov index satisfying a system of five *axioms*. The first four of these axioms are in a sense "minimal requirements" which should be part of any "reasonable" definition of the Maslov index; the fifth axiom is of a slightly subtler nature, and will be shortly commented in a moment.

In all that follows, the Greek capitals Σ, $\Sigma' \ldots$, will denote continuous paths $[0, 1] \longrightarrow Sp(n)$, and the calligraphic letters ℓ, $\ell' \ldots$ will denote a Lagrangian plane in the standard $2n$-dimensional symplectic space. We will define a *Maslov index of symplectic paths* to be an integer-valued function

$$(\Sigma, \ell) \longmapsto m(\Sigma, \ell)$$

satisfying the following system of axioms:

Axiom 1 (homotopy invariance): If the symplectic paths Σ and Σ' are homotopic with fixed endpoints, then

$$m(\Sigma, \ell) = m(\Sigma', \ell);$$

Axiom 2 (concatenation): If Σ and Σ' are two consecutive symplectic paths, then their concatenation $\Sigma * \Sigma'$ satisfies

$$m(\Sigma * \Sigma', \ell) = m(\Sigma, \ell) + m(\Sigma', \ell);$$

Axiom 3 (dimensional additivity): If Σ_1 is a path in $Sp(n_1)$ and Σ_2 a path in $Sp(n_2)$, then the path $\Sigma_1 \oplus \Sigma_2$ in $Sp(n_1 + n_2)$ satisfies

$$m_{1,2}(\Sigma_1 \oplus \Sigma_2, \ell) = m_1(\Sigma_1, \ell) + m_2(\Sigma_2, \ell);$$

Axiom 4 (zero in strata): If the symplectic path Σ is such that the intersection $\Sigma(t)\ell \cap \ell$ keeps constant dimension when t varies, then $m(\Sigma, \ell) = 0$;

Axiom 5 (spectral flow): Let $t \mapsto P(t)$ be a continuous mapping of the interval $[0, 1]$ in the space of all real symmetric $n \times n$ matrices. The matrices

$$\Sigma_P(t) = \begin{pmatrix} I_{n \times n} & 0_{n \times n} \\ -P(t) & I_{n \times n} \end{pmatrix} \tag{1.3}$$

are then symplectic, and the symplectic path $t \mapsto \Sigma_P(t)$ is such that

$$m(\Sigma_P, X) = \text{Inert } P(1) - \text{Inert } P(0). \tag{1.4}$$

The fifth axiom can be interpreted as follows: when t varies from 0 to 1, the number of negative eigenvalues of the symmetric matrix $P(t)$ also varies. The net variation of this number of eigenvalues is referred to as the "spectral" flow of the path of operators $P(\cdot)$. Axiom (5) then just says that this spectral flow is identical with the Maslov index $m(\Sigma_P, \ell)$.

2   Leray's index for pairs of Lagrangian paths

In this section we review and complement the theory of the Leray index.

Let $X = \mathbb{R}^n$; we endow the direct sum $X \oplus X^*$ with the canonical symplectic structure defined by

$$\omega(z, z') = \langle p, x' \rangle - \langle p', x \rangle$$

for $z = x + p$, $z' = x' + p'$. The associated symplectic group is denoted by $Sp(n)$: $s \in Sp(n)$ if and only if $\omega(sz, sz') = \omega(z, z')$ for all z, z'. We recall that $Sp(n)$ is a connected Lie group, homeomorphic to the product $U(n) \times \mathbb{R}^{n(2n+1)}$. In particular $\pi_1(Sp(n)) \equiv (\mathbb{Z}, +)$.

We denote by $\Lambda(n)$ the Lagrangian Grassmannian of $(X \oplus X^*, \omega)$: $\ell \in \Lambda(n)$ if and only if ℓ is an n-dimensional linear subspace of $X \oplus X^*$ on which the symplectic form ω vanishes identically. One proves that $\Lambda(n)$ is a connected and compact submanifold of the Grassmannian of all n-planes, and that $\pi_1(\Lambda(n)) \equiv (\mathbb{Z}, +)$. The symplectic group $Sp(n)$ acts transitively, not only on $\Lambda(n)$, but also on the sets

$$\Lambda_k^2(n) = \left\{ (\ell, \ell') \in \Lambda(n)^2 : \dim \ell \cap \ell' = k \right\}$$

where $0 \leq k \leq n$.

We will denote by $Sp_\infty(n)$ and $\Lambda_\infty(n)$ the universal coverings of $Sp(n)$ and $\Lambda(n)$.

One can identify the Lagrangian Grassmannian with the set $W(n, \mathbb{C})$ of all unitary symmetric $n \times n$ matrices. This identification goes as follows. Defining the action of $R = A + iB$ in $U(n, \mathbb{C})$ on Lagrangian planes by

$$R\ell = \begin{pmatrix} A & -B \\ B & A \end{pmatrix} \ell$$

(this identifies $U(n, \mathbb{C})$ with a subgroup of $Sp(n)$), the mapping

$$\ell = RX^* \longmapsto W(\ell) = RR^T \quad (R \in U(n, \mathbb{C}))$$

is a bijection $\Lambda(n) \longrightarrow W(n, \mathbb{C})$ (see [20]).

Remark 2.1. The Maslov index appearing in (1.2) is given by the formula

$$m(\gamma) = \frac{1}{2\pi i} \int_{\Gamma(\gamma)} \frac{d(\det W)}{\det W} \tag{2.1}$$

where $\Gamma(\gamma)$ is a loop in $\Lambda(n) \equiv W(n, \mathbb{C})$ defined as follows: to every point $\gamma(t)$ of the Lagrangian manifold V one associates the tangent Lagrangian plane $\ell(t) = T_{\gamma(t)} V$; one then chooses a unitary matrix $R(t) = A(t) + iB(t)$, depending smoothly on t, and such that $\ell(t)$ has the equation

$$A(t)x + B(t)p = 0. \tag{2.2}$$

The loop $\Gamma(\gamma)$ is then defined by $\Gamma(\gamma)(t) = R(t)^T R(t)$ (it is independent of the choice of the equation (2.2) representing $\ell(t)$).

Identifying the universal covering $\Lambda_\infty(n)$ of the Lagrangian Grassmannian $\Lambda(n)$ with the set

$$W_\infty(n, \mathbb{C}) = \left\{ (W, \theta) : W \in W(n, \mathbb{C}), \det W = e^{i\theta} \right\},$$

the *Leray index* is, by definition, the mapping

$$m : \Lambda_\infty(n) \times \Lambda_\infty(n) \longrightarrow \mathbb{Z}$$

defined in the following way:

1) Transversal case: If the projections ℓ and ℓ' of ℓ_∞ and ℓ'_∞ are transversal, that is if $\ell \cap \ell' = 0$, then

$$m(\ell_\infty, \ell'_\infty) = \frac{1}{2\pi} \left[\theta - \theta' + i \operatorname{Tr} \operatorname{Log}(-W(W')^{-1}) \right] + \frac{n}{2} \tag{2.3}$$

where $\operatorname{Tr} \operatorname{Log} M$ is the trace of the usual logarithm of a matrix M having no eigenvalues on the negative half-axis:

$$\operatorname{Log}(M) = \int_{-\infty}^{0} \left((\lambda I - M)^{-1} - (\lambda - 1)^{-1} I \right) d\lambda.$$

The term $\mathrm{Log}(-W(W')^{-1}$ in (2.3) is well defined because

$$\ell \cap \ell' = 0 \iff \det(W - W') \neq 0$$

and hence $-W(W')^{-1}$ cannot have ≤ 0 eigenvalues.

An essential property of m is that

$$m(\ell_\infty, \ell'_\infty) - m(\ell_\infty, \ell''_\infty) + m(\ell'_\infty, \ell''_\infty) = \mathrm{Inert}(\ell, \ell', \ell'') \qquad (2.4)$$

where $\mathrm{Inert}(\ell, \ell', \ell'')$ is the index of inertia of the triple (ℓ, ℓ', ℓ'') of pairwise transversal Lagrangian planes: it is (see [14]) the common index of inertia of the quadratic forms $z \longmapsto \omega(z, z')$, $z' \longmapsto \omega(z', z'')$, $z'' \longmapsto \omega(z'', z)$ on ℓ, ℓ', ℓ'', respectively, when $z + z' + z'' = 0$.

2) *General case*: To define $m(\ell_\infty, \ell'_\infty)$ in the general case, one first makes the following essential observation ([5, 7]): the index of inertia is given by the formula

$$\mathrm{Inert}(\ell, \ell', \ell'') = \frac{1}{2}(\sigma(\ell, \ell', \ell'') + n)$$

where $\sigma(\ell, \ell', \ell'')$ is the signature of the quadratic form

$$Q(z, z', z'') = \omega(z, z') + \omega(z', z'') + \omega(z'', z)$$

on the sum $\ell \oplus \ell' \oplus \ell''$. Now, $\sigma(\ell, \ell', \ell'') + n$ is not an even integer in general, but

$$\sigma(\ell, \ell', \ell'') + n - \dim(\ell \cap \ell') + \dim(\ell \cap \ell'') - \dim(\ell' \cap \ell'')$$

is. This allows us to define the index of inertia for a triple (ℓ, ℓ', ℓ'') in general position by the formula

$$\mathrm{Inert}(\ell, \ell', \ell'') = \frac{1}{2}(\sigma(\ell, \ell', \ell'') + n - \partial \dim(\ell, \ell', \ell'')) \qquad (2.5)$$

where we have set

$$\partial \dim(\ell, \ell', \ell'') = \dim(\ell \cap \ell') - \dim(\ell \cap \ell'') + \dim(\ell' \cap \ell'').$$

Now, the signature function σ is a cocycle: $\partial \sigma = 0$, and hence so is the index of inertia:

$$\partial \, \mathrm{Inert} = \frac{1}{2}(\partial \sigma + \partial n - \partial^2 \dim) = 0.$$

This cocycle property allows us to define $m(\ell_\infty, \ell'_\infty)$ for arbitrary pairs $(\ell_\infty, \ell'_\infty)$: choose ℓ''_∞ such that $\ell'' \cap \ell = \ell'' \cap \ell' = 0$, and set

$$m(\ell_\infty, \ell'_\infty) = m(\ell_\infty, \ell''_\infty) - m(\ell'_\infty, \ell''_\infty) + \mathrm{Inert}(\ell, \ell', \ell'') \qquad (2.6)$$

where $m(\ell_\infty, \ell''_\infty)$ and $m(\ell'_\infty, \ell''_\infty)$ are calculated using formula (2.3). This defines unambiguously $m(\ell_\infty, \ell'_\infty)$, because the relation $\partial \, \mathrm{Inert} = 0$ implies that the right-hand side of (2.6) is independent of the choice of ℓ''_∞ (see [5, 7] for an explicit proof). Remark that with this definition we have $m(\ell_\infty, \ell_\infty) = 0$.

The following theorem summarizes the properties of the Leray index:

Theorem 2.1. *The function m is the only function* $\Lambda_\infty(n) \times \Lambda_\infty(n) \longrightarrow \mathbb{R}$ *having the following properties:*

(1) *The coboundary of m is the index of inertia:*

$$\partial m(\ell_\infty, \ell'_\infty, \ell''_\infty) = \text{Inert}(\ell, \ell', \ell''); \qquad (2.7)$$

(2) *m is locally constant on each of the sets*

$$\pi^{-1}(\Lambda_k^2(n)) = \{(\ell_\infty, \ell'_\infty) : \dim \ell \cap \ell' = k\};$$

(3) *m is* $Sp_\infty(n)$*-invariant:*

$$m(s_\infty \ell_\infty, s_\infty \ell'_\infty) = m(\ell_\infty, \ell'_\infty) \ , \ \forall s_\infty \in Sp_\infty(n); \qquad (2.8)$$

(4) *the action of* $\pi_1(\Lambda(n))$ *on m is given by*

$$m(\beta^k \ell_\infty, \beta^{k'} \ell'_\infty) = m(\ell_\infty, \ell'_\infty) + k - k' \ , \ \forall k \in \mathbb{Z} \qquad (2.9)$$

(β *is the generator of* $\pi_1(\Lambda(n))$ *whose image in* $\mathbb{Z}$ *is* +1).

See M. de Gosson [5, 7] for a proof.

The Leray index moreover has a "dimensional additivity" property. Before we state and prove the result, let us introduce some notation: for any integers n_1, n_2 we set $X_1 = \mathbb{R}^{n_1}$, $X_2 = \mathbb{R}^{n_2}$. The symplectic forms ω_1, ω_2 on $X_1 \oplus X_1^*$ and $X_2 \oplus X_2^*$ are related to the symplectic form ω on

$$X \oplus X^* = (X_1 \oplus X_1^*) \oplus (X_2 \oplus X_2^*)$$

by the formula $\omega = \omega_1 \oplus \omega_2$, where, by definition:

$$(\omega_1 \oplus \omega_2)(z_1 \oplus z_2, z_1' \oplus z_2') = \omega_1(z_1, z_1') + \omega_2(z_2, z_2').$$

The direct sum

$$\Lambda(n_1) \oplus \Lambda(n_2) = \{\ell_1 \oplus \ell_2 : \ell_1 \in \Lambda(n_1), \ell_2 \in \Lambda(n_2)\} .$$

is clearly a submanifold of $\Lambda(n_1 + n_2)$; similarly

$$Sp(n_1) \oplus Sp(n_2) = \{s_1 \oplus s_2 : s_1 \in Sp(n_1), s_2 \in Sp(n_2)\}$$

can be identified with a subgroup of $Sp(n)$; by definition

$$(s_1 \oplus s_2)(z_1, z_2) = (s_1 z_1, s_2 z_2).$$

Theorem 2.2. *Let* m_1, m_2 *and m be the Leray indices on* $\Lambda_\infty(n_1)$, $\Lambda_\infty(n_2)$ *and* $\Lambda_\infty(n)$ $(n = n_1 + n_2)$, *respectively. We have*

$$m(\ell_{1,\infty} \oplus \ell_{2,\infty}, \ell'_{1,\infty} \oplus \ell'_{2,\infty}) = m_1(\ell_{1,\infty}, \ell'_{1,\infty}) + m_2(\ell_{2,\infty}, \ell'_{2,\infty}). \qquad (2.10)$$

Proof. Assume first that $\ell_1 \cap \ell_1' = 0$ and $\ell_2 \cap \ell_2' = 0$. Then

$$(\ell_1 \oplus \ell_2) \cap (\ell_1' \oplus \ell_2') = 0$$

and one checks that under the identifications $\ell_1 \equiv W_1$, $\ell_1' \equiv W_1'$, etc...,

$$\mathrm{Log}(W_1(W_1')^{-1} \oplus W_2(W_2')^{-1}) = \mathrm{Log}(W_1(W_1')^{-1} \oplus \mathrm{Log}(W_1(W_1')^{-1}$$

(see S. de Gosson [11]). The equality (2.10) readily follows in this case. To prove (2.10) in the general case, it suffices to proceed as in the construction of the Leray index in the nontransversal case outlined above, and to use the obvious relation

$$\mathrm{Inert}(\ell_1 \oplus \ell_2, \ell_1' \oplus \ell_2', \ell_1'' \oplus \ell_2'') = \mathrm{Inert}_1(\ell_1, \ell_1', \ell_1'') + \mathrm{Inert}_2(\ell_2, \ell_2', \ell_2'')$$

between the indices of inertia for triples of Lagrangian planes in $\Lambda(n_1)$, $\Lambda(n_2)$ and $\Lambda(n)$. $\qquad\square$

3   Construction of the Maslov index $m(\Sigma, \ell)$

We are going to see that a symplectic path intersection index verifying Axioms (1) to (5) can be easily constructed using the Leray index. Let us introduce the following notation: if $\Sigma : [0, 1] \longrightarrow Sp(n)$ we set $s_0 = \Sigma(0)$ and $s_1 = \Sigma(1)$. We denote by $s_{0,\infty}$ the homotopy class (with fixed endpoints) of an arbitrary continuous path γ_0 in $Sp(n)$ joining I (the identity) to s_0, and by $s_{1,\infty}$ that of the concatenation $\gamma_0 * \Sigma$. The homotopy classes $s_{0,\infty}$ and $s_{1,\infty}$ are thus elements of the universal covering group of $Sp(n)$: $s_{0,\infty}, s_{1,\infty} \in Sp_\infty(n)$. Leray shows in [14] that there is a canonical injection $\pi_1(Sp(n)) \longrightarrow \pi_1(\Lambda(n))$ which is multiplication by 2 on $\mathbb{Z}$; denoting by α (resp. β) the generator of $\pi_1(Sp(n))$ (resp. $\pi_1(\Lambda(n))$) whose image in $\mathbb{Z}$ is $+1$, it follows that the action of $Sp(n)$ on $\Lambda(n)$ can be lifted into a (transitive) action of $Sp_\infty(n)$ on $\Lambda_\infty(n)$ satisfying

$$(\alpha s_\infty)\ell_\infty = \beta^2(s_\infty \ell_\infty) = s_\infty(\beta^2 \ell_\infty) \tag{3.1}$$

for all $(s_\infty, \ell_\infty) \in Sp_\infty(n) \times \Lambda_\infty(n)$.

We claim that:

Theorem 3.1. *Let ℓ_∞ be any element of $\Lambda_\infty(n)$ with projection ℓ.*
  (1) *The difference*

$$m(s_{1,\infty}\ell_\infty, \ell_\infty) - m(s_{0,\infty}\ell_\infty, \ell_\infty) \tag{3.2}$$

only depends on Σ and ℓ;
  (2) *The formula*

$$m(\Sigma, \ell) = m(s_{1,\infty}\ell_\infty, \ell_\infty) - m(s_{0,\infty}\ell_\infty, \ell_\infty) \tag{3.3}$$

defines a Maslov index for symplectic paths which satisfies the Axioms (1) through (5).

Proof of (1). For $s_\infty \in Sp_\infty(n)$ and $\ell_\infty \in \Lambda_\infty(n)$ set

$$m_\ell(s_\infty) = m(s_\infty \ell_\infty, \ell_\infty)$$

(cf. [6]). We are going to show (as the notation suggests) that $m_\ell(s_\infty)$ only depends on the projection $\pi(\ell_\infty) = \ell$; it will follow that the difference (3.2) only depends on ℓ. Suppose that ℓ'_∞ has the same projection ℓ as ℓ_∞; then $\ell'_\infty = \beta^k \ell_\infty$ for some integer k and hence

$$
\begin{aligned}
m(s_\infty \ell'_\infty, \ell'_\infty) &= m(s_\infty(\beta^k \ell_\infty), \beta^k \ell_\infty) \\
&= m(s_\infty(\beta^k \ell_\infty), \ell_\infty) - k \\
&= m(\beta^k \ell_\infty, s_\infty^{-1} \ell_\infty) - k \\
&= m(\ell_\infty, s_\infty^{-1} \ell_\infty) \\
&= m(s_\infty \ell_\infty, \ell_\infty)
\end{aligned}
$$

by repeated use of the properties (2.9), (2.8). Let us next show that (3.2) is independent of the choice of $s_{0,\infty}$; it will follow that (3.2) only depends on Σ. Suppose we change the path γ_0 whose equivalence class is $s_{0,\infty}$ into another path γ'_0 joining I to s_0. Then $s_{0,\infty}$ will be changed into some $s'_{0,\infty} \in Sp_\infty(n)$ having the same projection s_0 on $Sp(n)$, and $s_{1,\infty}$ will be changed into $s'_{1,\infty}$, the homotopy class of the concatenation $\gamma'_0 * \Sigma$. We can thus find an integer k such that $s'_{0,\infty} = \alpha^k s_{0,\infty}$, $s'_{1,\infty} = \alpha^k s_{1,\infty}$, and hence, by (3.1):

$$
\begin{aligned}
&m(s'_{1,\infty} \ell_\infty, \ell_\infty) - m(s'_{0,\infty} \ell_\infty, \ell_\infty) \\
=&m((\alpha^k s_{1,\infty}) \ell_\infty, \ell_\infty) - m((\alpha^k s_{0,\infty}) \ell_\infty, \ell_\infty) \\
=&m(\beta^{2k}(s_{1,\infty} \ell_\infty), \ell_\infty) - m(\beta^{2k}(s_{0,\infty} \ell_\infty), \ell_\infty) \\
=&m(s_{1,\infty} \ell_\infty, \ell_\infty) - m(s_{0,\infty} \ell_\infty, \ell_\infty)
\end{aligned}
$$

which ends the proof of the first part of the theorem.

Proof of (2). Axioms (1), (2) and (3) are obviously satisfied: the first by the definition of a universal covering, the second is trivial, and the third by Theorem 2.2. Let us prove that $m(\Sigma, \ell)$ satisfies Axiom (4). Denoting by $s_{t,\infty}$ the homotopy class of $\gamma_0 * \Sigma(t)$, the condition that dim $\Sigma(t)\ell \cap \ell$ remains constant implies that the continuous function

$$t \longmapsto m(s_{t,\infty} \ell_\infty, \ell_\infty)$$

must also be constant, in view of the topological property (2) in Theorem 2.1 of the Leray index. There remains to check the "spectral flow formula" (1.4). Let us begin by calculating the Maslov index $m(\Sigma_P, X^*)$ where Σ_P is the path defined by (1.3). Since

$$
\begin{pmatrix} I_{n \times n} & 0_{n \times n} \\ -P(t) & I_{n \times n} \end{pmatrix} \begin{pmatrix} 0 \\ p \end{pmatrix} = \begin{pmatrix} 0 \\ p \end{pmatrix}
$$

we will have dim $\Sigma_P(t)X^* \cap X^* = n$ for all t and hence, using again the topological property (2), Theorem 2.1:

$$m(\Sigma_P, X^*) = m(s_{1,\infty}X^*_\infty, X^*_\infty) - m(s_{0,\infty}X^*_\infty, X^*_\infty) = 0. \tag{3.4}$$

In view of the cohomological property (1) (*ibid.*) we have, for every $s_\infty \in Sp_\infty(n)$:

$$m(s_\infty X_\infty, X_\infty) - m(s_\infty X^*_\infty, X^*_\infty) = \text{Inert}(sX, X, X^*) - \text{Inert}(sX, sX^*, X^*)$$

from which follows that

$$\begin{aligned} m(\Sigma_P, X) = {} & \text{Inert}(s_1 X, X, X^*) - \text{Inert}(s_0 X, X, X^*) \\ & - (\text{Inert}(s_1 X, s_1 X^*, X^*) - \text{Inert}(s_0 X, s_0 X^*, X^*)). \end{aligned}$$

Since $s_0 X^* = s_1 X^* = X^*$ we have, in view of the antisymmetry of the signature σ:

$$\text{Inert}(s_1 X, s_1 X^*, X^*) = \frac{1}{2}\sigma(s_1 X, X^*, X^*) = 0$$

and

$$\text{Inert}(s_0 X, s_0 X^*, X^*) = \frac{1}{2}\sigma(s_0 X, X^*, X^*) = 0.$$

The Maslov index is thus given by

$$m(\Sigma_P, X) = \text{Inert}(s_1 X, X, X^*) - \text{Inert}(s_0 X, X, X^*). \tag{3.5}$$

Set now $s_t = \Sigma_{P(t)}$; we have, using the fact that dim $X \cap X^* = 0$ and dim $s_t X \cap X^* = 0$:

$$\text{Inert}(s_t X, X, X^*) = \frac{1}{2}(\sigma(s_t X, X, X^*) + n - \dim(s_t X \cap X)).$$

A direct calculation shows that

$$\sigma(s_t X, X, X^*) = -\operatorname{sign} P(t) \ , \ \dim(s_t X \cap X)) = \operatorname{corank} P(t)$$

and hence

$$\text{Inert}(s_t X, X, X^*) = \frac{1}{2}(-\operatorname{sign} P(t) + \operatorname{rank} P(t)) = \text{Inert } P(t).$$

The spectral flow formula (1.4) follows in view of (3.5). $\qquad\square$

Remark 3.2. Using property (2.7) of the Leray index, one can rewrite definition (3.3) of the Maslov index as

$$m(\Sigma, \ell) = m(s_{0,\infty}^{-1}\ell_\infty, s_{1,\infty}^{-1}\ell_\infty) - \text{Inert}(\ell, s_0^{-1}\ell, s_1^{-1}\ell).$$

Remark 3.3. The following formula, which makes explicit the effect of a change of ℓ in the Maslov index, also follows from (3.3):

$$\begin{aligned} m(\Sigma, \ell) - m(\Sigma, \ell') = {} & \text{Inert}(s_1\ell, \ell, \ell') - \text{Inert}(s_1\ell, s_1\ell', \ell') \\ & - (\text{Inert}(s_0\ell, \ell, \ell') - \text{Inert}(s_0\ell, s_0\ell', \ell')). \end{aligned}$$

The relation between the Maslov index $m(\Sigma, \ell)$ and the index $m(\gamma)$ defined by Maslov is the following:

Corollary 3.2. *Let γ be a loop in an oriented Lagrangian manifold; then*

$$m(\gamma) = m(\alpha^{m(\gamma)/2}, \ell) \tag{3.6}$$

where α is the generator of $\pi_1(Sp(n))$ and ℓ an arbitrary Lagrangian plane.

Proof. Recall that $m(\gamma)$ is an even integer when γ is carried by an oriented Lagrangian manifold. By definition,

$$m(\alpha^{m(\gamma)/2}, \ell) = m((\alpha^{m(\gamma)/2} I_\infty)\ell_\infty, \ell_\infty) - m(\ell_\infty, \ell_\infty)$$

where I_∞ is the unit of $Sp_\infty(n)$. In view of (3.1) we have

$$m((\alpha^{m(\gamma)/2} I_\infty)\ell_\infty, \ell_\infty) = m(\beta^{m(\gamma)}\ell_\infty, \ell_\infty)$$

hence (3.6), using (2.9) in Theorem 2.1.   □

4 Conclusion

The Leray index and the universal covering group $Sp_\infty(n)$ are topological objects. Since our constructions involves only Leray's index and $Sp_\infty(n)$, it follows that the Maslov index $m(\Sigma, \ell)$ is itself a purely topological object. (The latter is certainly not clear from the constructions in [19], and cited by [16], since these authors use the properties of differential paths.)

We also remark that Booss-Bavnbek and Furutani have constructed in [3] a Maslov index for symplectic paths in infinite dimensional Hilbert spaces; their construction, however, deeply differs from the one given here in the finite dimensional case (in particular, these authors do not define an infinite-dimensional "Leray index").

References

[1] Arnold, V.I., *A characteristic class entering in quantization conditions*, Funkt. Anal. i. Priloz. **1**(1), 1–14 (in Russian) (1967); Funct. Anal. Appl. **1**, 1–14 (English translation) (1967).

[2] Brack, M. and Bhaduri R.K., *Semiclassical Physics*, Addison-Wesley, 1997.

[3] Booss-Bavnbek B. and Furutani K., *The Maslov Index: A Functional Analytical Definition and the Spectral Flow Formula*, Tokyo J. Math. **21**(1) (1998).

[4] Creagh, S.C., Robbins, J.M. and Littlejohn, R.G., *Geometrical properties of Maslov indices in the semiclassical trace formula for the density of states*, Phys. Rev. A, **42**(4), 1990.

[5] de Gosson, M., *La définition de l'indice de Maslov sans hypothèse de transversalité*, C.R. Acad. Sci., Paris, **309**, Série I, (1990), 279–281.

[6] de Gosson, M., *La relation entre Sp_∞, revêtement universel du groupe symplectique Sp et Sp $\times$ $\mathbb{Z}$*, C.R. Acad. Sci., Paris, **310**, Série I, (1990), 245–248.

[7] de Gosson, M., *The structure of q-symplectic geometry*, J. Math. Pures et Appl. **71** (1992), 429–453.

[8] de Gosson, M., *Lagrangian path intersections and the Leray index: Aarhus Geometry and Topology Conference*, Contemp. Math. **258**, Amer. Math. Soc., Providence, RI, 2000, pp. 177–184.

[9] de Gosson, M., de Gosson, S., *The Cohomological Meaning of Maslov's Lagrangian Path Intersection Index*. In Proceedings of the Conference in the Honor of Jean Leray, Karlskrona 1999, (M. de Gosson, ed.), Kluwer Acad. Publ., 2001.

[10] de Gosson, S., *Lagrangian Path Intersection Indices*, MSc Thesis, Kalmar, 1998.

[11] de Gosson, S., *Maslov Indices for Symplectic Paths and Applications to Trace Formulae*, Ph.D. Thesis, Växjö, 2002.

[12] Einstein, A., *Zum Quantensatz von Sommerfeld und Epstein*, Verhandlungen der Deutschen Phys. Ges., nr. 9/10 (1917).

[13] Keller, J.B., *Corrected Bohr–Sommerfeld Quantum Conditions for Nonseparable Systems*, Ann. of Physics **4** (1958), 180–188.

[14] Leray, J., *Lagrangian Analysis and Quantum Mechanics, a mathematical structure related to asymptotic expansions and the Maslov index*, MIT Press, Cambridge, Mass., 1981; translated from *Analyse Lagrangienne* RCP 25, Strasbourg Collège de France, 1976–1977.

[15] Long, Y., *Precise Iteration Formulae of the Maslov-type Index Theory and Ellipticity of Closed Characteristics*, Advances in Mathematics **154** (2000), 76–131.

[16] McDuff, D and Salamon, D., *Introduction to Symplectic Topology*, Oxford Science Publications, 1998.

[17] Maslov, V.P., *Théorie des Perturbations et Méthodes Asymptotiques*, Dunod, Paris, 1972; translated from Russian [original Russian edition 1965].

[18] Maslov, V.P. and Fedoriuk, M.V., *Semi-Classical Approximations in Quantum Mechanics*, Reidel, Boston, 1981.

[19] Robbin, J.W. and Salamon, D.A, *The Maslov index for paths*, Topology **32** (1993), 827–44.

[20] Souriau, J.-M., *Construction explicite de l'indice de Maslov*, Group Theoretical Methods in Physics, Lecture Notes in Physics **50**, Springer-Verlag, 1975, pp. 17–148.

[21] Souriau, J.-M., *Indice de Maslov des variétés lagrangiennes orientables*, C. R. Acad. Sci., Paris, Série A, **276** (1973), 1025–1026.

[22] Sugita, A., *Geometrical properties of Maslov indices in periodic-orbit theory*, Phys. Lett. A **266** (2000), 321–330.

Maurice de Gosson
Blekinge Institute of Technology
371 79 Karlsrona, Sweden
mdg@bth.se

Serge de Gosson Växjö
University MSI, Mathematics and System Engineering
351 95 Växjö, Sweden
sergedegosson@hotmail.com

A Global Cauchy–Kowalewski Theorem in Some Gevrey Classes

Daniel Gourdin and Mustapha Mechab

ABSTRACT We solve a global linear Cauchy–Kowalewski problem for partial differential equations with order m in two different situations.

The first situation extends both results of Treves (using Fourier–Borel isomorphism) and of Persson (using the Ovcyanikov theorem) and more recently those of Pongerard and Wagschal (in the class of entire functions with respect to (t, x)) in a class of functions which are entire with respect to t and Gevrey with respect to x with bounded index d.

The second situation generalises the theorem of Nagumo and both results of Choquet-Bruhat and Tarama (using hyperbolicity) to a class of functions which are m times differentiable with respect to t and Gevrey with respect to x with bounded index d.

1   Statement of results

We denote by t the points of $\mathbb{C}$ or $\mathbb{R}$, $x = (x_1, \ldots, x_n)$ those of $\mathbb{R}^n$, $\alpha = (\alpha_1, \ldots, \alpha_n) \in \mathbb{N}^n$, $|\alpha| = \displaystyle\sum_{i=1}^{n} \alpha_i$, $\alpha! = \displaystyle\prod_{i=1}^{n} (\alpha_i!)$, $D_x^\alpha = D_{x_1}^{\alpha_1} \ldots D_{x_1}^{\alpha_1}$, $D_{x_i}^{\alpha_i} = \dfrac{\partial^{\alpha_i}}{\partial x_i^{\alpha_i}}$.

Definition 1.1. Let $d \geq 1$ and Ω be an open set of $\mathbb{R}^n$. $\mathbf{G}^{(d)}(\Omega)$ is the set of functions $v \in C^\infty(\Omega)$ such that for all compact $K \subset \Omega$,

$$\forall h > 0, \ \exists C > 0, \ \forall \alpha \in \mathbb{N}^n, \quad \sup_{x \in K} \left| D_x^\alpha v(x) \right| \leq C h^{|\alpha|} (|\alpha|!)^d . \tag{1.1}$$

We consider the Cauchy problem

$$\begin{cases} D_t^m u(t, x) = \displaystyle\sum_{(j,\alpha) \in B} a_{j,\alpha}(t, x) D_x^\alpha D_t^j u(t, x) + f(t, x), & \text{in } \mathbb{R} \times \mathbb{R}^n \\[2mm] & \text{(resp. in } \mathbb{C} \times \mathbb{R}^n), \\[2mm] D_t^j u(0, x) = v_j(x) \text{ in } \mathbb{R}^n, \ j = 0, 1, \ldots, m-1 \end{cases} \tag{1.2}$$

where:

- B is a part of $\{(j, \alpha) \in \mathbb{N} \times \mathbb{N}^m, \ j < m, \ \alpha \neq 0, \ j + |\alpha| \leq m\}$.

- $a_{j,\alpha}(t, x) = \sum_{|\beta| < |\alpha|} a^{\beta}_{j,\alpha}(t) x^{\beta}.$

We say that v is in the projective Gevrey class with index d upon Ω.

Definition 1.2. $\mathbf{G}^{(0,d)}(\mathbf{R} \times \Omega)$ is the space of functions u defined on $\mathbf{R} \times \Omega$ with continuous derivatives $D^{\alpha}_x u$ such that for all $T > 0$ and K a compact subset of Ω,

$$\forall h > 0, \ \exists C > 0, \tag{1.3}$$

$$\forall (t, x) \in] - T, T[\times K, \ \forall \alpha \in \mathbf{N}^n, \quad \left| D^{\alpha}_x u(t, x) \right| \leq C h^{|\alpha|} (|\alpha|!)^d. \tag{1.4}$$

Definition 1.3. $\mathbf{G}^{(m,d)}(\mathbf{R} \times \Omega)$ is the space of functions u defined on $\mathbf{R} \times \Omega$ such that $D^j_t u \in \mathbf{G}^{(0,d)}(\mathbf{R} \times \Omega)$ for all $j = 0, \ldots, m$.

Definition 1.4.

- $C^{\omega,\infty}(\mathbb{C} \times \Omega)$ is the algebra of functions $u : \mathbb{C} \times \Omega \longrightarrow \mathbb{C}$ with derivatives $D^{\alpha}_x u$ continuous on $\mathbb{C} \times \Omega$ and holomorphic with respect to t.

- $\mathbf{G}^{(\omega,d)}(\mathbb{C} \times \Omega)$ is the algebra of functions $u \in C^{\omega,\infty}(\mathbb{C} \times \Omega)$ such that for all $T > 0$ and compact $K \subset \Omega$: $\forall h > 0, \ \exists C_h > 0,$

$$\forall t \in \mathbb{C}, \ |t| \leq T, \ \forall x \in K, \ \forall \alpha \in \mathbf{N}^n, \quad \left| D^{\alpha}_x u(t, x) \right| \leq C h^{|\alpha|} (|\alpha|!)^d. \tag{1.5}$$

Definition 1.5. The Cauchy problem (1.2) is well posed in $\mathbf{G}^{(\omega,d)}(\mathbb{C} \times \mathbf{R}^n)$ (resp. $\mathbf{G}^{(m,d)}(\mathbf{R} \times \mathbf{R}^n)$) if for all functions $v_0, \ldots, v_{m-1}$ in $\mathbf{G}^{(d)}(\mathbf{R}^n)$ and all functions $f \in \mathbf{G}^{(\omega,d)}(\mathbb{C} \times \mathbf{R}^n)$ (resp. $\mathbf{G}^{(m,d)}(\mathbf{R} \times \mathbf{R}^n)$) there exist unique functions $u \mathbf{G}^{(\omega,d)}(\mathbb{C} \times \mathbf{R}^n)$ (resp. $\mathbf{G}^{(m,d)}(\mathbf{R} \times \mathbf{R}^n)$) verifying (1.2).

Let

$$\delta = \sup\{s \geq 1; \ \forall (j, \alpha) \in B, \ j + s|\alpha| \leq m\}$$

Theorem 1.6. *For all entire functions* $a^{\beta}_{j,\alpha}$, $(j, \beta) \in B$, $|\beta| < |\alpha|$, *the Cauchy problem (1.2) is well posed in* $\mathbf{G}^{(\omega,d)}(\mathbb{C} \times \mathbf{R}^n)$, *for all* $d \in [1, \delta]$.

Theorem 1.7. *For all functions* $a^{\beta}_{j,\alpha} \in \mathcal{B}^0(\mathbf{R})$ *(i.e., continuous and bounded on* $\mathbf{R}$*),* $(j, \beta) \in B$ $|\beta| < |\alpha|$, *the Cauchy problem (1.2) is well posed in* $\mathbf{G}^{(m,d)}(\mathbf{R} \times \mathbf{R}^n)$, *for all* $d \in [1, \delta]$.

Remarks.

R1. Theorem 1.6 shows that the regularity $\mathbf{G}^{(d)}$ of the initial data is transmitted to the solution u of any time t.

R2. For equations of this kind

$$D_t^m u(t, x) = \sum_{j=0}^{m-1} \sum_{0<|\alpha|\leq m-j} a_{j,\alpha}(t) D_t^j D_x^\alpha u(t, x) + f(t, x). \qquad (1.6)$$

Theorem 1.6 solves the problem with complete initial data, verifying the same inequalities as in [7], and this is also proved in [7].

One difference with [7] is that for some operators, we can also solve an associated Cauchy problem, although initial data are not complete. For example, the Cauchy problem for

$$P(D) = D_t^4 + D_t^2 D_x$$

is well posed in the class of functions which are entire with respect to time and Gevrey with index $d \in [1, 2]$ with respect to spatial variable x, and this is not proved in [7].

R3. Theorem 1.7 is a generalization of the Nagumo Theorem. The same results are proved in [8] under the hyperbolicity condition. Our results are more "precise" than those of [8] in the sense that Theorem 1.6 solves the Cauchy problem for initial data in $G^{(d)}(\mathbf{R}^n)$. On the other hand, our technique permits us to solve the global Cauchy problem for all finite time (i.e., in $] - T, T[\times \mathbf{R}^n)$.

2 Sketch of the proof

We prove only Theorem 1.6,

2.1 Transformation of the problem

With the change of unknowns

$$v(t, x) = u(t, x) - \sum_{j=0}^{m-1} \frac{1}{j!} t^j v_j(x), \qquad (2.1)$$

denoting by $D_t^{-1} u$ the primitive of u with respect to t which vanishes for $t = 0$, and denoting by $D_t^m u = v$ (although u has been used before), the problem is equivalent to the equation

$$\mathcal{L}u = \sum_{(j,\alpha)\in B} a_{j,\alpha}(t, x) D_t^{j-m} D_x^\alpha u(t, x)$$

$$+ \sum_{(j,\alpha)\in B} \sum_{k=j}^{m-1} a_{j,\alpha}(t, x) \frac{1}{(k-j)!} t^{k-j} D_x^\alpha v_k(x) + f(t, x), \qquad (2.2)$$

which means that the solution is a fixed point of the application $\mathcal{L}$ defined by

$$\mathcal{L}u = \sum_{(j,\alpha) \in B} a_{j,\alpha}(t,x) D_t^{j-m} D_x^\alpha u(t,x) \tag{2.3}$$

$$+ \sum_{(j,\alpha) \in B} \sum_{k=j}^{m-1} a_{j,\alpha}(t,x) \frac{1}{(k-j)!} t^{k-j} D_x^\alpha v_k(x) + f(t,x).$$

2.2 The Banach algebras $\mathbf{G}^{\omega,d}_{\rho,T,\zeta}(\mathcal{U}_{\rho,T} \times \mathbb{R}^n)$.

For $T, \rho > 0$, we denote by $\mathcal{U}_T = \{t \in \mathbb{C}; |t| \leq T\}$, $\mathcal{U}_{\rho,T} = \{t \in \mathbb{C}; \rho|t| \leq T\}$, then $\mathcal{U}_{\rho,T} = \mathcal{U}_T$.

For $\zeta = (\zeta_1, \ldots, \zeta_n) \in (\mathbb{R}_+^*)^d$, $\mathbf{G}^{\omega,d}_{\rho,T,\zeta}(\mathcal{U}_{\rho,T} \times \mathbb{R}^n)$ is the algebra of functions $u \in C^{\omega,\infty}(\mathcal{U}_{\rho,T} \times \mathbb{R}^n)$ such that: $\exists C > 0$, $\forall \alpha \in \mathbb{N}^n$, $\forall k \in \mathbb{N}$,

$$\forall x \in \mathbb{R}^n, \quad \left| D_t^k D_x^\alpha u(0,x) \right| \leq C \zeta^\alpha \rho^k (|\alpha|!)^{d-1} D^{k+|\alpha|} \phi_T(0) \tag{2.4}$$

where

$$\phi_T(t) = \frac{1}{L}\theta\left(\frac{t}{T}\right), \qquad \theta(t) = \sum_{j=0}^{\infty} \frac{t^j}{(j+1)^2}$$

with L verifying $\theta^2 \ll L\theta$, (i.e., the module of each coefficient of the expansion θ^2 is less than L times the corresponding coefficient of the expansion θ).

This expression (2.4) (cf Wagschal [10]) is denoted by

$$u \ll C\Phi^{\omega,d}_{\rho,T,\zeta}$$

and with the norm

$$\|u\| = \inf \left\{ C > 0, \ u \ll C\Phi^{\omega,d}_{\rho,T,\zeta} \right\},$$

$\mathbf{G}^{\omega,d}_{\rho,T,\zeta}(\mathcal{U}_{\rho,T} \times \mathbb{R}^n)$ is a Banach algebra.

We have the following properties:

P1. For all ρ, $T > 0$ and $\zeta \in (\mathbb{R}_+^*)^n$,

$$\mathbf{G}^{\omega,d}_{\rho,T,\zeta}(\mathcal{U}_{\rho,T} \times \mathbb{R}^n) \subset \mathbf{G}^{\omega,d}(\mathcal{U}_{\rho,T} \times \mathbb{R}^n)$$

where $\mathbf{G}^{\omega,d}(\mathcal{U}_{\rho,T} \times \mathbb{R}^n)$ is the space of functions $u \in C^{\omega,\infty}(\mathcal{U}_{\rho,T} \times \mathbb{R}^n)$ such that there exists $C > 0$ verifying

$$\forall \alpha \in \mathbb{N}^n, \ \forall t \in \mathcal{U}_{\rho,T}, \ \forall x \in \mathbb{R}^n, \quad \left| D_x^\alpha u(t,x) \right| \leq C^{|\alpha|+1}(|\alpha|!)^d.$$

P2. For all $\zeta \in (\mathbf{R}_+^*)^n$, $C > 0$ and $u \in \mathbf{G}^{\omega,d}(\mathcal{U}_{\rho,T} \times \mathbf{R}^n)$ such that $u(0, .) = 0$, there exists $T_0 > 0$ such that

$$\forall T' \in \,]0, T_0[, \quad u \in \mathbf{G}^{\omega,d}_{\rho,T',\zeta}(\mathcal{U}_{\rho,T'} \times \mathbf{R}^n).$$

P3. For all $(k, \alpha) \in \mathbf{Z} \times \mathbf{N}^n$ such that $k + d|\alpha| \leq 0$, there exists $C_{k,\alpha} > 0$ such that $\forall \rho, T > 0$, $\forall \zeta \in (\mathbf{R}_+^*)^n$,

$$\forall u \in \mathbf{G}^{\omega,d}_{\rho,T,\zeta}(\mathcal{U}_{\rho,T} \times \mathbf{R}^n), \quad \|D_t^k D_x^\alpha u\| \leq C_{k,\alpha} \zeta^\alpha \rho^k T^{-k-|\alpha|}\|u\|. \qquad (2.5)$$

For all compact K of $\mathbf{R}^n$ such that $\overset{\circ}{K} \neq \emptyset$, we denote

$$K_T = \mathcal{U}_T \times \overset{\circ}{K}.$$

Adapting the computations of [10] (p. 333), and acting on h instead of T, we prove the following lemmas.

Lemma 2.1. *If $f \in \mathbf{G}^{\omega,d}(\mathbf{C} \times \mathbf{R}^n)$, then for all compact $K \subset \mathbf{R}^n$ ($\overset{\circ}{K} \neq \emptyset$) and for all T, $\rho > 0$ and $\zeta \in (\mathbf{R}_+^*)^n$, $f \in \mathbf{G}^{\omega,d}_{\rho,\rho T,\zeta}(K_T)$.*

Lemma 2.2. *Every entire function a is in $\mathbf{G}^{\omega,d}_{\rho,\rho T,\zeta}(K_T)$ (the norm of a depends only on T).*

Lemma 2.3. *Every function $v \in \mathbf{G}^{(d)}(\mathbf{R}^n)$ is in $\mathbf{G}^{\omega,d}_{\rho,\rho T,\zeta}(K_T)$.*

Lemma 2.4. *Let T, $\rho > 0$ and K be a compact subset of $\mathbf{R}^n$ such that $\overset{\circ}{K} \neq \emptyset$. All polynomials P of degree k are is in $\mathbf{G}^{0,d}_{\rho,\zeta,\rho T}\left(\overset{\circ}{K}_T\right)$ and there exists a constant $C_{P,K} > 0$, which does not depend on ρ, T and ζ, such that*

$$\|P\| \leq C_{P,K} \max_{|\gamma| \leq k} \left\{(\rho T)^{|\gamma|} \frac{1}{\zeta^\gamma}\right\}.$$

2.3 Fixed points of $\mathcal{L}$ defined on K_T

Proposition 2.1. *For all $T > 0$, $\zeta \in (\mathbf{R}_+^*)^n$ and for all compact $K \subset \mathbf{R}^n$ ($\overset{\circ}{K} \neq \emptyset$), there exist $\rho_0 > 0$ and $C \in \,]0, 1[$ such that for all $\rho \geq \rho_0$ there exist $r_\rho > 0$ satisfying*

$$\forall r \geq r_\rho, \quad \mathcal{L}(\mathcal{B}(0, r)) \subset \mathcal{B}(0, r) \subset \mathbf{G}^{\omega,d}_{\rho,\rho T,\zeta}(K_T), \qquad (2.6)$$

$$\forall u, u' \in \mathcal{B}(0, r) \subset \mathbf{G}^{\omega,d}_{\rho,\rho T,\zeta}(K_T), \quad \|\mathcal{L}u - \mathcal{L}u'\| \leq C\|u - u'\| \qquad (2.7)$$

where $\mathcal{B}(0, r)$ is the closed ball with center 0 and radius r in $\mathbf{G}^{\omega,d}_{\rho,\rho T,\zeta}(K_T)$.

Proof. We denote

$$\Psi(t, x) = \sum_{(j,\alpha)\in B} \sum_{k=j}^{m-1} a_{j,\alpha}(t, x)\frac{1}{(k-j)!}t^{j-m}D_x^\alpha v_k(x) + f(t, x).$$

As $G^{\omega,d}_{\rho,\rho T,\zeta}(K_T)$ are Banach algebras, thanks to Lemmas 2.1–2.4, we show that $\Psi \in G^{\omega,d}_{\rho,\rho T,\zeta}(K_T)$; then with (2.5) we get

$$\forall u \in G^{\omega,d}_{\rho,\rho T,\zeta}(K_T), \quad \|\mathcal{L}u\| \leq C_0 \sum_{(j,\alpha)\in B} \zeta^\alpha \rho^{-|\alpha|}T^{-(j-m)-|\alpha|}\|u\| + \|\Psi\| \quad (2.8)$$

where C_0 does not depend on ρ. As for each $(j, \alpha) \in B$ we have $\alpha \neq 0$, then if $\rho > 0$ is large enough, we have

$$0 < C_1 = C_0 \sum_{(j,\alpha)\in B} \zeta^\alpha \rho^{-|\alpha|}T^{-(j-m)-|\alpha|} < 1$$

and (2.6) is fullfilled as soon as $r_\rho \geq \dfrac{\|\Psi\|}{1-C_1}$.
In the same way (2.7) is also verified.

With straightforward computations, we prove the following lemma.

Lemma 2.5. *For all $\zeta_1 \leq \zeta_2$ and $\rho_1 \geq \rho_2$ we have*

$$G^{\omega,d}_{\rho_1,\rho_1 T,\zeta_1}(K_T) \subset G^{\omega,d}_{\rho_2,\rho_2 T,\zeta_2}(K_T).$$

For $p \in \mathbf{N}^*$, we denote $\zeta_p = \left(\dfrac{1}{p}, \ldots, \dfrac{1}{p}\right) \in (\mathbf{R}_+^*)^n$; with Proposition 2.1 and Lemma 2.5 we get:

Proposition 2.2. *We can choose an increasing sequence of positive numbers $(\rho_p)_p$ and a fixed point of $\mathcal{L}$:*

$$u_1 \in \bigcap_{p\in\mathbf{N}^*} G^{\omega,d}_{\rho_p,\rho_p T,\zeta_p}(K_T)$$

unique in $G^{\omega,d}_{\rho_1,\rho_1 T,\zeta_1}(K_T).$

2.4 Fixed point of $\mathcal{L}$ defined on $\mathcal{U}_T \times \mathbf{R}^n$

Let $T > 0$ and an exhaustive sequence $(K_p)_p$ of compacts such that $\overset{\circ}{K}_p \neq \emptyset$ and $\bigcup_p K_p = \mathbf{R}^n$.

From Proposition 2.2, by induction on $j \in \mathbf{N}^*$, we construct an increasing sequence of positive numbers $\rho_{p,j} = \rho_{p,K_j}$ such that

$$\rho_{1,1} < \rho_{2,2} > \cdots < \rho_{j-1,j-1} < \rho_{1,j} < \rho_{2,j} < \cdots < \rho_{p,j} < \cdots$$

and a fixed point $u_j \in \bigcap_{p \in \mathbf{N}^*} \mathbf{G}^{\omega,d}_{\rho_{p,j},\rho_{p,j}T,\zeta_p}(\mathcal{U}_T \times \overset{\circ}{K}_j)$ of $\mathcal{L}$ unique in each

$\mathbf{G}^{\omega,d}_{\rho_{p,j},\rho_{p,j}T,\zeta_p}(\mathcal{U}_T \times \overset{\circ}{K}_j)$.

From Lemma 2.5 and unicity of the fixed point of $\mathcal{L}$ in

$$\mathbf{G}^{\omega,d}_{\rho_{j-1,j-1},\rho_{j-1,j-1}T,\zeta_{j-1}}(\mathcal{U}_T \times \overset{\circ}{K}_{j-1})$$

we get $u_j\big|_{\overset{\circ}{K}_{j-1}} = u_{j-1}$. By gluing of u_j we define a function u on $\mathcal{U}_T \times \mathbf{R}^n$ which is a fixed point of $\mathcal{L}$ such that

$$\forall j \in \mathbf{N}^*, \quad u\big|_{\overset{\circ}{K}_j} \in \mathbf{G}^{\omega,d}_{\rho_{j,j},\rho_{j,j}T,\zeta_j}(\mathcal{U}_T \times \overset{\circ}{K}_j).$$

As the sequence $(K_j)_j$ is exhaustive in $\mathbf{R}^n$, for every compact K of $\mathbf{R}^n$, there exists $\rho_0 \in \mathbf{N}$ such that $K \subset K_p$ for all $p \geq p_0$ and

$$\forall p \geq p_0, \quad u \in \mathbf{G}^{\omega,d}_{\rho_{p,p},\rho_{p,p}T,\zeta}(\mathcal{U}_T \times \overset{\circ}{K}).$$

With the calculations of [10] p. 333 and [2] p. 577, we prove that u verifies (2.2) in $\mathcal{U}_{T'} \times \mathbf{R}^n$, so $u \in \mathbf{G}^{(\omega,d)}(\mathcal{U}_{T'} \times \mathbf{R}^n$ for all $T' \in]0, T[$.

2.5 End of the proof of Theorem 1.6

With property **P2**, we prove local unicity of the fixed point, as a holomorphic function in t. So we get uniqueness in $\mathcal{U}_T \times \mathbf{R}^n$ and by gluing we get a unique fixed point u of $\mathcal{L}$ in the space $\mathbf{G}^{(\omega,d)}(\mathbb{C} \times \mathbf{R}^n)$; then

$$v = D_t^{-m} + \sum_{j=0}^{m-1} \frac{1}{j!} t^j v_j(x)$$

is the solution of the problem (1.2).   □

References

[1] Y. Choquet-Bruhat, *Diagonalisation des systèmes quasi linéaires et hyperbolicité non stricte*, J. Maths Pures et Appl. **45** (1966), 371–386.

[2] D. Gourdin, M. Mechab, *Problème de Goursat non linéaire dans les espaces de Gevrey pour les équations de Kirchhoff généralisées*, J. Math. Pures Appl. **75** (1996), 569–593.

[3] D. Gourdin, M. Mechab, *Solutions globales du problème de Goursat non linéaire dans les classes de Gevrey*, Bull. Sci. Math. **121** (1997), 323–344.

[4] J. Leray, *Hyperbolic Differential Equations* (Chapter 73), Cours de Princeton, 1953.

[5] J. Leray, L. Waelbroeck, *Norme formelle d'une fonction composée (Préliminaire à l'étude des systèmes non linéaires, hyperboliques non stricts)*, Colloque de Liège, CBRM, 1964, pp. 145–152.

[6] J. Persson, *On the local and global non-characteristic Cauchy problem when the solutions are holomorphic functions or analytic functionals in space variables*, Arkiv for Mathematik **9** (1971), 171–180.

[7] P. Pongérard, Cl. Wagschal, *Problème de Cauchy dans des espaces de fonctions entières*, J. Math. Pures Appl. **75** (1996), 409–418.

[8] S. Tarama, *Sur les équations hyperboliques à coefficients analytiques par rapport aux variables spaciales*, J. Math. Kyoto Uni. **27**(3) (1987), 553–556.

[9] F. Trèves, *Linear partial differential equations with constant coeficients*, Gordon and Breach, New York, 1966.

[10] C. Wagschal, *Le problème de Goursat non linéaire*, J. Math. Pure et Appl. **58** (1979), 309–337.

Daniel Gourdin
Université Paris 6
U.F.R 920, 4 Place Jussieu
75252 Paris cedex 05, France
gourdin@math.jussieu.fr

Mustapha Mechab
Université Djilali Liabès
Laboratorie de Mathmatiques, B.P 89
22000 Sidi Bel Abbès, Algeérie

Sub-Riemannian Geometry and Subelliptic PDEs

Peter C. Greiner

ABSTRACT I shall discuss the connection between complex Hamiltonian mechanics and sub-Riemannian geometry on the Heisenberg group. Using these geometric concepts I shall describe the subelliptic heat kernel and its small time asymptotics. To extend this work to higher step operators I shall apply some of these concepts to a particular step 3 example.

Introduction

$X_1, \ldots, X_m$ are linearly independent vector fields on M_n,

$$\Delta_X = \frac{1}{2} \sum_{j=1}^{m} X_j^2,$$

$m = n \Rightarrow \Delta_X$ is elliptic and $m < n$ (+ conds.) $\Rightarrow \Delta_X$ is subelliptic. We want to construct: fundamental solutions (inverse kernel), heat kernels, and wave kernels.

Δ_X induces a geometry:

$X_1, \ldots, X_m$ are $\perp$ and have length 1; this yields a metric and a volume.

Given a curve $\gamma(s)$,

$$\gamma'(s) = \sum_{1}^{m} a_j(s) X_j\big(\gamma(s)\big),$$

whose length is given by

$$\ell(\gamma) = \int \sqrt{\sum_{1}^{m} a_j^2}.$$

The distance between two points is the minimum of the lengths of curves which connect the two points with tangents in the *horizontal* subspace $X = \{X_1, \ldots, X_m\}$ of TM_n. If X generates TM_n by bracketing, then by Chow's theorem there always exists a curve with horizontal tangents between every two points, and minimizing lengths we obtain a distance; Riemannian if $m = n$, sub-Riemannian if $m < n$ (Carnot–Caratheodory).

Research partially supported by NSERC Grant OGP0003017.

Note that if $m = n$, and

$$\int_{M_n} (X_j u) v = \int_{M_n} u(X_j^* v),$$

then

$$-\frac{1}{2} \sum_{j=1}^{n} X_j^* X_j$$

is the classical Laplace–Beltrami operator.

A *geodesic* is the projection of the Hamiltonian flow, or bicharacteristic, onto the base:

$$\dot{x}_j(s) = H_{\xi_j}, \quad \dot{\xi}_j(s) = -H_{x_j}, \quad j = 1, \ldots, m,$$

$$x(0) = 0, \quad x(\tau) = x.$$

In the Riemannian case, locally, there is a unique geodesic between every two points.

In the sub-Riemannian case, there exist points which are connected by an infinite number of geodesics, even locally.

Step 2.

$$X_1 = \frac{\partial}{\partial x_1} + 2ax_2 \frac{\partial}{\partial x_0},$$

$$X_2 = \frac{\partial}{\partial x_2} - 2ax_1 \frac{\partial}{\partial x_0}.$$

$$[X_1, X_2] = X_1 X_2 - X_2 X_1 = -4a \frac{\partial}{\partial x_0},$$

which is equivalent to the *Heisenberg uncertainty principle*. Set

$$\Delta_H = \frac{1}{2}(X_1^2 + X_2^2),$$

the *Heisenberg sub-Laplacian*. Note that

$$\Delta_H + \frac{1}{2} \frac{\partial^2}{\partial x_0^2}$$

is the Casimir operator, which is elliptic.

The Hamiltonian for Δ_H is

$$H = \frac{1}{2}(\xi_1 + 2ax_2\theta)^2 + \frac{1}{2}(\xi_2 - 2ax_1\theta)^2,$$

and $H = 0$ on a line in $T^* \mathbb{R}^3$ over every point $x \in \mathbb{R}^3$.

The heat kernel for Δ_H

$$P_H u(x, t) = e^{\Delta_H t} u = \int_{\mathbb{R}^3} P_H(y^{-1} \circ x, t) u(y) dy,$$

where

$$P_H(x, t) = \frac{1}{(2\pi t)^2} \int_{\mathbb{R}} e^{-f(x,\tau)/t} V(\tau) d\tau,$$

$$f(x, \tau) = -i\tau x_0 + a\tau(x_1^2 + x_2^2) \coth(2a\tau) = \tau g(x, \tau),$$

$$V(\tau) = \frac{2a\tau}{\sinh(2a\tau)}, \quad \text{the Van Vleck determinant.}$$

X_1, X_2, and therefore Δ_H, are left-invariant with respect to the *Heisenberg translation*:

$$x \circ x' = (x_0 + x_0' + 2a[x_2 x_1' - x_1 x_2'], x_1 + x_1', x_2 + x_2'),$$

with $x^{-1} = -x$. This explains the formula for the heat kernel. $\mathbb{R}^3$ with the Heisenberg translation is called the *Heisenberg group*.

Complex Hamiltonian mechanics

$$\dot{x}_j = \frac{\partial H}{\partial \xi_j}, \quad j = 1, 2, \ \dot{x}_0 = \frac{\partial H}{\partial \theta},$$

$$\dot{\xi}_j = -\frac{\partial H}{\partial x_j}, \ j = 1, 2, \ \dot{\theta} = -\frac{\partial H}{\partial x_0} = 0,$$

with

$$x_j(0) = 0, \quad x_j(\tau) = x_j, \quad j = 1, 2,$$
$$\theta(0) = -i, \quad x_0(\tau) = x_0.$$

Then

$$g(x, \tau) = \int_0^\tau (\xi_1 \dot{x}_1 + \xi_2 \dot{x}_2 + \theta \dot{x}_0 - H) ds - i x_0(0)$$

is a *complex action*, and $f = \tau g$ may be thought of as the square of a complex distance.

Note that g is the solution of a Hamilton–Jacobi equation:

$$\frac{\partial g}{\partial \tau} + H(x, \nabla_x g) = 0.$$

Sub-Riemannian geometry on H

The real Hamiltonian flow is

$$\dot{x}_j = H_{\xi_j}, \quad \dot{\xi}_j = -H_{x_j}, \quad j = 1, 2,$$
$$\dot{x}_0 = H_\theta, \quad \dot{\theta} = -H_{x_0} = 0 \Rightarrow \theta(s) = \theta(0) = \theta,$$

with boundary conditions

$$x(0) = 0, \quad x(1) = x.$$

This has a solution if and only if

$$x_0 = a\mu(2a\theta)(x_1^2 + x_2^2),$$

where

$$\mu(\varphi) = \frac{\varphi}{\sin^2 \varphi} - \cot \varphi$$

is a new transcendent.

$$a = \frac{1}{2}:$$

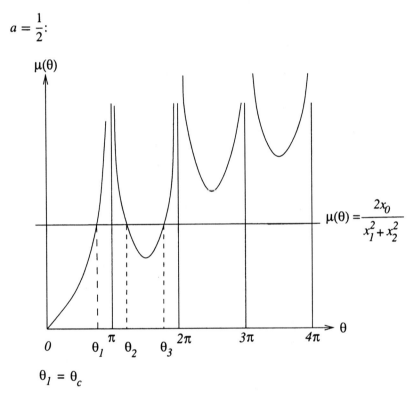

$$\theta_1 = \theta_c$$

Next we use the real action:

$$S(x, \theta) = \int_0^1 (\xi_1 \dot{x}_1 + \xi_2 \dot{x}_2 + \theta \dot{x}_0 - H) ds.$$

Theorem 1. (i) *There are finitely many geodesics which join the origin to x if and only if* $(x_1, x_2) \neq 0$. *These geodesics are parametrized by solutions* θ_j, $j = 1, 2, \ldots$ *of*

$$\mu(2a\theta_j) = \frac{x_0}{a(x_1^2 + x_2^2)},$$

and their lengths increase strictly with θ_j. The square of the length d_j^2 of such a geodesic associated to a θ_j is

$$d(x, \theta_j)^2 = 2S(x, \theta_j).$$

The shortest (Carnot–Caratheodory) distance is $d(x, \theta_c)$, $\theta_c = \theta_1 \in \left[0, \dfrac{\pi}{2a}\right)$.

   (ii) *On the x_0-axis one has lengths*

$$d_m(x_0, 0)^2 = \frac{m\pi |x_0|}{a}.$$

For each length d_m, the geodesics of that length are parametrized by the circle S^1. Again d_1 is the Carnot–Caratheodory distance.

Theorem 2. *The critical points, with respect to τ, of the complex action*

$$f(x, \tau) = -i\tau x_0 + a\tau(x_1^2 + x_2^2)\coth(2a\tau),$$

i.e., solutions $\tau_j(x)$ of

$$\frac{\partial f}{\partial \tau}(x, \tau_j(x)) = 0,$$

are given by $\tau_j(x) = i\theta_j(x)$, where

$$x_0 = a\mu(2a\theta_j)(x_1^2 + x_2^2).$$

Furthermore

$$f(x, \tau_j(x)) = \frac{1}{2}d(x, \theta_j)^2.$$

Theorem 3. (i) $(x_1, x_2) \neq 0$, $x_0 = \mu(2a\theta_c)(x_1^2 + x_2^2)$, $\theta_c \in \left[0, \dfrac{\pi}{2a}\right)$. *Then for small t:*

$$P_H(x, t) \sim \frac{1}{(2\pi t)^{3/2}} e^{-d(x,\theta_c)^2/2t} \Theta(x),$$

$$\Theta(x) = \frac{V(i\theta_c(x))}{\sqrt{f''(x, \tau_c(x))}}, \qquad f'' = \frac{\partial^2 f}{\partial \tau^2}.$$

   (ii) *At points $(x_0, 0)$, $x_0 \neq 0$,*

$$P_H(x_0, 0, t) = \frac{1}{4at^2} e^{-d(x_0,0)^2/2t} \left\{1 + O\left(e^{-d(x_0,0)^2/2t}\right)\right\}$$

as $t \to 0$.

Theorem 4. *The heat kernel $P_H(x, t)$ has the following sharp upper bound:*

$$P_H(x, t) \leq C\frac{e^{-d(x)^2/2t}}{t^2} \min\left\{1, \left(\frac{t}{\|x'\|d(x)}\right)^{1/2}\right\},$$

$x = (x_0, x') = (x_0, x_1, x_2) \in \mathbb{R}^3$.

Work in progress

(i) These results are being extended to step 2 Heisenberg manifolds.

(ii) We are working out the sub-Riemannian geodesics on spheres in $\mathbb{C}^n$; this is the first example of a compact, step 2, CR structure.

(iii) It is important to obtain results for higher step operators. For example, given

$$\left. \begin{array}{l} X = \dfrac{\partial}{\partial x} + y^2 \dfrac{\partial}{\partial t}, \\[2ex] Y = \dfrac{\partial}{\partial y} \end{array} \right\}$$

one has

$$[X, Y] = -2y\frac{\partial}{\partial t},$$

$$\big[[X, Y], Y \big] = 2\frac{\partial}{\partial t},$$

so

$$\frac{1}{2}(X^2 + Y^2)$$

is step 2 if $y \neq 0$, and step 3 at $y = 0$. In this case one has

Theorem 5. *The number of geodesics connecting* $(0, 0, 0)$ *to* (x, y, t) *is finite if and only if* $y \neq 0$.

(iv) The construction of the complex action seems to work for all subelliptic operators. The volume element V is a solution of a transport equation for step 2 operators. In the case of higher (> 2) step operators we have evidence which suggests that V may be the solution of an Euler–Poisson–Darboux equation.

References

[1] R. Beals, B. Gaveau and P. Greiner, Hamilton–Jacobi theory and the heat kernel on Heisenberg groups, *J. Math. Pures Appl.* **79**, 7 (2000), 633–689.

[2] P. Greiner and O. Calin, On sub-Riemannian geodesics, submitted.

Peter C. Greiner
Mathematics Department
University of Toronto
Toronto, Ontario, Canada
greiner@math.toronto.edu

On the Analytic Continuation of the Solution of the Cauchy Problem

Yûsaku Hamada

ABSTRACT In this article, we give some results on analytic continuations of the solution of the Cauchy problem for differential operators with coefficients of entire functions or polynomial coefficients in the complex domain.

1   Introduction and Results

J. Leray [L] and L. Gårding, T. Kotake and J. Leray [GKL] have studied singularities and analytic continuations of the solution of the Cauchy problem in the complex domain. [P], [PW] and [HLT] have studied analytic continuations in the case of differential operators with coefficients of entire functions or polynomial coefficients.

Let $x = (x_0, x')$ $[x' = (x_1, \ldots, x_n)]$ be a point of $\mathbf{C}^{n+1}$. We consider $a(x, D)$ a differential operator of order m with coefficients of entire functions on $\mathbf{C}^{n+1}$. We denote its principal part by $g(x, D)$ and assume that $g(x; 1, 0, \ldots, 0) = 1$.

Let S be the hyperplane $x_0 = 0$, therefore non-characteristic with respect to g. We study the Cauchy problem

$$a(x, D)u(x) = v(x), \quad D_0^h u(0, x') = w_h(x'), \quad 0 \le h \le m - 1, \qquad (1.1)$$

where $v(x)$, $w_h(x')$, $0 \le h \le m - 1$, are entire functions on $\mathbf{C}^{n+1}$ and $\mathbf{C}^n$ respectively.

The Cauchy–Kowalewski theorem asserts that there exists a unique holomorphic solution in a neighborhood of S in $\mathbf{C}^{n+1}$. How far can this local solution be continued analytically? In general, various complicated phenomena happen.

In [H1], by applying a result of L. Bieberbach and P. Fatou to this problem, we have constructed an example such that the domain of holomorphy of the solution has the nonempty exterior in $\mathbf{C}^{n+1}$, that is, it does not contain a ball in $\mathbf{C}^{n+1}$, for the differential operator with coefficients of entire functions. In fact, [B] and [F] have constructed entire functions on $\mathbf{C}^2$, $f_i(x_1, x_2)$, $i = 1, 2$, verifying the following conditions;

(i) The functional determinant $\dfrac{\partial(f_1, f_2)}{\partial(x_1, x_2)} = 1$ on $\mathbf{C}^2$, $f_i(0, 0) = 0$, $i = 1, 2$.

(ii) Let T be the mapping $u_1 = f_1(x_1, x_2)$, $u_2 = f_2(x_1, x_2)$, of $\mathbf{C}^2$ into $\mathbf{C}^2$.

The image $\Delta = T(\mathbf{C}^2)$ of $\mathbf{C}^2$ by T has an exterior point in $\mathbf{C}^2$. T is a biholomorphic mapping of $\mathbf{C}^2$ onto Δ.

Consider the Cauchy problem

$$[D_0 + D_2 f_2(x_1, x_2)D_1 - D_1 f_2(x_1, x_2)D_2]u_j(x) = 0,$$

$$u_j(0, x_1, x_2) = x_j, \quad j = 1, 2, \quad x = (x_0, x_1, x_2).$$

In [H1], we have shown that the domain of holomorphy of the solution has the nonempty exterior in $\mathbf{C}^2$. Now we consider the case of operators with polynomial coefficients. In [H2], we have given an example such that, roughly speaking, the domain of holomorphy of the ramified solution has an exterior point, for the differential operator with polynomial coefficients.

In order to explain this situation, we recall a result of [HLT].

Theorem 1.1 [HLT]. *Suppose that*

$$g(x, D) = D_0^m + \sum_{k=1}^m L_k(x, D_{x'})D_0^{m-k},$$

where $L_k(x, D_{x'})$, $1 \le k \le m$, *is of order* k *in* $D_{x'}$ *and polynomial in* x' *of degree* μk, *with an integer* μ (≥ 0). *Then there exists a constant* C ($0 < C \le 1$) *depending only on* $M(R)$ *such that the solution is holomorphic on*

$$\left\{ x \in C^{n+1}; \, | x_0 | \le C \min\left[(1+\| x' \|)^{-\max(\mu-1,0)}, R\right]\right\},$$

where $M(R)$ *is the maximum modulus on* $\{x_0; | x_0 | \le R\}$ *of coefficients of polynomials in* x' *of* $g(x, D)$ *and* $\|x'\| = \max_{1 \le i \le n} | x_i |$.

We recall that this theorem can be proved by using the following lemma.

Lemma 1.1 [HT]. *In the problem* (1.1), *we suppose that the operator* a *and the initial data* v, w_k *are holomorphic on* $X = \{x; | x_j | \le r_j, 0 \le j \le n\}$. *We denote by* $\|g(\cdot, \xi) - \xi_0^m\|_X$ *the spectral function of* $g(x, D) - D_0^m$ *on this domain;*

$$H_X(\xi) = \|g(\cdot, \xi) - \xi_0^m\|_X = \sum_{|\alpha|=m, \alpha_0 < m} \sup_{x \in X} | a_\alpha(x) | \, \xi^\alpha, \xi \in (R_+)^{n+1}.$$

Choose a number θ_0 ($\ge 1/r_0$) *satisfying*

$$H_X(1, \theta'/\theta_0) < 1 \ \textit{for} \ \theta' = (\theta_1, \ldots, \theta_n), \theta_j = 1/r_j, 1 \le j \le n.$$

Then the solution $u(x)$ *of the problem* (1.1) *is holomorphic on*

$$\{x; \sum_{j=0}^n \theta_j | x_j | < 1 - H_X(1, \theta'/\theta_0)\}.$$

Note. Proposition 8 in [HLT] is of a more precise form than this lemma.

We then have the following.

Corollary 1.1 [HLT]. *Let $\mathcal{D}$ be a domain in C and $0(\in \mathcal{D})$ be a base point. We denote by $\mathcal{R}(\mathcal{D})$ the universal covering space of $\mathcal{D}$. Suppose that in (1.1), the coefficients of operator $a(x, D)$ and the data $v(x)$, $w_k(x')$ can be continued analytically on $\mathcal{R}(\mathcal{D}) \times C^n$ and that its principal part $g(x, D)$, holomorphic on $\mathcal{R}(\mathcal{D}) \times C^n$, satisfies the condition of Theorem 1.1 with $\mu = 0, 1$. Then the problem (1.1) has a unique holomorphic solution on $\mathcal{R}(\mathcal{D}) \times C^n$. In particular, in the case of $\mathcal{D} = C$, the solution is an entire function on C^{n+1}. This has already been shown in $[P], [WP]$ and $[HLT]$.*

Note. A result of [HLT] is of a more precise form than this corollary.

Proof. Let γ be an arbitrary path of origin 0 in $\mathcal{D}$; $\gamma = \{x_0 = \gamma(t), t \in [0, 1]\}$, where $\gamma(t)$ is a continuous function on $[0, 1]$. First, we note that according to Theorem 1.1, the solution $u(x)$ is holomorphic in $V(0) \times C^n$, $V(0)$ being a neighborhood of 0 in C. Set $\tau_0 = \sup \tau$, where, for $\gamma_\tau = \{x_0 = \gamma(t); t \in [0, \tau]$, $0 \le \tau \le 1\}$, there exists a neighborhood $V(\gamma_\tau)$ of γ_τ in C such that the solution $u(x)$ of problem (1.1) is holomorphic on $V(\gamma_\tau) \times C^n$. In a neighborhood of $x_0 = \gamma(\tau_0)$, all coefficients $a_{\alpha,\beta'}(x_0)$ in x' and ξ of $g(x, \xi)$ are holomorphic and bounded. Then for τ $(0 < \tau < \tau_0)$, near τ_0, the $D_0^k u(\gamma(\tau), x'), 0 \le k \le m-1$, are holomorphic on C^n and by Theorem 1.1, $u(x)$ is holomorphic on $V(\gamma(\tau_0)) \times C^n$. Therefore we see that $\tau_0 = 1$ and that the solution $u(x)$ can be continued analytically on $V(\gamma) \times C^n$, $V(\gamma)$ being a neighborhood of γ in C. By Propositions 7.1 and 7.2 in [HLT], the solution $u(x)$ can be continued analytically on $\mathcal{R}(\mathcal{D}) \times C^n$. This completes the proof. $\square$

By employing Lemma 1.1, this corollary is generalized as follows.

Corollary 1.2. *Write z a point in C^ℓ. Let $a(x, z, D_x, D_z)$ be a holomorphic differential operator on $\mathcal{R}(\mathcal{D}) \times C^n \times C^\ell$. Suppose that its principal part $g(x, D_x, D_z)$ is independent of z and of the form $g(x, D_x, D_z) = g_0(x, D_x) + g_1(x, D_x, D_z)$, where $g_0(x, \xi)$ is a polynomial in ξ, x' satisfying the condition of Theorem 1.1 with $\mu = 0, 1$ and $g_1(x, \xi, \eta) = 0$ for $\eta = 0$.*
Consider the Cauchy problem

$$a(x, z, D_x, D_z)u(x, z) = v(x, z), \qquad (1.2)$$

$$D_0^k u(0, x', z) = w_k(x', z), 0 \le k \le m - 1,$$

where $v(x, z)$, $w_k(x', z)$ are holomorphic on $\mathcal{R}(\mathcal{D}) \times C^n \times C^\ell$. Then this problem (1.2) has a unique holomorphic solution on $\mathcal{R}(\mathcal{D}) \times C^n \times C^\ell$.

Note. It is not necessary that the coefficients of $g_1(x, D_x, D_z)$ be polynomials in x'.

In [H2], [H3], by employing the results of a modular function and its differential equation, and differential equations of Darboux–Halphen and of Chazy [C], we have given examples such that the domain of holomorphy of the solution has the nonempty exterior for the differential operators with polynomial coefficients.

In fact, J. Chazy [C] has studied ordinary differential equations of third order and Darboux–Halphen's system of ordinary differential equations. (Also see [AF]). We employ these results.

Consider the Cauchy problems

$$\{D_0 + \sum_{i=1}^{3} H_i(x')D_i\}U_{1,j}(x) = 0, \ U_{1,j}(0, x') = x_j, \ 1 \le j \le 3, \qquad (1.3)$$

$$[x = (x_0, x'), x' = (x_1, x_2, x_3)]$$

where

$$H_1(x') = \frac{1}{2}[(x_2 + x_3)x_1 - x_2x_3],$$

$$H_2(x') = \frac{1}{2}[(x_1 + x_3)x_2 - x_1x_3], \qquad (1.4)$$

$$H_3(x') = \frac{1}{2}[(x_1 + x_2)x_3 - x_1x_2].$$

This concerns Darboux–Halphen's system of ordinary differential equations ([C]).

Consider the Cauchy problems

$$\{D_0 + x_2D_1 + x_3D_2 + (2x_1x_3 - 3x_2^2)D_3\}U_{2,j}(x) = 0, \qquad (1.5)$$
$$U_{2,j}(0, x') = x_j, \ 1 \le j \le 3.$$

This concerns Chazy's ordinary differential equation. ([C], [AF]).

J. Leray [L] and L. Gårding, T. Kotake and J. Leray [GKL] have studied the Cauchy problem in the case where the initial surface has characteristic points. In [H2], we have studied an exceptional case in [L] and [GKL] and here we give a complement to the results of [H2].

Consider the Cauchy problems

$$\left\{ \sum_{i=0}^{3} A_i(x')D_i \right\} U_{3,j}(x) = 0, \ U_{3,j}(0, x') = x_j, \ 1 \le j \le 3, \qquad (1.6)$$
$$[x = (x_0, x'), x' = (x_1, x_2, x_3)]$$

where

$$A_0(x') = 2x_1^2(1 - x_1)^2 x_2,$$
$$A_1(x') = A_0(x')x_2, \qquad (1.7)$$
$$A_2(x') = A_0(x')x_3,$$
$$A_3(x') = 3x_1^2(1 - x_1)^2 x_3^2 - (1 - x_1 + x_1^2)x_2^4.$$

This concerns the ordinary differential equation of a modular function. ([C], [Hi], [AF]).

In [H3], we proved the following.

Proposition 1.1. *The domains of holomorphy* $\mathcal{D}_i$, $1 \leq i \leq 3$, *of the solutions* $U_{i,j}(x)$, $1 \leq i, j \leq 3$, *of the problems* (1.3), (1.5) *and* (1.6) *are domains in* C^4. *They have the nonempty exteriors in* C^4.

By a birational mapping and an algebraic mapping, problems (1.3) and (1.5) are transformed to problems (1.6) respectively. In these examples, the dimension n of space $x' = (x_1, ..., x_n)$ is always $n \geq 2$.

Ferruccio Colombini has posed the question as to whether, in the case of $n = 1$, the domain of holomorphy of the solution can have or not have the nonempty exterior.

In this paper, we observe the case of $n = 1$, by giving some examples for certain differential operators of principal part with polynomial coefficients.

2   Some Examples

We study the Cauchy problem

$$(D_0^m - \alpha x_0^p x_1^q D_1^m)u(x) = b(x, D)u(x) + v(x), \tag{2.1}$$
$$D_0^k u(0, x_1) = w_k(x_1), 0 \leq k \leq m - 1, x = (x_0, x_1) \in C^2,$$

where α is a constant, $p(\geq 0), q(\geq m + 1)$ are integers and $b(x, D)$ is an operator of order $m - 1$. The coefficients of $b(x, D)$, $v(x)$ and $w_k(x_1), 0 \leq k \leq m - 1$, are holomorphic on C^2.

Proposition 2.1. *The Cauchy problem* (2.1) *has a unique holomorphic solution on* $\mathcal{R}[C^2 \setminus \cup_{i=0}^r K_i]$, *where the* K_i, $0 \leq i \leq r$, *are characteristic surfaces for* $D_0^m - \alpha x_0^p x_1^q D_1^m$, *r being an integer* (≥ 1).

Note. In the case of $0 \leq q \leq m$, or $\alpha = 0$ in the problem (2.1), by Corollary 1.1, the solution is holomorphic on C^2.

Proof. By making the change of variable $X_1 = 1/x_1$, the problem (2.1) is transformed into the problem

$$[D_0^m - \alpha x_0^p X_1^{-q}(-X_1^2 D_{X_1})^m]U(x_0, X_1)$$
$$= b(x_0, 1/X_1, D_0, -X_1^2 D_{X1})U(x_0, X_1) + v(x_0, 1/X_1), \tag{2.2}$$
$$D_0^k U(0, X_1) = w_k(1/X_1), 0 \leq k \leq m - 1,$$

where $U(x_0, X_1) = u(x_0, 1/X_1)$. Moreover, make the change of variables in a neighborhood of a point $(x_0, X_1) = (0, X_1^{(0)})$, $X_1^{(0)} \neq 0$, fixing a branch of X_1^λ at the point $X_1^{(0)}$, if λ is a rational number;

$$t = x_0/X_1^\lambda, s = X_1, \lambda = (q - m)/(p + m). \tag{2.3}$$

We then have

$$D_0 = s^{-\lambda} D_t, \quad D_{X_1} = (1/s)(-\lambda t D_t + s D_s). \tag{2.4}$$

The problem (2.2) is transformed into the problem

$$[D_t^m + G(t, D_t, s D_s)]\tilde{U}(t, s)$$
$$= B(t, s, D_t, D_s)\tilde{U}(t, s) + \tilde{V}(t, s), \tag{2.5}$$
$$D_t^k \tilde{U}(0, s) = \tilde{W}_k(s), 0 \le k \le m - 1,$$

where $G(t, \eta_0, \eta_1)$, $B(t, s, \eta_0, \eta_1)$ are polynomials of degree m and $m - 1$, respectively, in η_0, η_1 with coefficients continued analytically on $\mathcal{R}[t \in \mathbf{C}; \alpha\lambda^m t^{p+m} \ne 1] \times \mathcal{R}[s \in \mathbf{C}; s \ne 0]$, and $G(t, \eta_0, \eta_1)$ does not contain η_0^m.

Setting $s = e^y$, the problem (2.5) is transformed into the problem

$$[D_t^m + G(t, D_t, D_y)]\hat{U}(t, y)$$
$$= \hat{B}(t, y, D_t, D_y)\hat{U}(t, y) + \hat{V}(t, y), \tag{2.6}$$
$$D_t^k \hat{U}(0, y) = \hat{W}(y), 0 \le k \le m - 1,$$

where $\hat{B}$ is a differential operator order $m - 1$, and its coefficients, $\hat{V}(t, y)$ and $\hat{W}(y)$ are holomorphic functions on $\mathcal{R}[t \in \mathbf{C}; \alpha\lambda^m t^{p+m} \ne 1] \times \{y \in \mathbf{C}\}$.

In view of Corollary 1.1, the solution $\hat{U}(t, y)$ of (2.6) is continued analytically on $\mathcal{R}[t \in \mathbf{C}; \alpha\lambda^m t^{p+m} \ne 1] \times \{y \in \mathbf{C}\}$.

Let $\tilde{K}_i$, $1 \le i \le r$, be the irreducible components of $\{(x_0, X_1) \in \mathbf{C}^2; X_1^{q-m} = \alpha\lambda^m x_0^{p+m}\}$ and $\tilde{K}_0 = \{(x_0, X_1) \in \mathbf{C}^2; X_1 = 0\}$. The solution $U(x_0, X_1)$ of (2.2) is continued analytically on $\mathcal{R}[\mathbf{C}^2 \setminus \cup_{i=0}^r \tilde{K}_i]$, so the solution $u(x)$ of problem (2.1) is continued analytically on $\mathcal{R}[\mathbf{C}^2 \setminus \cup_{i=0}^r K_i]$, where K_i, $0 \le i \le r$, are the irreducible components of $\{x \in \mathbf{C}^2; \alpha\lambda^m x_0^{p+m} x_1^{q-m} = 1\}$, $K_0 = \{x_1 = 0\}$. This proves Proposition 2.1. □

Using Corollary 1.2, this result is easily generalized as follows.
Consider the Cauchy problem

$$[D_0^m - \alpha x_0^p x_1^q D_1^m + h(x_0, x_1, D)]u(x) = b(x, D)u(x) + v(x),$$
$$D_0^k u(0, x') = w_k(x'), 0 \le k \le m - 1, \tag{2.7}$$
$$x = (x_0, x'), \quad x' = (x_1, x''), \quad x'' = (x_2, ..., x_n) \in \mathbf{C}^{n-1},$$

where α is a constant, $p\ (\ge 0)$, $q\ (\ge m + 1)$ are integers, $h(x_0, x_1, D)$ and $b(x, D)$ are operators of order m and $m - 1$, respectively, holomorphic on $\mathbf{C}^{n+1}$, and $h(x_0, x_1, \xi_0, \xi_1, \xi'') = 0$, for $\xi'' = (\xi_2, \cdot, \cdot, \cdot, \xi_n) = 0$. $v(x)$, $w_k(x')$ are holomorphic functions on $\mathbf{C}^{n+1}$.

Then we have the following.

Corollary 2.1. *The Cauchy problem (2.7) has a unique holomorphic solution on* $\mathcal{R}[\mathbf{C}^2 \setminus \cup_{i=0}^r K_i] \times \mathbf{C}^{n-1}$, *where* K_i, $0 \le i \le r$, *are characteristic surfaces of Proposition 2.1.*

This method is applicable to the following Cauchy problems.
Consider the problem

$$(D_0 - 2x_0 x_1^2 D_1) D_0 u(x) = b(x, D)u(x) + v(x), \tag{2.8}$$
$$D_0^k u(0, x_1) = w_k(x_1), 0 \le k \le 1, \quad x = (x_0, x_1) \in \mathbf{C}^2,$$

where $b(x, D)$ is of order 1, and its coefficients, $v(x)$ and $w_k(x_1)$ are holomorphic on $\mathbf{C}^2$. The solution of (2.8) is holomorphic on $\mathcal{R}[\mathbf{C}^2 \setminus K]$, $K = \{x \in \mathbf{C}^2; x_0^2 x_1 = 1\} \cup \{x_1 = 0\}$.
Next, consider the problem for an operator of Clairaut type;

$$[D_0^3 + (x_0 D_0 - x_1 D_1)(-x_1^2 D_1)^2]u(x) = b(x, D)u(x) + v(x), \tag{2.9}$$
$$D_0^k u(0, x_1) = w_k(x_1), 0 \le k \le 2, x = (x_0, x_1) \in \mathbf{C}^2,$$

where $b(x, D)$ is of order 2 and its coefficients, $v(x)$ and $w_k(x_1)$ are holomorphic on $\mathbf{C}^2$.
By the change of variable $X_1 = 1/x_1$, problem (2.9) becomes

$$[D_0^3 + (x_0 D_0 + X_1 D_{X_1}) D_{X_1}^2]U(x_0, X_1)$$
$$= \tilde{b}(x_0, X_1, D_0, D_{X_1})U(x_0, X_1) + \tilde{v}(x_0, X_1), \tag{2.10}$$
$$D_0^k U(0, X_1) = \tilde{w}_k(X_1), 0 \le k \le 2,$$

where $\tilde{b}$ is of order 2 and its coefficients, $\tilde{v}(x_0, X_1)$ and $\tilde{w}_k(X_1)$ are holomorphic on $[\mathbf{C}^2 \setminus \{X_1 = 0\}]$. The principal part of the operator in (2.10) is a particular operator in [D]. The equation of bicharacteristic curves is

$$X_1 = x_0 \frac{dX_1}{dx_0} + \left(\frac{dX_1}{dx_0}\right)^3,$$

which is a differential equation of Clairaut type.
Making the change of variables

$$t = x_0/X_1^{2/3}, \quad s = X_1$$

and using the method of this section, we can see that the solution of (2.9) is holomorphic on $\mathcal{R}[\mathbf{C}^2 \setminus K]$, $K = \{x \in \mathbf{C}^2; 1 + (4/27)x_0^3 x_1^2 = 0\}$, $\cup \{x_1 = 0\}$.

References

[B]   L. Bieberbach, *Beispiel zweier ganzer Funktionen zweier komplexer Variablen, welche eine schlichite volumtreue Abbildung der R_4 auf einen Teil seiner selbst vermitteln*, S. B. preuss. Akad. Wiss. (1933), 476–479.

[AF]   M. J. Ablowitz and A. S. Fokas, *Complex Variables: Introduction and Applications*, Cambridge Texts in Applied Mathematics, Cambridge University Press, 1997.

[C]   J. Chazy, *Sur les équations différentielles du troisième ordre et d'ordre supérieur dont l'intégrale générale a ses points critiques fixes*, Acta Math. **34** (1911), 317–385.

[D]   S. Delache, *Les solutions élémentaires hyperboliques d'opérateurs de Tricomi-Clairaut*, Bull. Soc. Math. France **97** (1969), 5–79.

[F]   P. Fatou, *Sur les fonctions méromorphes de deux variables*, C. R. Acad. Sc. Paris **175** (1922), 862–865; *Sur certaines fonctions uniformes de deux variables*, C. R. Acad. Sc. Paris **175** (1922), 1030–1033.

[Fo]   R. Forsyth, *Theory of Differential Equations*, Dover, 1958.

[GKL] L. Gårding, T. Kotake et J. Leray, *Uniformisation et développement asymptotique de la solution du problème de Cauchy linéaire à données holomorphes; analogue avec la théorie des ondes asymptotiques et approchées*, Bull. Soc. Math. France **92** (1964), 263–361.

[HLT] Y. Hamada, J. Leray and A. Takeuchi, *Prolongements analytiques de la solution du problème de Cauchy linéaire*, J. Math. Pures Appl. **64** (1985), 257–319.

[H1]   Y. Hamada, *Une remarque sur le domaine d'existence de la solution du problème de Cauchy pour l'opérateur différentiel à coefficients des fonctions entières*, Tohoku Math. J. **50** (1998), 133–138.

[H2]   Y. Hamada, *Une remarque sur le problème de Cauchy pour l'opérateur différentiel de partie principale à coefficients polynomiaux*, Tohoku Math. J. **52** (2000), 79–94.

[H3]   Y. Hamada, *Une remarque sur le problème de Cauchy pour l'opérateur différentiel de partie principale à coefficients polynomiaux* II, Tohoku Math. J. **54** (2002), 294–307.

[HT]   Y. Hamada et A. Takeuchi, *Sur le prolongement analytique de la solution du problème de Cauchy*, C.R. Acad. Sc. Paris, t.295, Série I (1982), 329–332. Observation du présenteur.

[Hi]   E. Hille, *Ordinary Differential Equations in the Complex Domain*, John Wiley, 1976.

[L]   J. Leray, *Uniformisation de la solution du problème linéaire analytique de Cauchy près de la variété qui porte les données de Cauchy (Problème de Cauchy I)*, Bull. Soc. Math. France **85** (1957), 389–429.

[N]   T. Nishino, *Theory of Functions of Several Complex Variables* [Tahensu Kansu Ron] (in Japanese), Univ. of Tokyo Press, 1996.

[P]   J. Persson, *On the local and global non-characteristic Cauchy problem when the solutions are holomorphic functions or analytic functionals in the space variables*, Ark. Mat. **9** (1971), 171–180.

[PW]   P. Pongérard and C. Wagschal, *Problème de Cauchy dans des espaces de fonctions entières*, J. Math. Pures Appl. **75** (1996), 409–418.

Yûsaku Hamada
61-36 Tatekura-Cho
Shimogamo, Sakyo-Ku
Kyoto, 606-0806, Japan
yhamada@oak.ocn.ne.jp

Strong Gevrey Solvability for a System of Linear Partial Differential Equations

Kunihiko Kajitani and Sergio Spagnolo

ABSTRACT We consider a class of linear systems whose principal symbol satisfies a certain condition of semi-hyperbolicity, and we prove the local surjectivity in suitable Gevrey spaces.

0   Introduction

We investigate the local solvability for $m \times m$ systems of type

$$\left[D_t + \sum_{j=1}^{n} A_j(t, x) D_{x_j} + B(t, x) \right] u = f(t, x), \qquad (0.1)$$

where $D_{x_j} = -i \partial_{x_j}$, $D_t = -i \partial_t$. Here $u(t, x)$ and $f(t, x)$ are $\mathbf{C}^m$-valued functions on $\mathbf{R}^{1+n}$, while the $A_j(t, x)$'s are $m \times m$ matrix functions, *uniformly analytic* in $\mathbf{R}^{1+n}$ in the sense that there is some $C_0 > 0$ for which

$$|\partial_t^k \partial_x^\alpha A_j(t, x)| \le C_0^{1+k+|\alpha|} (k + |\alpha|)! \qquad \forall k, \alpha. \qquad (0.2)$$

We denote by $\tau_1(t, x, \xi), \ldots, \tau_m(t, x, \xi)$ the eigenvalues of the matrix

$$A(t, x, \xi) = \sum_{j=1}^{n} A_j(t, x) \xi_j, \qquad \xi \in \mathbf{R}^n, \qquad (0.3)$$

repeated following their multiplicities, and define

$$p(t, x, z, \xi) = \det(zI - A(t, x, \xi)) = \prod_{h=1}^{m} (z - \tau_h(t, x, \xi)).$$

We do not assume $\tau_h \ne \tau_k$ for $h \ne k$, hence we cannot apply the classical theory of C^∞-solvability of Hörmander, Nirenberg and Treves for equations of principal type. Consequently, we expect solvability only in suitable Gevrey classes.

Our main assumption is that, for each fixed $\xi \in \mathbf{R}^n$, the imaginary parts of all the τ_h's keep the same sign when (t, x) runs in $\mathbf{R}^{1+n}$, that is

$$\forall \xi \in \mathbf{R}^n: \quad \text{either} \quad \Im \tau_h(t, x, \xi) \ge 0 \quad \forall h, t, x,$$
$$\text{or} \quad \Im \tau_h(t, x, \xi) \le 0 \quad \forall h, t, x. \qquad (0.4)$$

Under this assumption, it was proved by Spagnolo ([7]) that (0.1) is locally solvable in each Gevrey class γ^d with $1 < d < m/(m-1)$, provided the coefficients $A_j \equiv A_j(t)$ depend only on the time variable. This result does not extend to the general case where $A_j \equiv A_j(t, x)$; for example, the scalar equation

$$[D_t + x_2 D_{x_1} + i D_{x_2}]u = f$$

fulfills condition (0.4) but is not locally solvable in any γ^s with $s > 1$. Here $\gamma^d(\mathbf{R}^n)$ denotes the class of functions $u \in C^\infty(\mathbf{R}^n)$ which satisfy an estimate like

$$|D_x^\alpha u(x)| \le C \rho^{-|\alpha|} |\alpha|!^d, \qquad \forall \alpha \in \mathbf{N}^n, \ \forall x \in \mathbf{R}^n.$$

Here, we confine ourselves to the special case when the leading coefficients depend only on some of the space variables. More precisely, we assume that

$$A_j = A_j(t, x'), \quad j = 1, \dots, n, \qquad \text{where} \quad x' = (x_1, \dots, x_l), \qquad (0.5)$$

for some $l \in \{1, \dots, n\}$, and, correspondingly, that

$$A_1(t, x), \ \dots, \ A_l(t, x) \quad \text{are hyperbolic matrices.} \qquad (0.6)$$

We recall that a matrix is called *hyperbolic* when all its eigenvalues are real.

Remark 1 (*borderline cases*). The case considered in [7], i.e., $A_j \equiv A_j(t)$ but no hyperbolicity condition is required, corresponds to $\{(0.5)–(0.6)\}$ with $l = 0$. On the other hand, the case when (0.1) is a hyperbolic system corresponds to $l = n$: in such a case condition (0.5) becomes trivial, while $\{(0.4)–(0.6)\}$ simply means that the matrix $A(t, x, \xi)$ in (0.3) is hyperbolic. Indeed, each of the hyperbolic matrices A_j has a trace with imaginary part equal to zero, thus, taking (0.4) into account, we have

$$|\Im \tau_h(t, x, \xi)| \le |\sum_{k=1}^m \Im \tau_k(t, x, \xi)|$$

$$= |\Im \operatorname{tr}(A(t, x, \xi))| = |\sum_{j=1}^l \Im \operatorname{tr}(A_j(t, x)) \xi_j| = 0.$$

Remark 2 ($m = 1$). The scalar operators satisfying (0.5)–(0.6) have a principal part of the form

$$p(t, x, D_t, D_x) = D_t + \sum_1^n a_j(t, x_1, \dots, x_l) D_{x_j}$$

where the coefficients $a_1, \dots, a_l$ are real functions. Thus (0.4) is stronger than the condition (P) of Nirenberg and Treves: indeed, $\Im p(t, x, \tau, \xi)$ is independent of $\xi_1, \dots, \xi_l$, while $\xi_{l+1}(s), \dots, \xi_n(s)$ keep constant along each bicharacteristic of $\Re p(t, x, \tau, \xi)$. Hence, we have the local solvability in C^∞.

The purpose of this paper is to prove the following:

Theorem 0.1. *Consider an $m \times m$ system of type* (0.1) *with coefficients satisfying* (0.2). *Assume* (0.4), (0.5) *and* (0.6). *Therefore, for all $f \in C^\infty(\mathbf{R}; \gamma^d(\mathbf{R}^n)) \cap C_0^\infty(\mathbf{R}^{1+n})$ there is a solution $u \in C^\infty(\mathbf{R}; \gamma^d(\mathbf{R}^n))$, provided*

$$1 < d < \min\left\{2, \frac{m}{m-1}\right\}. \tag{0.7}$$

It should be noted that in this theorem there are no conditions on the lower order terms, that is, the operators are solvable in the Gevrey classes in a strong sense. We recall that Gramchev and Rodino [3] treated the case of operators with characteristics of constant multiplicity, while Gramchev [4] and Popivanov [6] investigated the necessary conditions on lower order terms for local solvability.

Examples. If $n = 2$, the operator

$$D_t + H_1(t, x_1)D_{x_1} + i\, H_0(t, x_1)D_{x_2}$$

satisfies the conditions (0.4), (0.5) and (0.6) whenever the H_j's are Hermitian matrices, and $H_0 \geq 0$. More generally, any operator of the form

$$D_t + \sum_1^n H_j(t, x_1, \ldots, x_l)D_{x_j} + i\, H_0(t, x_1, \ldots, x_l) \sum_{l+1}^n c_j D_{x_j}$$

with H_j, H_0 as above and $c_j \in \mathbf{R}$ satisfies (0.4), (0.5), (0.6).

A similar result to Theorem 0.1 holds true for a class of scalar equations of the form

$$p(t, x, D_t, D_x)u = \left[D_t^m + \sum_{|\alpha|+h=m} a_{\alpha h}(t, x')D_x^\alpha D_t^h \right.$$

$$\left. + \sum_{|\alpha|+h<m} a_{\alpha h}(t, x)D_x^\alpha D_t^h \right]u = f, \tag{0.8}$$

where $x' = (x_1, \ldots, x_l)$. To state our result, let us introduce the functions $a_j(t, x)$, $j = 1, \ldots, n$, by writing the principal part of the equation in the form

$$p_m(t, x', D_t, D_x) = D_t^m + \sum_{j=1}^n a_j(t, x')D_{x_j} D_t^{m-1}$$

$$+ \sum_{h=0}^{m-2} \sum_{|\alpha|=m-h} a_{\alpha h}(t, x')D_x^\alpha D_t^h.$$

Here, the eigenvalues $\{\tau_h\}$ of the matrix (0.3) are replaced by the roots of

$$p_m(t, x', \tau, \xi) = 0,$$

and the condition (0.6) is replaced by the following:

$$a_1(t, x), \ldots, a_l(t, x) \quad \text{are real}. \tag{0.9}$$

Therefore, we have

Theorem 0.2. *Under the assumptions* (0.4) *and* (0.9), *the equation* (0.8), *with uniformly analytic leading coefficients, is locally solvable in* γ^d *for* $1 < d < m/(m-1)$.

Notation.

$$\langle \xi \rangle = \sqrt{1 + |\xi|^2}, \quad \langle \xi \rangle_h = \sqrt{h^2 + |\xi|^2}, \quad (h \geq 1), \quad \kappa = \frac{1}{d} \qquad (0.10)$$

$$x' = (x_1, \ldots, x_l), \quad x'' = (x_{l+1}, \ldots, x_n), \quad \xi' = (\xi_1, \ldots, \xi_l), \quad \xi'' = (\xi_{l+1}, \ldots, \xi_n).$$

1   Semi-hyperbolic polynomials

When the polynomial

$$p(t, x, z, \xi) = \det [zI + \sum A_j(t, x)\xi_j] \qquad (1.1)$$

is hyperbolic in z, that is, the imaginary part of its roots $\tau_h(t, x, \xi)$ vanishes identically, we can construct a solution of (0.1) by solving the Cauchy problem with zero initial data (see [2] and [5]). Therefore, we assume that p is not hyperbolic. We start with two lemmas which make more explicit the assumptions (0.4) and (0.6), or (0.9). These lemmas apply more generally to any homogeneous polynomial

$$p(t, x, z, \xi) = z^m + \sum_{h=0}^{m-1} \sum_{|\alpha|=m-h} a_{h\alpha}(t, x)\, \xi^\alpha z^h$$

$$\equiv z^m + z^{m-1} \sum_{j=1}^{n} a_j(t, x)\xi_j + \cdots \qquad (1.2)$$

with uniformly bounded coefficients on $\mathbf{R}^{1+n}$.

In the case of systems, i.e., when p is given by (1.1), we have

$$a_j(t, x) = \operatorname{tr}(A_j(t, x)). \qquad (1.3)$$

In such a case, (0.9) is equivalent to (0.6): indeed if, for some matrix A, the eigenvalues of A have imaginary parts with the same sign, then $\Im \operatorname{tr}(A) = 0$ iff A is hyperbolic.

Lemma 1.1. *If the roots* $z = \tau_h(t, x, \xi)$, $h = 1, \ldots, m$, *of the polynomial* (1.2) *satisfy* (0.4), *then we have necessarily*

$$\Im a_j(t, x) = c_j\, \beta(t, x) \quad \text{with} \quad \beta(t, x) \geq 0 \qquad (j = 1, \cdots, n) \qquad (1.4)$$

for some real constants c_j, *which can be assumed to satisfy* $c_1^2 + \cdots + c_n^2 = 1$. *Introducing the linear function*

$$\theta(\xi) = c_1\xi_1 + \cdots + c_n\xi_n \qquad (\textstyle\sum c_j^2 = 1), \qquad (1.5)$$

(0.4) *reads*

$$\Im\tau_h(t, x, \xi) \geq 0 \quad (resp. \leq 0) \quad \forall h, t, x \quad if \quad \theta(\xi) \geq 0 \quad (resp. \leq 0),$$
(1.6)

while (0.9) *means that* $c_1 = \cdots = c_l = 0$, *i.e.*,

$$\theta = \theta(\xi'') \quad with \quad \xi'' = (\xi_{l+1}, \ldots, \xi_n).$$
(1.7)

We note that $\theta(\xi)$ is uniquely defined (except when p is hyperbolic). Once fixed such a function, we define, for $\varepsilon \geq 0$, the following subsets of $\mathbf{C} \times \mathbf{R}^n$:

$$\begin{aligned} W_\varepsilon^- &= \{(z, \xi) : \Im z > 0, \ \theta(\xi) \leq \varepsilon \Im z\}, \\ W_\varepsilon^+ &= \{(z, \xi) : \Im z < 0, \ \theta(\xi) \geq \varepsilon \Im z\}, \end{aligned}$$
(1.8)

$$W_\varepsilon = W_\varepsilon^- \cup W_\varepsilon^+ .$$
(1.9)

Condition (1.6) can be expressed by saying that

$$|p(t, x, z, \xi)| > 0 \qquad \forall (z, \xi) \in W_0 ;$$

the following lemma provides some estimates of p on the set W_ε for (small) $\varepsilon > 0$.

Lemma 1.2. *Under the same assumptions of Lemma 1.1, it is possible to find some* $\bar\varepsilon > 0$ *such that, for all* $(z, \xi) \in W_\varepsilon$ *with* $\varepsilon \leq \bar\varepsilon$, *it results that*

$$|p(t, x, z, \xi)| \geq \frac{1}{2} |\Im z|^m ,$$
(1.10)

$$|p(t, x, \bar z, \xi)| \leq C |p(t, x, z, \xi)| ,$$
(1.11)

$$\frac{|\partial_{z,\xi}^\alpha D_{t,x}^\beta p(t, x, z, \xi)|}{|p(t, x, z, \xi)|} \leq C \langle\xi\rangle^{|\beta|} |\Im z|^{-1} \quad for \quad |\alpha| + |\beta| = 1.$$
(1.12)

Proof of Lemma 1.1. By (1.2) we have

$$\sum_{h=1}^m \Im\tau_h(t, x, \xi) = \sum_{j=1}^n \beta_j(t, x)\xi_j \quad where \quad \beta_j(t, x) = \Im a_j(t, x).$$

Condition (0.4) implies that, for each ξ, the function $(t, x) \mapsto \sum \beta_j(t, x)\xi_j$ does not change sign on $\mathbf{R}^{1+n}$, but this is possible only if $\beta_j(t, x) = c_j\beta(t, x)$ for some c_j and β, as one can easily see. Therefore we have

$$\sum_{h=1}^m \Im\tau_h(t, x, \xi) = \sum_{j=1}^n \beta_j(t, x)\xi_j = \beta(t, x)\theta(\xi).$$
(1.13)

Since $\Im\tau_1, \ldots, \Im\tau_n$ have the same sign, for each fixed ξ, we get (1.6). $\qquad\square$

Proof of Lemma 1.2. In order to prove (1.10) and (1.11), let us write

$$p(t, x, z, \xi) = \prod_{h=1}^{m} (z - \tau_h(t, x, \xi)), \qquad \overline{p(t, x, \overline{z}, \xi)} = \prod_{h=1}^{m} (z - \overline{\tau_h(t, x, \xi)}).$$

Assume that $(z, \xi) \in W_\varepsilon^-$, i.e., $\Im z > 0$ and $\theta(\xi) \leq \varepsilon \Im z$. We distinguish two cases:
a) If $\theta(\xi) \leq 0$, we know by (1.6) that $\Im \tau_h \leq 0$ for all h, thus we have

$$|p(t, x, z, \xi)| = \prod_{h=1}^{m} |z - \tau_h| \geq \prod_{h=1}^{m} |\Im z - \Im \tau_h| \geq \prod_{h=1}^{m} |\Im z| = |\Im z|^m .$$

Moreover we have $|z - \overline{\tau_h}| \leq |z - \tau_h|$, since $\Im z \cdot \Im \tau_h \leq 0$, hence we conclude that

$$|p(t, x, z, \xi)| \geq |p(t, x, \overline{z}, \xi)| .$$

b) If $0 < \theta(\xi) \leq \varepsilon \Im z$, we have, by (1.13),

$$0 \leq \Im \tau_h \leq \sum_{k=1}^{m} \Im \tau_k = \beta(t, x)\theta(\xi) \leq M\varepsilon \Im z, \qquad (1.14)$$

where $M = \sup_{\mathbf{R}^{1+n}} \beta(t, x)$, hence $|\Im z - \Im \tau_h| \geq |\Im z| / \sqrt[m]{2}$ for ε sufficiently small with respect to M and m. Thus we find the estimate

$$|p(t, x, z, \xi)| \geq \prod_{h=1}^{m} |\Im z - \Im \tau_h| \geq |\Im z|^m / 2.$$

As to (1.11), we note that, by (1.14), we have $0 \leq \Im \tau_h \leq |\Im z - \Im \tau_h|$ for $\varepsilon \leq 1/(2M)$. Hence we get

$$|z - \overline{\tau_h}| \leq |z - \tau_h| + 2|\Im \tau_h| \leq |z - \tau_h| + 2|\Im z - \Im \tau_h| \leq 3|z - \tau_h|$$

which gives (1.11) with $C = 3^m$. This proves that (1.10) and (1.11) hold on the set W_ε^-, for small ε. In a similar way we prove that these estimates hold on W_ε^+.

In order to prove (1.12), following S. Wakabayashi ([8], Theorem 2), we effect the *Hermite decomposition*

$$(1+i) p(t, x, z, \xi) = p_1(t, x, z, \xi) + i \, p_2(t, x, z, \xi),$$

where $p_k(t, x, z, \xi)$, $k = 1, 2$ are polynomials in z with real coefficients. This means that

$$p_1 = \frac{1}{2}[(1+i)p + (1-i)\tilde{p}], \qquad p_2 = \frac{1}{2}[(1+i)p - (1-i)\tilde{p}] \quad (1.15)$$

where

$$\tilde{p}(t, x, z, \xi) = \overline{p(t, x, \overline{z}, \xi)} .$$

In particular, by (1.11) and (1.15), we have, for $\varepsilon \leq \bar{\varepsilon}$,

$$|p_k(t, x, z, \xi)| \leq 3\sqrt{2}\,|p(t, x, z, \xi)|, \ \forall\,(z, \xi) \in W_\varepsilon^- \cup W_\varepsilon^+ \ (k = 1, 2). \quad (1.16)$$

Now we can easily see (Hermite's Theorem) that the polynomials p_1 and p_2 are hyperbolic as soon as p is *upper semi-hyperbolic*, in the sense that its roots have a nonnegative imaginary part, or when p is *lower semi-hyperbolic* (imaginary part of the roots ≤ 0). Our assumption (0.4) implies that, for each ξ, $p(t, x, z, \xi)$ is either upper or lower semi-hyperbolic, thus we conclude that the polynomials $p_k(t, x, z, \xi)$, $k = 1, 2$, are always hyperbolic. But we know from Bronshtein [1] that the roots of a hyperbolic polynomial are Lipschitz continuous; consequently, $p_1(t, x, z, \xi)$ and $p_2(t, x, z, \xi)$ will satisfy (1.12) for all (t, x, z, ξ). Therefore, by (1.16), $p(t, x, z, \xi)$ satisfies (1.12) on W_ε for $\varepsilon \leq \bar{\varepsilon}$.   □

2   Proof of Theorem 1

Step 1 (*reduction to a system with scalar principal part*)
Let us go back to our system

$$[D_t + \sum A_j(t, x')D_j + B(t, x)]\,v$$
$$\equiv [L_1(t, x', D_t, D_x) + B(t, x)]\,v = f(t, x) \quad (2.1)$$

where $x' = (x_1, \ldots, x_l)$. Denoting by L_1^{co} the cofactor matrix of L_1, we get

$$[L_1(t, x, D_t, D_x) + B(t, x)] \circ L_1^{co}(t, x, D_t, D_x)$$
$$= p(t, x', D_t, D_x)I + Q(t, x, D_t, D_x),$$

where

$$p(t, x', \tau, \xi) = \det L_1(t, x', \tau, \xi) = \tau^m + \sum_{\substack{|\alpha|+h=m \\ h<m}} a_{\alpha h}(t, x')\,\xi^\alpha \tau^h, \quad (2.2)$$

$$Q(t, x, \tau, \xi) = \left[\sum_{|\alpha|+h\,\leq m-1} q_{\alpha h}^{ij}(t, x)\,\xi^\alpha \tau^h\right]_{i,j=1,\ldots,m}. \quad (2.3)$$

By the assumptions (0.4) and (0.6), the roots $\{\tau_h(t, x', \xi)\}$ of p satisfy (1.6) for some linear function $\theta(\xi'')$ depending only on $\xi'' = (\xi_{l+1}, \ldots, \xi_n)$.

In order to exhibit a solution to (2.1), it will be sufficient to find a solution of

$$L(t, x, D_t, D_x)\,u \equiv [p(t, x', D_t, D_x)I + Q(t, x, D_t, D_x)]\,u = f(t, x), \quad (2.4)$$

and then take $v = L_1^{co}(t, x', D_t, D_x)u$. Note that (2.4) is an $m \times m$ system of partial differential equations of order m with principal part of scalar type.

Step 2 (*conjugation with Fourier Integral Operators*)

We define $\kappa = 1/d$, so that by (0.7) we have

$$\max \left\{ \frac{m-1}{m}, \frac{1}{2} \right\} < \kappa = \frac{1}{d} < 1. \tag{2.5}$$

Let us fix two functions χ_+ and χ_- in $\gamma^d(\mathbf{R})$, with $0 \leq \chi_\pm(t) \leq 1$, such that

$$\chi_+(t) = \begin{cases} 0 & \text{for } t \leq -1, \\ 1 & \text{for } t \geq -1/2, \end{cases} \qquad \chi_-(t) = \begin{cases} 0 & \text{for } t \geq 1, \\ 1 & \text{for } t \leq 1/2, \end{cases} \tag{2.6}$$

hence, in particular,

$$|\chi_\pm^{(h)}(t)\, t^k| \leq C^{h+1} h!^d, \qquad \forall h \geq 1, \forall k \geq 0. \tag{2.7}$$

Let us fix two parameters ε_j (to be choosen later: see the proof of Lemma 2.4) with

$$0 < \varepsilon_0 < \varepsilon_1. \tag{2.8}$$

Then, for $j = 0, 1$, let us define the cutoff functions

$$\Phi_j^+(\xi) = \chi_+(\varepsilon_j^{-1}\theta(\xi'')\langle\xi\rangle_h^{-\kappa}), \qquad \Phi_j^-(\xi) = \chi_-(\varepsilon_j^{-1}\theta(\xi'')\langle\xi\rangle_h^{-\kappa}), \tag{2.9}$$

where $\theta \equiv \theta(\xi'')$ is the function of Lemma 1.1. Since $\chi_\pm(\varepsilon_0^{-1}t) \leq \chi_\pm(\varepsilon_1^{-1}t)$, we have

$$\Phi_0^\pm(\xi) \leq \Phi_1^\pm(\xi), \qquad \text{supp}(\Phi_j^\pm) \subseteq \left\{ \xi : \pm\theta(\xi'') \geq -\varepsilon_j \langle\xi\rangle_h^\kappa \right\}. \tag{2.10}$$

In particular $\text{supp}(\Phi_0^\pm) \subset\subset \text{supp}(\Phi_1^\pm)$, more precisely:

$$\forall \xi \in \text{supp}(\Phi_0^\pm), \ |\eta| \leq (\varepsilon_1 - \varepsilon_0)\langle\xi\rangle_h^\kappa \implies \xi + \eta \in \text{supp}(\Phi_1^\pm). \tag{2.11}$$

Next, we define the *phase functions*

$$\Lambda_\pm^0(\xi) = \Phi_1^\pm(\xi)\, \langle\xi\rangle_h^\kappa, \qquad \Lambda_\pm(t, \xi) = \rho_\pm(t)\, \Lambda_\pm^0(\xi), \tag{2.12}$$

where

$$\rho_-(t) = \rho_0 - \mu t, \qquad \rho_+(t) = \rho_0 + \mu t, \qquad t \in [-T, T], \tag{2.13}$$

for some $T, \rho_0, \mu > 0$, to be determined later, such that $\rho_\pm(t) > 0$ on $[-T, T]$.
Taking (2.7) and (2.10) into account, and recalling that $\kappa < 1$, we can see that

$$|\partial_\xi^\alpha \Phi_j^\pm(\xi)| \leq C_0\, \rho_0^{-|\alpha|}|\alpha|!^d \langle\xi\rangle_h^{-|\alpha'|-\kappa|\alpha''|}, \tag{2.14}$$

for all $\alpha = (\alpha', \alpha'')$ in $\mathbf{N}^l \times \mathbf{N}^{n-l}$, hence also

$$|\partial_\xi^\alpha \Lambda_\pm^0(\xi)| \leq C_0'\, \rho_0^{-|\alpha|}|\alpha|!^d \langle\xi\rangle_h^{\kappa-|\alpha'|-\kappa|\alpha''|}. \tag{2.15}$$

Note that the Fourier integral operators

$$e^{\Lambda_\pm(t,D)} u(x) = (2\pi)^{-n} \int e^{ix\xi + \Lambda_\pm(t,\xi)} \hat{u}(\xi)\, d\xi$$

operate on the Gevrey class

$$L_\kappa^2 = \{v \in L^2(\mathbf{R}^n) : e^{\rho\,(\xi)_h^\kappa}\,\hat{v} \in L^2(\mathbf{R}^n) \quad \forall\rho > 0\}. \tag{2.16}$$

Now, we conjugate the matrix operator L in (2.4) by these integral operators. Let $u(t, x)$ be a Gevrey solution of (2.4), then the functions

$$u_+(t, x) = e^{\Lambda_+(t,D)} u(t, x), \qquad u_-(t, x) = e^{\Lambda_-(t,D)} u(t, x), \tag{2.17}$$

satisfy the equations

$$L_{\Lambda_\pm}(t, x, D_t, D_x) u_\pm = f_\pm(t, x), \tag{2.18}$$

where

$$L_{\Lambda_\pm} = e^{\Lambda_\pm(t,D)} L(t, x, D_t, D_x) e^{-\Lambda_\pm(t,D)}, \tag{2.19}$$

$$f_\pm = e^{\Lambda_\pm(t,D)} f(t, x). \tag{2.20}$$

Since $L = pI + Q$, we get (writing for brevity Λ in place of $\Lambda_\pm$)

$$L_\Lambda(t, x, D_t, D_x) = p_\Lambda(t, x', D_t, D_x) I + Q_\Lambda(t, x, D_t, D_x),$$

where

$$p_\Lambda = e^{\Lambda(t,D)} p(t, x', D_t, D_x) e^{-\Lambda(t,D)},$$
$$Q_\Lambda = e^{\Lambda(t,D)} Q(t, x, D_t, D_x) e^{-\Lambda(t,D)}.$$

Now we have

$$e^{\Lambda(t,D)} D_t^h e^{-\Lambda(t,D)} = (D_t + i\,\partial_t\Lambda(t, D))^h, \qquad \partial_t\Lambda = \rho'(t)\Lambda^0(\xi),$$

hence, by (2.2) and (2.3), we get

$$Q_{\Lambda_\pm}(t, x, D_t, D_x) = \left[\sum_{|\alpha|+h<m} (q_{\alpha h}^{ij})_{\Lambda_\pm}(t, x, D_x) D_x^\alpha (D_t + i\partial_t\Lambda_\pm(t, D))^h \right]_{ij},$$
$$\tag{2.21}$$

$$p_{\Lambda_\pm}(t, x', D_t, D_x) = \sum_{|\alpha|+h=m} a_{\alpha h \Lambda_\pm}(t, x', D_x) D_x^\alpha (D_t + i\,\partial_t\Lambda_\pm(t, D))^h,$$
$$\tag{2.22}$$

with $a_{0m\Lambda_\pm} \equiv 1$. Here, $(q_{\alpha h}^{ij})_{\Lambda_\pm}(t, x, D_x)$ and $a_{\alpha h \Lambda_\pm}(t, x', D_x)$ are pseudodifferential operators of order 0 and type $(\kappa, 0)$, which will be made explicit in the following lemma.

Lemma 2.1. i) *Let* $a(x)$ *be an analytic function such that*

$$|D_x^\alpha a(x)| \le C\rho_0^{-|\alpha|}|\alpha|!, \qquad \forall x \in \mathbf{R}^n, \ \forall \alpha \in \mathbf{N}^n. \qquad (2.23)$$

Let $\Lambda^0(\xi)$ *be a symbol satisfying* (2.15) *and let us define*

$$\omega(\xi, \eta) = \int_0^1 \nabla_\xi \Lambda^0(\xi + s\eta) \, ds. \qquad (2.24)$$

Then there is some $C_0 > 0$ *for which, if* $|\rho| \le C_0\rho_0$, *the pseudodifferential operator*

$$a_\Lambda(x, D) = e^{\Lambda(D)}a(x)e^{-\Lambda(D)}, \qquad where \qquad \Lambda(\xi) = \rho\Lambda^0(\xi),$$

has a symbol such that, for all integer $N \ge 1$,

$$a_\Lambda(x, \xi) = \sum_{|\alpha| < N} \frac{1}{\alpha!} \partial_\eta^\alpha D_x^\alpha a(x + i\rho\omega(\xi, \eta))\Big|_{\eta=0} + r_N(x, \xi), \qquad (2.25)$$

where $r_N(x, \xi)$ *satisfies, for all* $\alpha = (\alpha', \alpha'') \in \mathbf{R}^l \times \mathbf{R}^{n-l}$ *and* $\beta \in \mathbf{R}^n$,

$$\left|\partial_\xi^\alpha D_x^\beta r_N(x, \xi)\right| \le C_{\alpha\beta}\langle\xi\rangle_h^{-\kappa N - |\alpha'| - \kappa|\alpha''|}. \qquad (2.26)$$

ii) *In the special case when* $a = a(x')$, $x' = (x_1, \ldots, x_l)$, *we have*

$$a_\Lambda(x', \xi) = \sum_{|\alpha'| < N} \frac{1}{\alpha!} [\partial_{\eta'}^{\alpha'} D_{x'}^{\alpha'} a(x' + i\rho\omega'(\xi, \eta'))]\Big|_{\eta'=0} + r_N(x', \xi), \qquad (2.27)$$

where

$$\omega'(\xi, \eta') = \int_0^1 \nabla_{\xi'}\Lambda^0(\xi' + s\eta', \xi'') \, ds, \qquad (2.28)$$

$$|\partial_\xi^\alpha D_{x'}^{\beta'} r_N(x', \xi)| \le C_{\alpha\beta'}\langle\xi\rangle_h^{-N - |\alpha'| - \kappa|\alpha''|}, \qquad \forall \alpha \in \mathbf{N}^n, \ \beta' \in \mathbf{N}^l. \qquad (2.29)$$

In particular, for $N = 1$ *we have*

$$\left|\partial_\xi^\alpha D_x^\beta [a_\Lambda(x, \xi) - a(x + i\rho\omega(\xi))]\right| \le C_{\alpha\beta}\langle\xi\rangle_h^{-\kappa - |\alpha'| - \kappa|\alpha''|} \qquad (2.30)$$

and, in case that $a \equiv a(x')$,

$$\left|\partial_\xi^\alpha D_{x'}^{\beta'} [a_\Lambda(x', \xi) - a(x' + i\rho\omega'(\xi))]\right| \le C_{\alpha\beta'}\langle\xi\rangle_h^{-1 - |\alpha'| - \kappa|\alpha''|} \qquad (2.31)$$

where

$$\omega(\xi) = \nabla_\xi\Lambda^0(\xi), \qquad \omega'(\xi) = \nabla_{\xi'}\Lambda^0(\xi). \qquad (2.32)$$

The proof of this lemma will be given in Section 3.

Step 3 (*Hypoellipticity estimates and parametrix*).

From Lemma 2.1, with

$$\Lambda_\pm(t,\xi) = \rho_\pm(t)\Lambda_\pm^0(\xi), \qquad \rho_\pm(t) = \rho_0 \pm \mu t, \qquad \Lambda_\pm^0(\xi) = \Phi_1^\mp(\xi)\langle\xi\rangle_h^\kappa,$$

it follows that the operator $L_{\Lambda_\pm} = e^{\Lambda_\pm}(pI + Q)e^{-\Lambda_\pm}$ (see (2.2) and (2.3)) has the form

$$L_{\Lambda_\pm}(t, x, D_t, D_x) = p_\pm(t, x', D_t, D_x)I + B_\pm(t, x, D_t, D_x), \qquad (2.33)$$

where $p_\pm$ is the scalar symbol

$$p_\pm(t, x', \lambda, \xi) = p(t, \ x' + i\,\rho_\pm(t)\omega_\pm'(\xi), \ \lambda \pm i\mu\Phi_1^\mp(\xi)\langle\xi\rangle_h^\kappa, \ \xi), \qquad (2.34)$$

with $\omega_\pm'(\xi) = \nabla_{\xi'}\Lambda_\pm^0(\xi)$, while $B_\pm(t, x, \lambda, \xi)$ is a matrix symbol of order $< m$.

More precisely, by applying (2.30) and (2.31) to the coefficients of (2.21) and (2.22) respectively, we get the estimates

$$|\partial_\lambda^j \partial_\xi^\alpha D_t^k D_x^\beta B_\pm(t, x, \lambda, \xi)| \le C_{\alpha\beta kj} \, (|\lambda| + \langle\xi\rangle_h)^{m-1-j} \, \langle\xi\rangle_h^{-|\alpha'|-\kappa|\alpha''|}, \qquad (2.35)$$

for all $j \in \mathbf{N}$, $\alpha = (\alpha', \alpha'') \in \mathbf{N}^n$, $k \in \mathbf{N}$, $\beta \in \mathbf{N}^n$, and all $t \in [-T, T]$.

The following lemma, which is a consequence of the results of §1, says that the operator $p_\pm(t, x, D_t, D_x)$ is microlocally $(\kappa, \ 1 - \kappa)$-hypoelliptic in a suitable zone of the dual space. Note that by (2.5) we have $1/2 < \kappa < 1 - \kappa < 1$.

Lemma 2.2. *Let $\Phi_0^\pm(\xi)$ and $\Phi_1^\pm(\xi)$ be the cutoff functions defined in (2.9). Assume that the roots of the polynomial $p(t, x', \lambda, \xi)$ in (2.2) satisfy (1.6), with $\theta = \theta(\xi'')$, and let $p_\pm$ be the symbol defined in (2.34). Therefore, we can find T, ρ_0, μ in such a way that one has, for all $\xi \in \mathrm{supp}\,(\Phi_0^\pm)$,*

$$|p_\pm(t, x', \lambda, \xi)| \ge c_0 \,(|\lambda| + \langle\xi\rangle_h)^{\kappa m} \qquad (c_0 > 0), \qquad (2.36)$$

$$\frac{|\partial_{\lambda,\xi}^\alpha D_{t,x'}^\beta \, p_\pm(t, x', \lambda, \xi)|}{|p_\pm(t, x', \lambda, \xi)|} \le C_{\alpha\beta} \, \langle\xi\rangle_h^{-\kappa|\alpha|+(1-\kappa)|\beta|} \qquad (2.37)$$

for $t \in [-T, T]$, $x' \in \mathbf{R}^l$, $\lambda \in \mathbf{R}$, and for all $\alpha \in \mathbf{N}^{1+n}$, $\beta \in \mathbf{N}^{1+l}$.

The proof of this lemma will be given in Section 4.

By virtue of the estimates (1.10), (1.11) and (1.12), we are now in a position to construct a (partial) parametrix for the operator $L_{\Lambda_\pm}(t, x, D_t, D_x)$ (see (2.33)). To this end, for a given function $\chi_T(t) \in \gamma^d(\mathbf{R})$ with $\mathrm{supp}\,(\chi_T) \subseteq [-T, T]$, let us define the scalar symbol

$$s_\pm(t, x', \lambda, \xi) = \frac{\chi_T(t)\,\Phi_0^\pm(\xi)}{p_\pm(t, x', \lambda, \xi)}. \qquad (2.38)$$

Then we prove:

Lemma 2.3. *Under the same assumptions of Lemma 2.2, the symbol* (2.38) *is well defined and satisfies*

$$|\partial_{\lambda,\xi}^{\alpha} D_{t,x'}^{\beta} s_{\pm}(t, x', \tau, \xi)| \le C_{\alpha\beta} \frac{\langle \xi \rangle^{-\kappa|\alpha| + (1-\kappa)|\beta|}}{|p_{\pm}(t, x', \lambda, \xi)|}. \tag{2.39}$$

Moreover we have

$$s_{\pm}(t, x', D_t, D_x) L_{\Lambda\pm}(t, x, D_t, D_x) = \chi_T(t)\Phi_0^{\pm}(D_x) I + R_{\pm}(t, x, D_t, D_x), \tag{2.40}$$

where $R_{\pm}$ *satisfies, with h as in* (0.10),

$$|\partial_{\lambda,\xi}^{\alpha} D_{t,x}^{\beta} R_{\pm}(t, x, \lambda, \xi)| \le C_{\alpha\beta} (h^{(1-\kappa)m-1} + h^{1-2\kappa}) \langle \xi \rangle_h^{-\kappa|\alpha| + (1-\kappa)|\beta|}. \tag{2.41}$$

This lemma will be proved in Section 4.

Step 4 (*Conclusion of the proof of Theorem* (0.1))

In order to prove the existence of a local solution $u(t, x)$ to the equation $Lu = f$, with $L = p(t, x', D_t, D_x)I + Q(t, x, D_t, D_x)$ (see (2.2)–(2.3)), we shall prove an a priori estimate for the adjoint operator L^*. More precisely we prove that, for all $v \in C_0^{\infty}(\mathbf{R}; \gamma^d(\mathbf{R}^n))$ with supp $(v) \subset (-T, T) \times \mathbf{R}^n$, one has (see (2.12))

$$\|e^{\Lambda_+(t,D_x)} L^*v\|_{L^2} + \|e^{\Lambda_-(t,D_x)} L^*v\|_{L^2}$$
$$\ge c_0 \|e^{\Lambda_+(t,D_x)} v\|_{L^2} + \|e^{\Lambda_-(t,D_x)} v\|_{L^2} \tag{2.42}$$

where $c_0 > 0$ and the L^2-norm is effected on $\mathbf{R}^{n+1}$. The local existence in γ_d will follow from (2.42) by an easy argument of duality. As a matter of fact, observing that the adjoint L^* has the same form of the direct operator L, indeed $L^* = (-1)^m pI + Q^*$, we can prove (2.42) for L. Hence, our goal will be an estimate of

$$\|e^{\Lambda_+(t,D_x)} Lu\|_{L^2} + \|e^{\Lambda_-(t,D_x)} Lu\|_{L^2}$$
$$\ge c_0 \|e^{\Lambda_+(t,D_x)} u\|_{L^2} + \|e^{\Lambda_-(t,D_x)} u\|_{L^2} \tag{2.43}$$

for all $u \in C_0^{\infty}(\mathbf{R}; L_\kappa^2)$, where L_κ^2 is the Gevrey class defined in (2.16).

Let u be a solution to $Lu = f$ with supp$(u) \subset (-T, T) \times \mathbf{R}^n$: we choose $\chi_T \in \gamma_0^d(\mathbf{R})$ such that $\chi_T(t) \equiv 1$ on supp(u). If $u_{\pm}$, $f_{\pm}$ are the functions defined in (2.17)–(2.20), from (2.18), (2.33) and (2.40), we get the equations

$$\chi_T(t) \Phi_0^{\pm}(D_x) u_{\pm} = R_{\pm}(t, x, D_t, D_x) u_{\pm} + s_{\pm}(t, x', D_t, D_x) f_{\pm}. \tag{2.44}$$

Summing up, we find

$$\Phi u = R u + S f, \tag{2.45}$$

where

$$\Phi = \Phi_0^+(D_x)e^{\Lambda_+(t,D_x)} + \Phi_0^-(D_x)e^{\Lambda_-(t,D_x)}, \tag{2.46}$$

$$R = R_+(t, x, D_t, D_x)e^{\Lambda_+(t,D_x)} + R_-(t, x, D_t, D_x)e^{\Lambda_-(t,D_x)}, \tag{2.47}$$

$$S = s_+(t, x, D_t, D_x) e^{\Lambda_+(t,D_x)} + s_-(t, x, D_t, D_x)e^{\Lambda_-(t,D_x)}. \tag{2.48}$$

The cutoff functions $\Phi_0^{\pm}(\xi)$ and $\Phi_1^{\pm}(\xi)$, defined in (2.9), can be chosen in such a way that the operator (2.46) has a coercive symbol. More precisely, we have:

Lemma 2.4. *Given ρ_0, $\mu > 0$, let us choose $T < \rho_0/\mu$. Therefore the functions $\rho_+(t) = \rho_0 + \mu t$ and $\rho_-(t) = \rho_0 - \mu t$, are positive on $[-T, T]$, and we can choose the parameters ε_0, ε_1 (see (2.8), (2.9)) in such a way that, for some $\delta_0 > 0$, it results that*

$$\Phi_0^{\pm}(\xi) \leq \delta_0 \quad \Longrightarrow \quad \Phi_1^{\mp}(\xi) \leq \frac{\rho_0 - \mu T}{\rho_0 + \mu T} \equiv \inf_{|t| \leq T} \frac{\rho_{\pm}(t)}{\rho_{\mp}(t)}. \tag{2.49}$$

Consequently we get, for some $c_0 > 0$,

$$\Phi_0^+(\xi) \, e^{\Lambda_+(t,\xi)} + \Phi_0^-(\xi) \, e^{\Lambda_-(t,\xi)} \geq c_0 \, (e^{\Lambda_+(t,\xi)} + e^{\Lambda_-(t,\xi)}). \tag{2.50}$$

Proof. The first assertion follows directly from the definitions (2.6) and (2.9), observing that

$$\chi_{\pm}\left(\frac{\varepsilon_0}{\varepsilon_1} t\right) \to \chi_{\pm}(t) \quad \text{uniformly on } \mathbf{R} \text{ when } \quad \frac{\varepsilon_0}{\varepsilon_1} \to 1^-.$$

Assume that (2.49) holds for some $\delta_0 < 1$. By the definition (2.6) it follows that $\chi_-(t) \equiv 1$ where $\chi_+(t) < 1$, hence, recalling (2.9), we see that $\Phi_0^-(\xi) \equiv \Phi_1^-(\xi) \equiv 1$ on the set $I_+ = \{\xi : \Phi_0^+(\xi) \leq \delta_0 < 1\}$. Thus we have, by (2.49),

$$\Lambda_-(t, \xi) \equiv \rho_-(t)\Phi_1^-(\xi)\langle\xi\rangle_h^\kappa$$
$$\equiv \rho_-(t)\langle\xi\rangle_h^\kappa \geq \rho_+(t)\Phi_1^+(\xi)\langle\xi\rangle_h^k \equiv \Lambda_+(t, \xi) \quad \text{on} \quad I_+.$$

This implies (2.50), with $c_0 = 1/2$, on I_+. In a similar way we prove (2.50) on the set $I_- = \{\Phi_0^-(\xi) \leq \delta_0\}$. Finally, on the set $\{\Phi_0^+(\xi) > \delta_0\} \cap \{\Phi_0^-(\xi) > \delta_0\}$, (2.50) holds with $c_0 = \delta_0$. $\qquad\square$

Now we conclude the proof of (2.43). Fixed T, δ_0 as in Lemma 2.4, and denoting by $\|\cdot\|$ the L_2 norm in $\mathbf{R}^{n+1}$, we have, by (2.50),

$$\|\Phi(t, D_x)u\| \geq c_0 \, (\|e^{\Lambda_+(t,D)}u\| + \|e^{\Lambda_-(t,D)}u\|). \tag{2.51}$$

On the other hand, by (2.41) it follows that

$$\|R_{\pm}(t, x, D_t, D_x)w\| \leq C \, h^{-\sigma}\|w\|,$$

with $\sigma = \min\{1 - (1 - \kappa)m, \, 2k - 1\} > 0$, while by (2.39) we have

$$\|s_{\pm}(t, x, D_t, D_x)w\| \leq C \, \|w\|.$$

In conclusion, by (2.47) and (2.51), we find

$$\|Ru\| \leq C \, h^{-\sigma}(\|e^{\Lambda_+(t,D)}u\| + \|e^{\Lambda_-(t,D)}u\|) \leq C' \, h^{-\sigma}\|\Phi(t, D_x)u\|,$$

hence, going back to (2.45) and (2.48), and choosing h sufficiently large, we obtain

$$\|\Phi(t, D_x)u\| \leq C \, \|Sf\| \leq (\|e^{\Lambda_+(t,D)}f\| + \|e^{\Lambda_-(t,D)}f\|). \tag{2.52}$$

From (2.51) and (2.52) it follows that (2.43) holds for all u with support in $(-T, T) \times \mathbf{R}^n$.

For the adjoint operator L^* of L, we get the same estimate, that is (2.42), hence the conclusion of Theorem 0.1 follows by the standard technique of duality.

3   Proof of Lemma 2.1

The symbol of the operator $e^{\Lambda(D)} a(x) e^{-\Lambda(D)}$, where $\Lambda(\xi) = \rho \Lambda_0(\xi)$, is given by

$$
\begin{aligned}
a_\Lambda(x, \xi) &= \iint e^{-iy\eta + \Lambda(\xi+\eta) - \Lambda(\xi)} \, a(x+y) \, dy \, d\eta \\
&= \iint e^{-i\eta \, (y - i\rho\omega(\xi, \eta))} \, a(x+y) \, dy \, d\eta,
\end{aligned}
$$

with $\omega(\xi, \eta) = \int_0^1 \nabla_\xi \Lambda_0(\xi + s\eta) \, ds$. Using the inequality $\langle \xi + s\eta \rangle_h^{-1} \le C \, \langle \xi \rangle_h^{-1} \langle \eta \rangle$, we derive from (2.15)

$$
|\partial_\xi^\alpha \partial_\eta^\beta \omega(\xi, \eta)| \le C_{\alpha\beta} \, \langle \xi \rangle_h^{-|\alpha'+\beta'| - \kappa|\alpha''+\beta''|} \, \langle \eta \rangle^{|\alpha'+\beta'| + \kappa|\alpha''+\beta''|}. \tag{3.1}
$$

By (2.23), we know that $a(z)$ is holomorphic in the complex strip $\{|\Im z| < \rho_0\}$; hence, for $|\rho| \le C_0 \rho_0$, we can write

$$
a_\Lambda(x, \xi) = \iint_{\mathbf{R}^n \times (\mathbf{R}^n - i\rho\omega)} e^{-iz\eta} \, a(x + z + i\rho\omega(\xi, \eta)) \, dz \, d\eta.
$$

Therefore, using Stokes' formula, we get

$$
\begin{aligned}
a_\Lambda(x, \xi) &= \iint_{\mathbf{R}^n \times \mathbf{R}^n} e^{-iy\eta} \, a(x + y + i\rho\omega(\xi, \eta)) \, dy \, d\eta \\
&= \sum_{|\beta| < N} \frac{1}{\beta!} \partial_\eta^\beta \{ D_x^\beta a(x + i\rho\omega(\xi, \eta)) \} \Big|_{\eta=0} + r_N(x, \xi)
\end{aligned}
$$

where

$$
\begin{aligned}
r_N(x, \xi) = \sum_{|\beta|=N} \frac{N}{\beta!} \int_0^1 \iint_{\mathbf{R}^{2n}} (1-s)^{N-1} e^{-iy\eta} \\
\times \partial_\eta^\beta \{ D_x^\beta a(x + sy + i\rho\omega(\xi, \eta)) \} \, dy \, d\eta \, ds. \tag{3.2}
\end{aligned}
$$

Let l_0, l_1 be two (sufficiently large) integers: integrating by parts, we find

$$
\begin{aligned}
&\partial_\xi^\alpha D_x^\gamma r_N(x, \xi) \\
&= \sum_{|\beta|=N} \frac{N}{\beta!} \int_0^1 \iint \langle \eta \rangle^{-l_0} \langle D_y \rangle^{l_0} \langle y \rangle^{-l_1} \langle D_\eta \rangle^{l_1} \\
&\qquad\qquad \times \partial_\xi^\alpha D_x^\gamma \partial_\eta^\beta \{ D_x^\beta a(x + sy + i\rho\omega(\xi, \eta)) \} \, \psi \, dy \, d\eta \, ds \\
&= \sum_{|\beta|=N} \frac{N}{\beta!} \int_0^1 \iint F_{\alpha\beta\gamma}(x, \xi, y, \eta, s) \, \psi \, dy \, d\eta \, ds,
\end{aligned}
$$

with $\psi = (1-s)^{N-1} e^{-iy\eta}$, and

$$F_{\alpha\beta\gamma}(x, \xi, y, \eta, s)$$
$$= \langle\eta\rangle^{-l_0} \langle D_y\rangle^{l_1} \langle y\rangle^{-l_1} \langle D_\eta\rangle^{l_1} D_x^\gamma \partial_\xi^\alpha \partial_\eta^\beta \{D_x^\beta a(x + sy + i\rho\omega(\xi, \eta))\}.$$

By (3.1) we have, since $-|\beta'| - \kappa|\beta''| \le -\kappa|\beta| = -\kappa N$,

$$\sum_{|\beta|=N} |F_{\alpha\beta\gamma}(x, \xi, y, \eta, s)|$$

$$\le C_{\beta\gamma N} \langle y\rangle^{-l_1} \langle\eta\rangle^{-l_0 + |\alpha'+\beta'| + \kappa|\alpha''+\beta''|} \langle\xi\rangle_h^{-\kappa N - |\alpha'| - \kappa|\alpha''|},$$

hence, taking $l_0 > n/2 + |\alpha' + \beta'| + \kappa |\alpha'' + \beta''| + 1$, and $l_1 > n/2 + 1$, we get (2.26) from (3.2).

When $a = a(x')$ is independent of $(x_l, \ldots, x_n)$, we can write

$$a_\Lambda(x', \xi) = \iint_{\mathbf{R}^{2l}} e^{-iy'\eta'} a(x' + y' + i\rho\omega'(\xi, \eta')) \, dy' \, d\eta'$$

$$= \sum_{|\beta'|<N} \frac{1}{\beta'!} \partial_{\eta'}^{\beta'} D_{x'}^{\beta'} a(x' + i\rho\omega'(\xi, \eta'))\Big|_{\eta'=0} + r_N(x', \xi),$$

where

$$r_N(x', \xi) = \sum_{|\beta'|=N} \frac{N}{\beta'!} \int_0^1 \iint (1-s)^{N-1} e^{-iy'\eta'}$$
$$\times \partial_{\eta'}^{\beta'} \{D_{x'}^{\beta'} a(x' + sy' + i\rho\omega'(\xi, \eta'))\} \, dy' \, d\eta' \, ds$$

and

$$\omega'(\xi, \eta') = \int_0^1 \nabla_{\xi'} \Lambda_0(\xi' + s\eta', \xi'') \, ds.$$

Using again (2.15), we have now

$$|\partial_\xi^\alpha \partial_{\eta'}^{\beta'} \omega'(\xi, \eta')| \le C_{\alpha\beta'} \langle\xi\rangle_h^{\kappa - 1 - |\alpha'+\beta'| - \kappa|\alpha''|} \langle\eta'\rangle^{1 - \kappa + |\alpha'+\beta'| + \kappa|\alpha''+\beta''|},$$

hence we get (2.29) similarly to (2.26). □

4 Proof of Lemma 2.2

In order to prove that the symbol (2.34) satisfies the estimates (2.36) and (2.37), we shall firstly prove that these estimates are fulfilled by the simpler symbol

$$\tilde{p}_\pm(t, x', \lambda, \xi) = p(t, x', \lambda \mp i\mu\Phi_1^\pm(\xi)\langle\xi\rangle_h^\kappa, \xi). \qquad (4.1)$$

To this end, let us observe that, by the definition (2.9), we have $\pm\theta(\xi) \ge -\varepsilon_0\langle\xi\rangle_h^\kappa$ for $\xi \in \mathrm{supp}(\Phi_0^\pm)$; hence, if we take ε_0 sufficiently small, and, $\xi \in \mathrm{supp}(\Phi_0^\pm)$, we see that the point $(\lambda \mp i\mu\Phi_1^\pm(\xi)\langle\xi\rangle_h^\kappa, \xi)$ belongs to the set W_ε of Lemma 1.2. Consequently we get, by (1.10),

$$|\widetilde{p}_\pm(t, x', \lambda, \xi)| \ge c_1\langle\xi\rangle_h^{\kappa m} \ge c_2\,(|\lambda| + \langle\xi\rangle_h)^{\kappa m}\,,$$

as soon as $|\lambda| \le M\langle\xi\rangle_h$ for some M. On the other hand, if $|\lambda| \gg \langle\xi\rangle_h$, taking into account that the roots of p satisfy $|\tau_j| \le C|\xi|$, we have

$$|p(t, x', \lambda + i\rho, \xi)| = \prod_{j=1}^m \left|(\lambda - \Re\tau_j)^2 + (\rho - \Im\tau_j)^2\right|^{1/2}$$

$$\ge \prod(||\lambda| - |\tau_j||)^m \ge c_0\,(|\lambda| + \langle\xi\rangle_h)^m.$$

Summing up, we have proved that, for all $\xi \in \mathrm{supp}\,(\Phi_0^\pm)$ and all $\lambda \in \mathbf{R}$, one has

$$|\widetilde{p}_\pm(t, x', \lambda, \xi)| \ge c_1(|\lambda| + \langle\xi\rangle_h)^{\kappa m}\,. \qquad (4.2)$$

Applying again Lemma 1.2, we get from (1.12) that

$$\frac{|\partial_{\lambda,\xi}^\alpha D_{t,x'}^\beta \widetilde{p}_\pm(t, x', \lambda, \xi)|}{|\widetilde{p}_\pm(t, x', \lambda, \xi)|} \le C\langle\xi\rangle_h^{-\kappa|\alpha|+(1-\kappa)|\beta|}\,, \qquad (4.3)$$

for all $(\lambda, \xi) \in \mathbf{R} \times \mathrm{supp}(\Phi_0^\pm)$, provided $|\alpha| + |\beta| = 1$.

Next we shall prove that, if $\xi \in \mathrm{supp}(\Phi_1^\pm)$, then (4.3) is valid for all α, β. We recall that $\mathrm{supp}(\Phi_1^\pm) \subset\subset \mathrm{supp}(\Phi_0^\pm)$ (see (2.9), (2.10), (2.11)). If $|\lambda| \gg \langle\xi\rangle_h$, (4.3) is trivial because p is elliptic. So, we prove (4.3) for $|\lambda| \le M\langle\xi\rangle_h$.

Defining

$$f(t, x', \lambda, \xi) = 2C\,\widetilde{p}_\pm(t, x, \lambda, \xi)\,(|\lambda| + \langle\xi\rangle_h)^{-\kappa m}\,,$$

(4.2) and (4.3) yield, for $|\alpha| + |\beta| = 1$,

$$\frac{|\partial_{\lambda,\xi}^\alpha D_{t,x'}^\beta f(t, x', \lambda, \xi)|}{|f(t, x', \lambda, \xi)|} \le C\,\langle\xi\rangle_h^{-\kappa|\alpha|+(1-\kappa)|\beta|}\,. \qquad (4.4)$$

Now we define the positive function

$$M(t, x', \lambda, \xi) = \log|f(t, x', \lambda, \xi)|\,,$$

and we rewrite (4.4) as

$$|\partial_{\lambda,\xi}^\alpha D_{t,x'}^\beta M(t, x', \lambda, \xi)| \le C\,\langle\xi\rangle_h^{-\kappa|\alpha|+(1-\kappa)|\beta|}\,, \qquad |\alpha| + |\beta| = 1. \quad (4.5)$$

Next, for any fixed $(y, \eta) \in \mathbf{R}^{1+l} \times \mathbf{R}^{1+n}$ with $|y| + |\eta|$ sufficiently small, we put $\widetilde{y} = y\langle\xi\rangle_h^{\kappa-1}$, $\widetilde{\eta} = \eta\langle\xi\rangle_h^\kappa$. Then, for $\xi \in \mathrm{supp}(\Phi_1^\pm)$, we have by (4.5)

$$|M((t, x') + \widetilde{y}, (\lambda, \xi) + \widetilde{\eta}) - M(t, x', \lambda, \xi)|$$

$$= \left|\sum_{|\alpha|+|\beta|=1} \int_0^1 \partial_{\lambda,\xi}^\alpha D_{t,x'}^\beta M((t, x') + s\widetilde{y}, (\lambda, \xi) + s\widetilde{\eta})\,\widetilde{y}^\beta \widetilde{\eta}^\alpha\,ds\right|$$

$$\le C(|y| + |\eta|),$$

provided $|\eta| \le \varepsilon_1 - \varepsilon_0$, and hence, by virtue of (2.11), $(\lambda, \xi) + s\tilde{\eta} \in \mathbf{R} \times \text{supp}(\Phi_0^{\pm})$. In conclusion, we have proved that

$$| M((t, x') + \tilde{y}, (\lambda, \xi) + \tilde{\eta}) - M(t, x', \lambda, \xi) | \le C,$$

and consequently

$$| f((t, x') + \tilde{y}, (\lambda, \xi) + \tilde{\eta}) | \le e^C |f(t, x', \lambda, \xi)|,$$

for all $(\lambda, \xi) \in \mathbf{R} \times \text{supp}(\Phi_0^{\pm})$. On the other hand, recalling that $f \in S_{1,0}^{m - \kappa m}$, we have

$$| f((t, x') + \tilde{y}, (\lambda, \xi) + \tilde{\eta}) |$$

$$= \left| \sum_{|\alpha| + |\beta| < N} \partial_{\lambda, \xi}^{\alpha} D_{t, x'}^{\beta} f(t, x', \lambda, \xi) \frac{\tilde{y}^{\beta} \tilde{\eta}^{\alpha}}{\alpha! \beta!} \right.$$

$$\left. + N \sum_{|\alpha| + |\beta| = N} \int_0^1 \partial_{\lambda, \xi}^{\alpha} D_{t, x'}^{\beta} f((t, x') + s\tilde{y}, (\lambda, \xi) + s\tilde{\eta}) \frac{\tilde{y}^{\beta} \tilde{\eta}^{\alpha}}{\alpha! \beta!} \, ds \right|$$

$$\ge \left| \sum_{|\alpha| + |\beta| < N} \partial_{\lambda, \xi}^{\alpha} D_{t, x'}^{\beta} f(t, x', \lambda, \xi) \tilde{y}^{\beta} \tilde{\eta}^{\alpha} / \alpha! \beta! \right|$$

$$- C_N (|y| + |\eta|)^N \langle \xi \rangle_h^{(1-\kappa)(m-N)}.$$

Hence, taking $N = m$ and $|y| + |\eta| \le 1$, and noting that $|f(t, x', \lambda, \xi)| > 1$, we get

$$\left| \sum_{0 < |\alpha| + |\beta| < N} \partial_{\lambda, \xi}^{\alpha} D_{t, x'}^{\beta} f(t, x', \lambda, \xi) \tilde{y}^{\beta} \tilde{\eta}^{\alpha} / \alpha! \beta! \right| \le C |f(t, x', \lambda, \xi)|. \quad (4.6)$$

But we can choose the vectors $y_\beta \in \mathbf{R}^{n+1-l}$ and $\eta_\alpha \in \mathbf{R}^{n+1}$ in such a way that the matrix $[y_\beta^\beta \eta_\alpha^\alpha]_{\{|\alpha| + |\beta| \le m-1\}}$ is not singular, hence (4.6) implies

$$| \partial_{\lambda, \xi}^{\alpha} D_{t, x'}^{\beta} f(t, x', \lambda, \xi) | \langle \xi \rangle_h^{\kappa |\alpha| - (1-\kappa)|\beta|} \le C |f(t, x, \lambda, \xi)|,$$

that is (4.4), for $0 < |\alpha| + |\beta| \le m - 1$. For $|\alpha + \beta| \ge m$, (4.4) is trivial since $f \in S_{1,0}^{(1-\kappa)m}$ and $|f| > 1$.

Thus, we have proved (4.4) for $(\lambda, \xi) \in \mathbf{R} \times \text{supp}(\Phi_0^{\pm})$, and every $\alpha \in \mathbf{N}^{1+n}$, $\beta \in \mathbf{N}^{1+l}$. By (4.4), we can easily derive that (4.3) holds for all α, β.

Finally, we show that (2.36) and (2.37) are valid also for the symbol (2.34). This can be written in the form

$$p_\pm(t, x', \lambda, \xi) = \tilde{p}_\pm(t, x' + i \rho(t)\omega'(\xi), \lambda, \xi) \quad (4.7)$$

with $\tilde{p}_\pm$ as in (4.1), $\rho(t) = \rho_0 \mp \mu t$, $\omega'(\xi) = \nabla_{\xi'} \Lambda_\pm^0(\xi) = \nabla_{\xi'}(\Phi_1^\pm(\xi)\langle \xi \rangle_h^\kappa)$. Note that, by (2.15) with $|\alpha'| = 1$, $|\alpha''| = 0$, it follows that

$$|\omega'(\xi)| \le C \langle \xi \rangle_h^{-(1-\kappa)}. \quad (4.8)$$

Thus, by Taylor expansion, we have

$$p_{\pm}(t, x', \lambda, \xi) = \sum_{|\beta| < N} \frac{1}{\beta!} \partial_{x'}^{\beta} \widetilde{p}_{\pm}(t, x', \lambda, \xi) \, \rho(t)^{|\beta|} \omega'(\xi)^{\beta} \; + \; R_N(t, x', \lambda, \xi),$$

(4.9)

where the remainder satisfies

$$|R_N(t, x', \lambda, \xi)| \le c_N \, (|\lambda| + \langle \xi \rangle_h)^m \, \langle \xi \rangle_h^{-(1-\kappa)N} \, .$$

Hence, by (4.2), we get

$$|R_N(t, x', \lambda, \xi)| \le C_N \, \langle \xi \rangle_h^{(m-N)(1-\kappa)} \, \widetilde{p}_{\pm}(t, x', \lambda, \xi) \, .$$

From (4.3), (4.8) and (4.9) it follows that

$$\frac{| p_{\pm}(t, x', \lambda, \xi) - \widetilde{p}_{\pm}(t, x', \lambda, \xi) |}{| \widetilde{p}_{\pm}(t, x', \lambda, \xi) |} \le \sum_{j=1}^{N-1} C_j \rho(t)^j \; + \; C_N \, \langle \xi \rangle_h^{(m-N)(1-\kappa)N}$$

$$\le C_N' \left[(\rho_0 + \mu T)^N + h^{-(1-\kappa)(N-m)} \right] \le \frac{1}{2} \, ,$$

provided $N > m$, and that $\rho_0 + \mu T + h^{-1}$ is small enough.

From the above inequality and (4.2), we get (2.36) for $p_{\pm}$. Moreover from (4.2), (4.3), and (4.9), we have, for $|\alpha| + |\beta| = 1$,

$$\frac{| \partial_{\lambda, \xi}^{\alpha} D_{t, x'}^{\beta} \, p_{\pm} |}{| p_{\pm} |}$$

$$\le \sum_{|\gamma| < N} \frac{1}{\alpha!} \frac{| (\partial_{\lambda, \xi}^{\alpha} D_{t, x'}^{\beta} \partial_{x'}^{\gamma} \widetilde{p}_{\pm}) \, \omega'(\xi)^{\alpha} + \partial_{\lambda, \xi}^{\alpha} D_{t, x'}^{\beta} R_N(t, x', \lambda, \xi) |}{| \widetilde{p}_{\pm} |} \cdot \frac{| \widetilde{p}_{\pm} |}{| p_{\pm} |}$$

$$\le C \sum_{|\gamma| < N} \langle \xi \rangle_h^{-\kappa|\alpha| + (1-\kappa)|\beta| - (1-2\kappa)|\gamma|} + C \, \langle \xi \rangle_h^{m-(1-\kappa)N - \kappa m - \kappa|\alpha|}$$

$$\le C \, \langle \xi \rangle_h^{-\kappa|\alpha| + (1-\kappa)|\beta|} \, .$$

This proves that $p_{\pm}$ satisfies (2.37), if we take $N = m$. It follows from (4.2) that

$$\frac{(|\lambda| + \langle \xi \rangle_h)^m \langle \xi \rangle_h^{-1}}{| \widetilde{p}_{\pm} |} \le C \, \langle \xi \rangle_h^{m-1-\kappa m} \le C \, h^{m-1-\kappa m} \; << \; 1,$$

since $\kappa > (m-1)/m$. This concludes the proof of Lemma 2.2.   □

5   Proof of Lemma 2.3

First, we prove that the symbol

$$s_{\pm}(t, x', \lambda, \xi) = \frac{\chi_T(t) \Phi_0^{\pm}(\xi)}{p_{\pm}(t, x', \lambda, \xi)}$$

(5.1)

satisfies, for all $\alpha, \beta \in \mathbf{N}^{1+n}$,

$$|\partial_{\lambda,\xi}^{\alpha} D_{t,x}^{\beta} s_{\pm}(t, x', \lambda, \xi)| \leq C_{\alpha\beta} \frac{\langle\xi\rangle_h^{-|\alpha|+(1-\kappa)|\beta|}}{|p_{\pm}(t, x', \lambda, \xi)|}. \tag{5.2}$$

In fact, differentiating the equality $s_{\pm}(t, x, \lambda, \xi) \, p_{\pm}(t, x, \lambda, \xi) = \chi_T(t)\Phi_0^{\pm}(\xi)$, we find

$$\partial_{\lambda,\xi}^{\alpha} D_{t,x}^{\beta} s_{\pm}$$
$$= -\frac{1}{p_{\pm}}\Big[\sum_{\alpha^1 < \alpha, \, \beta^1 < \beta} \binom{\alpha}{\alpha^1}\binom{\beta}{\beta^1} s_{\pm(\beta^1)}^{(\alpha^1)} p_{\pm(\beta-\beta^1)}^{(\alpha-\alpha^1)} - \partial_{\lambda,\xi}^{\alpha} D_{t,x}^{\beta}(\chi_T(t)\Phi_0^{\pm}(\xi)) \Big],$$

hence, recalling (2.14) and using Lemma 2.2 we can prove inductively (5.2), whence (2.39) follows immediately.

In view of (2.40), (2.41), recalling (2.33) we define the operators

$$R_{\pm} = s_{\pm}(t, x', D_t, D_x) [\, p_{\pm}(t, x', D_t, D_x)I$$
$$+ B_{\pm}(t, x', D_t, D_x)] - \chi_T(t)\Phi_0^{\pm}(D_x)I,$$
$$R_{\pm}^1 = [s_{\pm}(t, x', D_t, D_x) \, p_{\pm}(t, x', D_t, D_x) - \chi(t)\Phi_0^{\pm}(D_x)]\, I,$$
$$R_{\pm}^2 = s_{\pm}(t, x', D_t, D_x) \, B_{\pm}(t, x', D_t, D_x),$$

so that $R_{\pm} = R_{\pm}^1 + R_{\pm}^2$. We show that the symbol of the first summand satisfies

$$|\partial_{\lambda,\xi}^{\alpha} D_{t,x}^{\beta} R_{\pm}^1(t, x', \lambda, \xi)| \leq C_{\alpha\beta} (|\lambda| + \langle\xi\rangle_h)^{1-2\kappa-\kappa|\alpha|+(1-\kappa)|\beta|}. \tag{5.3}$$

Indeed we can write, recalling (2.38),

$$R_{\pm}^1(t, x', \lambda, \xi)$$
$$= \iint e^{-iy\eta} (s_{\pm} \cdot p_{\pm})(t, x', \lambda + \eta_0, (\xi' + \eta', \xi'')) \, dy \, d\eta - \chi(t)\Phi_0^{\pm}(\xi)$$
$$= \sum_{|\gamma|=1} \int_0^1 \iint e^{-iy\eta} \, \partial_{\lambda,\xi''}^{\gamma} s_{\pm}(t, x', \lambda + \eta_0, (\xi' + \eta', \xi''))$$
$$\times D_{t,x'}^{\gamma} p_{\pm}(t + \theta y_0, x' + \theta y', \lambda, \xi) \, dy \, d\eta \, d\theta.$$

Hence we can estimate

$$\partial_{\lambda,\xi}^{\alpha} D_{t,x}^{\beta} R_{\pm}^1(t, x', \lambda, \xi)$$
$$= \sum \binom{\alpha}{\alpha^1}\binom{\beta}{\beta^1} \iiint e^{-iy\eta} \, \partial_{\lambda,\xi''}^{\gamma+\alpha-\alpha^1} D_{t,x'}^{\beta-\beta^1} s_{\pm}(t, x', \lambda + \eta_0, (\xi', \xi'' + \eta''))$$
$$\times \partial_{\lambda,\xi}^{\alpha^1} D_{t,x'}^{\gamma+\beta^1} p_{\pm}(t + \theta y_0, x' + \theta y'', \lambda, \xi) \, dy \, d\eta \, d\theta$$

$$= \sum \binom{\alpha}{\alpha^1}\binom{\beta}{\beta^1} \iiint e^{-iy\eta} \langle y \rangle^{-l_0} \langle D_\eta \rangle^{l_0} \partial_{\lambda,\xi'}^{\gamma+\alpha-\alpha^1} D_{t,x'}^{\beta-\beta^1}$$

$$\times s_\pm(t, x', \lambda + \eta_0, (\xi' + \eta', \xi'' + \eta'')) \langle \eta \rangle^{-l_0} \langle D_y \rangle^{l_0}$$

$$\times \partial_{\lambda,\xi}^{\alpha^1} D_{t,x'}^{\gamma+\beta^1} p_\pm(t + \theta y_0, x' + \theta y', \lambda, \xi)\, dy\, d\eta\, d\theta.$$

Therefore, from (5.2) and (2.37) we can easily derive (5.3). On the other hand, by (5.2) and (2.35), we see that $R_\pm^2 = s_\pm(t, x', D_t, D_x) B_\pm(t, x, D_t, D_x)$ satisfies

$$|D_{t,x}^\beta \partial_{\lambda,\xi}^\alpha R_\pm^2(t, x, \lambda, \xi)| \le C_{\alpha\beta} h^{(1-\kappa)m-1} \langle \xi \rangle_h^{-\kappa|\alpha|+(1-\kappa)|\beta|}.$$

Thus we have proved that $R_\pm = R_\pm^1 + R_\pm^2$ satisfies (2.41).   □

Acknowledgements. The authors are indebted to Giovanni Taglialatela for his careful and critical reading of the manuscript.

References

[1] M.D. Bronshtein, *Smoothness of polynomials depending on parameters*, Sib. Mat. Zh. **20** (1979), 493–501.

[2] M.D. Bronshtein, *The Cauchy problem for hyperbolic operators with multiple variable characteristics*, Trudy Moskow Mat. Obsc. **41** (1980), 83–99; Trans. Moscow Math. Soc. **1** (1982), 87–103.

[3] T. Gramchev and L. Rodino, *Gevrey solvability for semilinear partial differential equations with multiple characteristics*, Boll. Un. Mat. It. **2-B** (1999), 65–120.

[4] T. Gramchev, *On the critical index of Gevrey Solvability for some linear partial differential equations*, Ann. Univ. Ferrara Sez. Sci. Mat., Suppl. **14** (1999), 139–153.

[5] K. Kajitani, *Local solution of the Cauchy problem for hyperbolic systems in Gervrey classes*, Hokkaido Math. J. **12** (1983), 434–460.

[6] P.R. Popivanov, *Local solvability of some classes of linear differential operators with multiple characteristics*, Ann. Univ. Ferrara Sez. Sci. Mat., Suppl. **24** (1999), 263–274.

[7] S. Spagnolo, *Local and semi-global solvability for systems of non-principal type*, Comm. Part. Diff. Equat. **25** (2000), 1115–1141.

[8] S. Wakabayashi, *Remarks on hyperbolic polynomials*, Tsukuba J. Math. **10** (1986), 17–28.

Kunihiko Kajitani
Institute of Mathematics
University of Tsukuba
Tsukuba, Japan

Sergio Spagnolo
Department of Mathematics,
University of Pisa,
Pisa, Italy
spagnolo@dm.unipi.it

Spherically Symmetric Solutions of the Compressible Euler Equation

Kiyoshi Mizohata

ABSTRACT We shall discuss the existence theorem of global weak solutions with spherical symmetry of the Euler equation and of the relativistic Euler equation.

1 Introduction

In this article we shall show some results of the compressible Euler equation for the case of spherical symmetry and related topics.

The compressible Euler equation for an isentropic gas in $\mathbf{R}^n$ is given by

$$
\begin{cases}
\rho_t + \sum_{j=1}^{n} \dfrac{\partial}{\partial x_j} \left(\rho u_j \right) = 0 \,, \\[2ex]
(\rho u_i)_t + \sum_{j=1}^{n} \dfrac{\partial}{\partial x_j} \left(\rho u_i u_j + \delta_{ij} P \right) = \rho f_i \,, \quad (i = 1, 2, \cdots, n)
\end{cases}
\tag{1.1}
$$

with the equation of state

$$
p = a^2 \rho^\gamma \,,
\tag{1.2}
$$

where density ρ, velocity $\vec{u}$ and pressure p are functions of $x \in \mathbf{R}^n$ and $t \geq 0$, while $a > 0$ and $\gamma \geq 1$ are given constants.

For the one-dimensional case ($n = 1$), the Cauchy problem for (1.1) with (1.2) has been studied by many authors. Nishida [12] established the existence of global weak solutions, for the first time, for the case $\gamma = 1$ with arbitrary initial data, by using Glimm's method [4]. DiPerna [3] extended the latter result to the case of large initial data, using the theory of compensated compactness under the restriction $\gamma = 1 + 2/(2m + 1)$, $m \geq 2$ integers. Ding et al. [1], [2] removed this restriction and established the existence of global weak solutions for $1 < \gamma \leq 5/3$.

On the other hand, little is known for the case $n \geq 2$. No global solutions have been known to exist, but only local classical solutions (see [7]).

T. Makino, K. Mizohata and S. Ukai have presented global weak solutions first for the case $n \geq 2$ in [8] for $\gamma = 1$. They constructed global weak solutions with spherical symmetry outside a ball with radius 1. We shall state this result and related topics in Section 2.

Next, we will extend this result to the relativistic equation. The relativistic Euler equation in $\mathbf{R}^3$ is given by

$$\begin{cases} \dfrac{\partial}{\partial t}\left(\dfrac{\rho c^2 + P}{c^2}\dfrac{1}{1-\frac{v^2}{c^2}} - \dfrac{P}{c^2}\right) + \sum_{j=1}^{3}\dfrac{\partial}{\partial x_j}\left(\dfrac{\rho c^2 + P}{c^2}\dfrac{v_j}{1-\frac{v^2}{c^2}}\right) = 0, \\[4mm] \dfrac{\partial}{\partial t}\left(\dfrac{\rho c^2 + P}{c^2}\dfrac{v_i}{1-\frac{v^2}{c^2}}\right) + \sum_{j=1}^{3}\dfrac{\partial}{\partial x_j}\left(\dfrac{\rho c^2 + P}{c^2}\dfrac{v_i v_j}{1-\frac{v^2}{c^2}} + \delta_{ij}P\right) = 0, \end{cases} \tag{1.3}$$

$i = 1, 2, 3$, with the equation of state

$$p = \sigma^2 \rho, \tag{1.4}$$

where the density ρ, the velocity $\mathbf{v} =^t (v_1, v_2, v_3)$ and the pressure p are functions of $x \in \mathbf{R}^3$ and $t \geq 0$, while c is the speed of light and σ is the speed of sound, which are constants. According to the relativistic theory, σ never exceeds c. Especially, the case $\sigma^2 = \frac{c^2}{3}$ has been important in the context of physics. If $c \to \infty$, (1.3) reduces to the classical compressible Euler equation (1.1).

In 1993, Smoller and Temple [13] constructed uniformly bounded weak solutions for the one-dimensional case by using Glimm's method [4]. After constructing approximate solutions, they showed that the total variation of $\log \rho$ of the approximate solutions is monotone decreasing.

Recently, K. Mizohata has constructed global weak solutions for (1.3) with spherical symmetry. We shall state this result in Section 3.

2   Spherically symmetric solutions

We look for solutions of the form

$$\rho = \rho(t, |x|), \quad \vec{u} = \frac{x}{|x|} \cdot u(t, |x|). \tag{2.1}$$

Then, denoting $r = |x|$, (1.1) becomes

$$\begin{aligned} \rho_t + \frac{1}{r^{n-1}}(r^{n-1}\rho u)_r &= 0, \\ \rho(u_t + u u_r) + p_r &= 0. \end{aligned} \tag{2.2}$$

This equation has a singularity at r=0. To avoid the difficulty caused by this singularity, we simply deal with the boundary value problem for (2.2) in the domain $1 \leq r < \infty$ (the exterior of a sphere) with the boundary condition $u(t, 1) = 0$, which is identical, under the assumption (2.1), to the usual boundary condition $\vec{n} \cdot \vec{u} = 0$ for (1.1) where $\vec{n}$ is the unit normal to the boundary.

Put $\tilde{\rho} = r^{n-1}\rho$. Then we get from (2.2)

$$\begin{aligned} \tilde{\rho}_t + (\tilde{\rho} u)_r &= 0, \\ u_t + u u_r + \frac{a^2 \gamma \tilde{\rho}_r}{\tilde{\rho}^{2-\gamma} r^{(n-1)(\gamma-1)}} &= \frac{a^2 \gamma (n-1)\tilde{\rho}^{\gamma-1}}{r^n \cdot r^{(n-1)(\gamma-2)}}. \end{aligned} \tag{2.3}$$

Introduce the Lagrangian mass coordinates

$$\tau = t, \quad \xi = \int_1^r \tilde{\rho}(t, r) \, dr . \qquad (2.4)$$

Then $\xi > 0$ as long as $\tilde{\rho} > 0$ for $r > 1$, and (2.3) is reformulated as

$$\tilde{\rho}_\tau + \tilde{\rho}^2 u_\xi = 0,$$
$$u_\tau + \frac{a^2 \gamma \tilde{\rho}_\xi}{\tilde{\rho}^{1-\gamma} r^{2\gamma-2}} = \frac{a^2 \gamma (n-1) \tilde{\rho}^{\gamma-1}}{r^n \cdot r^{(n-1)(\gamma-2)}}. \qquad (2.5)$$

Put $v = 1/\tilde{\rho}$ and note that the inverse transformation to (2.4) is given by

$$t = \tau, \quad r = 1 + \int_0^\xi v(\zeta, t) d\zeta. \qquad (2.6)$$

Then after changing τ to t and ξ to x, (2.5) is written as

$$v_t - u_x = 0,$$
$$u_t + \left(\frac{a^2}{v^\gamma}\right)_x \cdot \frac{1}{r^{(n-1)(\gamma-1)}} = \frac{a^2 \gamma (n-1) v^{1-\gamma}}{r^n \cdot r^{(n-1)(\gamma-2)}}, \qquad (2.7)$$

where r is now defined by $r = 1 + \int_0^x v(t, \zeta) d\zeta$.
Now we restrict ourselves to the case $\gamma = 1$. Then (2.7) becomes

$$v_t - u_x = 0,$$
$$u_t + \left(\frac{a^2}{v}\right)_x = \frac{K}{1 + \int_0^x v(t, \zeta) d\zeta} \qquad (2.8)$$

where $K = a^2(n-1)$.
Let us consider the initial boundary value problem for (2.8) in $t \geq 0$, $x \geq 0$ with the boundary and initial conditions

$$u(0, x) = u_0(x), \quad v(0, x) = v_0(x), \quad for \ x > 0, \qquad (2.9)$$

$$u(t, 0) = 0, \quad for \ t > 0. \qquad (2.10)$$

Let $BV(\mathbf{R}_+)$ denote the space of functions of bounded variation on $\mathbf{R}_+ = (0, \infty)$. Our result is as follows.

Theorem 1 ([8]). *Suppose that $u_0(x)$, $v_0(x) \in BV(\mathbf{R}_+)$, and that $v_0(x) \geq \delta_0 > 0$ for all $x > 0$ with some positive constant δ_0. Then (2.8), (2.9) and (2.10) have a global weak solution which belongs to the class*

$$u, \ v \ \in L^\infty(0, T; \ BV(\mathbf{R}_+)) \cap Lip([0, T]; \ L^1_{loc}(\mathbf{R}_+))$$

for any $T > 0$.

This theorem can be proved by following Nishida's argument [12] based on Glimm's method. Indeed this can be seen from the following two simple observations.

First, the homogeneous equation corresponding to (2.8),

$$
\begin{aligned}
v_t - u_x &= 0 , \\
u_t + \left(\frac{a^2}{v} \right)_x &= 0 ,
\end{aligned}
\tag{2.11}
$$

is just the same equation as solved by Nishida [12] using Glimm's method both on the Cauchy problem and the initial boundary value problem. Note that if $\gamma > 1$, the homogeneous equation for (2.7) has a variable coefficient and hence does not coincide with the one-dimensional Euler equation.

The second observation is that, as long as $v \geq 0$, the right-hand side of (2.8),

$$
\frac{K}{1 + \int_0^x v(t, \zeta) d\zeta} ,
\tag{2.12}
$$

is monotone decreasing in x and has an a priori estimate

$$
T. V. \left(\frac{K}{1 + \int_0^x v(t, \zeta) d\zeta} \right) \leq K ,
\tag{2.13}
$$

independent of v.

These observations allow us to use Nishida's ideas [12] to construct global weak solutions to (2.8), (2.9) and (2.10). For details, see [8].

Note that the class of the initial data of this result is not wide enough to include the stationary solutions. It is clear that the equation (1.1) admits the stationary solutions $\rho = \tilde{\rho} = constant > 0, u = 0$. The corresponding stationary solutions for (2.8) are

$$
v = \frac{1}{\tilde{\rho} \left(1 + \frac{n}{\tilde{\rho}} x \right)^{1 - \frac{1}{n}}} , \quad u = 0 .
\tag{2.14}
$$

For these stationary solutions, we have inf $v = 0$, i.e., v is not bounded away from zero. In [9], we extended this result to a wider class of the initial data so that it includes the stationary solutions by using a modified Glimm's scheme.

In [10], we also dicussed the equivalence of Eulerian and Lagrangian weak solutions of the compressible Euler equation with spherical symmetry.

Recently, Cheng-Hsiung Hsu and T. Makino have presented, under some conditions on the initial data, global weak solutions with spherical symmetry for the case $\gamma > 1$. For details, see [6].

3   Extended results to the relativistic Euler equation

The relativistic Euler equation in $\mathbf{R}^3$ is given by

$$\frac{\partial}{\partial t}\left(\frac{\rho c^2 + P}{c^2}\frac{1}{1-\frac{v^2}{c^2}}-\frac{P}{c^2}\right)+\sum_{j=1}^{3}\frac{\partial}{\partial x_j}\left(\frac{\rho c^2+P}{c^2}\frac{v_j}{1-\frac{v^2}{c^2}}\right)=0,$$

$$\frac{\partial}{\partial t}\left(\frac{\rho c^2 + P}{c^2}\frac{v_i}{1-\frac{v^2}{c^2}}\right)+\sum_{j=1}^{3}\frac{\partial}{\partial x_j}\left(\frac{\rho c^2+P}{c^2}\frac{v_i v_j}{1-\frac{v^2}{c^2}}+\delta_{ij}P\right)=0,$$

(3.1)

$i = 1, 2, 3$, with the equation of state

$$p = \sigma^2 \rho, \tag{3.2}$$

where the density ρ, the velocity $\mathbf{v} =^t (v_1, v_2, v_3)$ and the pressure p are functions of $x \in \mathbf{R}^3$ and $t \geq 0$, while c is the speed of light and σ is the speed of sound, which are constants. According to the relativistic theory, σ never exceeds c. Especially, the case $\sigma^2 = \frac{c^2}{3}$ has been important in the context of physics.

In 1993, Smoller and Temple [13] constructed uniformly bounded weak solutions for the one-dimensional case by using Glimm's method [4]. They have found that the structure of (3.1) for the one-dimensional case is very similar to Nishida's famous example [12] although (3.1) seems to be very complicated. After constructing approximate solutions, they showed that the total variation of $\log \rho$ of the approximate solutions is monotone decreasing.

Put $r = \sqrt{x_1^2 + x_2^2 + x_3^2}$. We look again for solutions of the form

$$\rho = \rho(t, |x|), \quad \vec{u} = \frac{x}{|x|} \cdot u(t, |x|). \tag{3.3}$$

Then (3.1) becomes

$$\frac{\partial}{\partial t}\left(\frac{\rho c^2+P}{c^2}\frac{1}{1-\frac{v^2}{c^2}}-\frac{P}{c^2}\right)+\frac{\partial}{\partial r}\left(\frac{\rho c^2+P}{c^2}\frac{v}{1-\frac{v^2}{c^2}}\right)$$
$$+\frac{\rho c^2+P}{c^2}\frac{1}{1-\frac{v^2}{c^2}}\frac{2v}{r}=0,$$

$$\frac{\partial}{\partial t}\left(\frac{\rho c^2+P}{c^2}\frac{v}{1-\frac{v^2}{c^2}}\right)+\frac{\partial}{\partial r}\left(\frac{\rho c^2+P}{c^2}\frac{v^2}{1-\frac{v^2}{c^2}}+P\right)$$
$$+\frac{\rho c^2+P}{c^2}\frac{v}{1-\frac{v^2}{c^2}}\frac{2v^2}{r}=0.$$

(3.4)

This equation has a singularity at $r = 0$. To avoid the difficulty caused by this singularity, we simply deal with the boundary value problem for (3.4) in the domain $1 \leq r < \infty$ (outside a unit ball) with the boundary value condition $v(t, 1) = 0$.

Introduce the Lagrangian transformation

$$x = \int_1^r \rho r^2 J\, dr , \qquad (3.5)$$

where

$$J = 1 + \frac{(c^2 + \sigma^2)v^2}{c^2(c^2 - v^2)} . \qquad (3.6)$$

Putting

$$W = \frac{1}{Jq}, \quad V = \frac{\frac{c^2+\sigma^2}{c^2-\sigma^2}}{J}v = \frac{(c^2 + \sigma^2)c^2}{c^4 + \sigma^2 v^2}v , \quad q = r^2\rho . \qquad (3.7)$$

Then (3.4) becomes

$$W_t - V_x = 0 ,$$
$$V_t + \left(\frac{\sigma^2}{J^2 W}\right)_x = \frac{2\sigma^2}{Jr} , \qquad (3.8)$$

where r is now defined by $r = 1 + \int_0^x W\, dx$.

Let us consider the initial boundary value problem for (3.8) in $t \geq 0$, $x \geq 0$ with the initial and boundary conditions

$$q(0, x) = q_0(x), \quad v(0, x) = v_0(x), \quad x > 0 , \qquad (3.9)$$
$$v(t, 0) = 0 . \qquad (3.10)$$

Then the following result holds, which is an extension of Theorem 1 to the relativistic equation.

Theorem 2 ([11]). *Suppose that* $\log q_0(x)$ *and* $\log \frac{c+v_0(x)}{c-v_0(x)}$ *are of bounded variation. Then there exists a global weak solution for (3.8), (3.9) and (3.10) satisfying*

$$|v| < c , \quad q > 0 . \qquad (3.11)$$

Let us explain the outline of the proof briefly. We first consider the Riemann problem for the homogeneous equation corresponding to (3.8) which is given by

$$W_t - V_x = 0 ,$$
$$V_t + \left(\frac{\sigma^2}{J^2 W}\right)_x = 0 . \qquad (3.12)$$

Then we can show that all shock curves have the same figure in the plane of Riemann invariants. Fortunately, their figure is similar to the figure of shock curves of the classical case. Thus we can use Nishida's idea similarly.

Next, we must prepare some auxiliary estimates for the approximate solutions by analyzing the waves in the plane of Riemann invariants z_1 and z_2 where they are given by

$$z_1 = \log q + \frac{c^2 + \sigma^2}{2\sigma c}\log\frac{c + v}{c - v} ,$$
$$z_2 = -\log q + \frac{c^2 + \sigma^2}{2\sigma c}\log\frac{c + v}{c - v} . \qquad (3.13)$$

By using the geometry of shock waves, we shall estimate the total variation of $\log q$. This is the main idea of Smoller and Temple [13]. But in this case, we must be more careful since there is an inhomogeneous term in (3.8). To obtain our desired uniform estimates, we use the transformation

$$v = c \tanh u = c \frac{e^u - e^{-u}}{e^u + e^{-u}}. \tag{3.14}$$

Instead of estimating v itself, we estimate u. Fortunately, it follows that

$$\log \frac{c + v}{c - v} = 2u. \tag{3.15}$$

The transformation (3.14) plays a crucial role in our paper. By using these estimates, we can derive uniform estimates of the total variation of our approximate solutions. Thus we can show, by using Glimm's theory, Theorem 2. For details, see [11].

Recently, Cheng-Hsiung Hsu, S. S. Lin and T. Makino have presented, under some conditions on the initial data, global weak solutions with spherical symmetry where the pressure p is a more general case. For details, see [5].

References

[1] X. Ding, G. Chen and P. Luo, *Convergence of the Lax–Friedrichs scheme for isentropic gas dynamics*, (I), (II), Acta Math. Sci. **5** (1985), 483–500; 501–540.

[2] X. Ding, G. Chen and P. Luo, *Convergence of the Lax–Friedrichs scheme for isentropic gas dynamics*, (III), Acta Math. Sci. **6** (1986), 75–120.

[3] R. DiPerna, *Convergence of the viscosity method for isentropic gas dynamics*, Commun. Math. Phys. **91**, (1983), 1–30.

[4] J. Glimm, *Solutions in the large for nonlinear hyperbolic systems of equations*, Comm. Pure Appl. Math. **18** (1965), 697–715.

[5] C. H. Hsu, S. S. Lin and T. Makino, *On spherically symmetric solutions of the relativistic Euler equation*, preprint.

[6] C. H. Hsu and T. Makino, *Spherically Symmetric Solutions to the Compressible Euler equation with an asymptotic γ-law*, preprint.

[7] P. D. Lax, *Hyperbolic Systems of Conservation Laws and the Mathematical Theory of Shock Waves*, SIAM Reg. Conf. Lecture 11, Philadelphia, 1973.

[8] T. Makino, K. Mizohata and S. Ukai, *The global weak solutions of the compressible Euler equation with spherical symmetry*, Japan J. Indust. Appl. Math. **9**, (1992), 431–449.

[9] T. Makino, K. Mizohata and S. Ukai, *The global weak solutions of the compressible Euler equation with spherical symmetry II*, Japan J. Indust. Appl. Math. **11**, 1994, 417–426.

[10] K. Mizohata, *Equivalence of Eulerian and Lagrangian weak solutions of the compressible Euler equation with spherical symmetry*, Kodai Mathematical Journal **17**, (1994), 69–81.

[11] K. Mizohata, *Global solutions to the relativistic Euler equation with spherical symmetry*, Japan J. Indust. Appl. Math. **14**, (1997), 125–157.

[12] T. Nishida, *Global solutions for an initial boundary value problem of a quasilinear hyperbolic system*, Proc. Japan Acad. **44** (1968), 642–646.

[13] J. Smoller and B. Temple, *Global solutions of the relativistic Euler equations*, Comm. Math. Phys. **156** (1993), 67–99.

Kiyoshi Mizohata
Faculty of Engineering
Doshisha University
Kyotababe, Kyoto 610-0394, Japan
kmizoha@mail.doshisha.ac.jp

Hyperbolic Cauchy Problem Well Posed in the Class of Gevrey

Yujiro Ohya

First of all, let us recall the two classical results of references [1] and [2]. These two papers are not related directly to the classical concept of hyperbolicity by J. Hadamard; that is, the well-posed Cauchy problem in the class of infinitely differentiable functions. Therefore, we were obliged to discuss the well-posedness in the class of Gevrey. In other words, we understand hyperbolicity when there is the influence (or dependence) domain extending the traditional hyperbolicity of differential operators.

Definition 0.1. Functions of Gevrey classes are defined by $\gamma^s(\mathbf{R}^l) = \{f(x);$ $f(x) \in C^\infty\}$ and for any multi-index α, any compact set K of $\mathbf{R}^l$, there exist C and A such that $\sup_K |D_x^\alpha f(x)| \leq CA^{|\alpha|}(|\alpha|!)^s\}$ and $\gamma_0^s(\mathbf{R}^l) = \gamma^s(\mathbf{R}^l) \cap C_0^\infty(\mathbf{R}^l)$.

We consider the following Cauchy problem:

$$(\alpha) \quad \begin{cases} P(t, x, D_t, D_x)u(t, x) = f(t, x) \quad on \quad \Omega = [0, T] \times \mathbf{R}^l, \\ \\ D_t^{j-1}u(0, x) = \varphi_j(x) \quad j = 1, 2, \ldots, m, \end{cases}$$

where

$$D_t = \frac{1}{i}\frac{\partial}{\partial t}, \quad D_x = \frac{1}{i}\frac{\partial}{\partial x_j} \quad (j = 1, 2, \cdots, l)$$

and

$$P(t, x, D_t, D_x) = D_t^m + \sum_{j+|v|\leq m} a_{jv}(t, x)D_t^j D_x^v \quad for \quad j < m.$$

i.e., P is the following two assumptions of hyperbolicity (see P.D. Lax [3], S. Mizohata[4]).

(a) Let $p_m(t, x, \tau, \xi)$ be the characteristic polynomial;

$$p_m(t, x, \tau, \xi) = \tau^m + \sum_{j+|v|\leq m} a_{jv}(t, x)\tau^j \xi^v.$$

The roots of $p_m(t, x, \tau, \xi) = 0$ for any $(t, x, \xi) \in \Omega \times \mathbf{R}^l$ have only real roots.

(b) (constant multiplicity). Let

$$p_m(t, x, \tau, \xi) = \prod_{j=1}^{k}(z - \lambda_j(t, x, \xi))^{\nu_j}, \quad \sum_{j=1}^{k}\nu_j = m.$$

In general, ν_j depends on (t, x, ξ), but we assume $\nu_j = constant$ for any (t, x, ξ) Let max $\nu_j = r$. Then we can consider the form $p_m = a_1 a_2 \cdots a_r$ where $a_j(t, x, \tau, \xi)$ has real and distinct roots; a_j is a strictly hyperbolic polynomial. We will denote the pseudo-differential operator $A_j(t, x, D_t, D)$ of order m with respect to $(D_t.D)$ in particular polynomial of D_t.

In place of (α) we solve the Cauchy problem

$$(\beta) \quad \begin{cases} A_1 A_2 \cdots A_r(t, x, D_t, D)u(t, x) = (A_1 A_2 \cdots A_r - P)u + f, \\ D_t^{j-1}u(0, x) = \varphi_j(x), \quad j = 1, 2, \cdots, m. \end{cases}$$

Clearly, the order of $A_1 A_2 \cdots A_r - P$ is at most order $m - 1$, but we analyze this more precisely.

Definition 0.2.

$$\| D^n f(t) \| = \sup_{k+s \le n} \| D_t^k f(t) \|_s$$

where

$$\| f(t) \|_s^2 = \int \int \cdots \int \left|(1 + |\xi|^2)^{\frac{s}{2}} \hat{f}(t, \xi)\right|^2 d\xi_1 d\xi_2 \cdots d\xi_l.$$

Then, from well-known results for the strictly hyperbolic Cauchy problem, we get the energy inequality for a strictly hyperbolic operator A of order m,

$$\| D^{n+m-1}u(t) \| \le \gamma \| D^{n+m-1}u(0) \| + C \int_0^t \| D^n f(s) \| \, ds$$

where $Au = f$.

Then we confirm to get for $A_1 A_2 \cdots A_r u = f$ with zero Cauchy data that

$$\| D^{n+m-1}u(t) \| \le C \int_0^t \| D^n f(s) \| \, ds.$$

Because of this property, we denote the order of $A_1 A_2 \cdots A_r - P$ by $m - r + q$ $(0 \le q \le r - 1)$. When we can choose (fortunately) $q = 0$, there is no difference from the treatment of the strictly hyperbolic Cauchy problem (A. Lax [6], M. Yamaguti [7]).

This leads to the prototype of the condition by E.E. Levi [8] in 1909.

Theorem 0.3 (Leray–Ohya [2]). *Suppose that $a_{j\nu}(t, x)$ and $f(t, x)$ belong to $\gamma^s(\Omega)$. $\varphi_j(x) \in \gamma^s(\mathbf{R}^l)$. Then, if $1 \le s \le \frac{r}{q}$, there exists a unique solution $u(t, x)$ belonging to $\gamma^s(\Omega)$.*

Remark 0.4. When $r = m$, $q = m - 1$, this theorem was obtained earlier in Ohya [1].

The opinion of J. Leray concerning this theorem is quoted as follows;

... Non, tout d'abord parce qu il est souvent déçu. Si l'on n'a réussi à établir l'existence des solution régissant la rélativité générale que durant un court intervalle de temps, pour y réussir en magnétohydrodynamique rélativiste, il nous faut en outre supposer indéfiniment différentiables les fonctions physiques données; certes, elles appartiennent à une classe de Gevrey, non quasi-analytique, ce que conseve un sens d'Einstein: aucune influence ne doit se propager plus vite que la lumière; et ce principe est effectivement respecté. Mais pourquoi nos théorèmes mathématiques ne valent-ils que sous des conditions d'nue étroitenesse si décevantes? ...

(extracted from "La mathématique et ses applications" in Accademia Nationale dei Lincei, Adananze Ataordinarie per il Conferimento dei Premi A. Fertinelli, pp. 191–197, 1972.)

Even though the assumption of constant multiplicity was too strong, this theorem explicitly illustrated the property of solutions for the Cauchy problems with multiple characteristics. Actually, J. Leray proved nonuniqueness which means the necessity of Gevrey assumption $1 \le s \le \frac{r}{q}$ referring to the paper of E. De Giorgi [9].

As another example, one night at the end of 1963, I asked him his opinion about the theory of singular integral operators. He answered me simply, at the corner of rue St. Jacques and rue Pierre et Marie Curie, "Ce n'est que la transformation de Fourier". Since the Cauchy problem was solved in this form, this has attracted many young researchers. Some of them wanted to extend this to variable multiplicity and others to discuss systems. Anyway, the progress of the theory of pseudo-differential operators about 10 years since 1965 (see L. Nirenberg [10]) implied quick resolution of the final (in some sense) target, that is, Theorem 1 without any assumption on multiplicity. It was obtained essentially by M.D. Bronshtein [11] in 1976. The following theorem explains what we understand by his result.

Theorem 0.5 (Ohya–Tarama [12]). *Let r be the maximal multiplicity of characteristic roots. Suppose that $a_{j\nu}(t, x) \in C^k([0, T]; \gamma^s(\mathbf{R}^l))$ and $f(t, x) \in C^0([0, T]; \gamma^s(\mathbf{R}^l))$. If,*

$$1 \le s \le \min\left(1 + \frac{\kappa}{r}, \frac{r}{r-1}\right),$$

then there exists one and only one solution $u(t, x)$ belonging to $C^0([0, T]; \gamma^s(\mathbf{R}^l))$ for $0 < \kappa \le 2$.

Remark 0.6. Concerning the regularity of characteristic roots, we do not discuss the topic here more precisely (see M.D. Bronshtein [11], S. Wakabayashi [13], S. Tarama [14]).

Sketch of Proof. First restrict the root of $p(t, x, \tau, \xi) = 0$ to be in $|\tau| \le C_1\langle\xi\rangle$, $|\Im\tau| \le C_2\langle\xi\rangle^{\frac{r-1}{r}}$ where $\langle\xi\rangle = (1 + |\xi|^2)^{\frac{1}{2}}$.

Taking into account this property, we define $\delta = \frac{r-1}{r}$,

$$q(t, x, \tau - iH < \xi >^\delta, \xi) = \int_{-\infty}^{\infty} \frac{\rho X(\rho(t-s))}{p_m(s, \kappa, \tau - iH < \xi >^\delta, \xi)} ds.$$

$X(x)$ belongs to $C_0^\infty(\mathbf{R})$ such that $\int X(s)ds = 1$. As usual, we associate the pseudo-differential operator $Q(t, x, D_t, D)$ by the symbol $q(t, x, \tau - iH\langle\xi\rangle^\delta, \xi)$. We choose q (or Q) such that $PQ = I + R$. This means, putting $u = Qv$,

$$r(t, x, \tau - iH < \xi >^\delta, \xi) = \sum \frac{1}{\alpha!} \partial_\xi^\alpha p(t, x, \tau - iH < \xi >^\delta, \xi) D_x^\alpha q - 1,$$

by the symbolic calculus of pseudo-differential operators.

Lemma 0.7 (see L. Hörmander [17]). *Suppose that $\varphi(x)$ belongs to $\gamma_0^s(\mathbf{R}^l)$. If we define*

$$\tilde{\varphi}(s) = (2\pi)^{-\frac{l}{2}} \int \int \cdots \int e^{-i<s,x>} \varphi(x) dx_1 dx_2 \cdots dx_l, \quad \zeta = \xi + i\eta,$$

then we obtain

$$|\tilde{\varphi}(\zeta)| \leq Ce^{A|\Im\zeta|}e^{-B|\Re\zeta|^\delta}.$$

Beginning from the definition of pseudo-differential operators, we get

$$R(t, x, D_t, D)v(t, x) = (2\pi)^{-\frac{l+1}{2}} \int \int \cdots \int e^{it(\tau - iH<\xi>^\delta)}$$

$$\times e^{i<x,\xi>} r(t, x, \tau - iH < \xi >^\delta, \xi)\tilde{v}(t, x, \tau - iH < \xi >^\delta, \xi)d\tau d\xi_1 d\xi_2 \cdots d\xi_l.$$

Substituting

$$\tilde{v}(t, x, \tau - iH < \xi >^\delta, \xi) = \frac{1}{\sqrt{2\pi}} \int e^{-it_1(\tau - iH<\xi>^\delta)}\hat{v}(t, \xi)dt_1$$

we get

$$Rv = (2\pi)^{-\frac{l+1}{2}} \int \int \cdots \int e^{H<\xi>^\delta(t-t_1)+i<x,\xi>}$$

$$\times \frac{1}{\sqrt{2\pi}} \int e^{(t-t_1)\tau} r(t, x, \tau - iH < \xi >^\delta, \xi)d\xi_1 d\xi_2 \cdots d\xi_l \hat{v}(t, \xi)d\tau$$

by the theorem of Fubini–Tonelli. Finally, we use the notation

$$Rv(t, x) = \frac{1}{\sqrt{2\pi}} \int_0^t \tilde{R}(t, x, t - t_1)v(t_1, x)dt_1$$

where

$$\tilde{R}(t, x, t - t_1) = (2\pi)^{-\frac{l}{2}} \int \int \cdots \int e^{(t-t_1)H<\xi>^\delta} \tilde{r}\hat{v}d\xi_1 d\xi_2 \cdots d\xi_l.$$

Taking into account the preceding lemma, if we define $e^{(h-tH)}\langle D\rangle^\delta v(t,x) = w(t,x) \in l^2(\mathbf{R}^l)$ for $t \in (0, \frac{h}{2H})$, then we can verify via the lemma of Sobolev $(m > \frac{l}{2})$

$$|v(t,x)| \leq const. \quad \| <\xi>^m v(t,\xi)\|$$
$$\leq C \sup_\xi | <\xi>^m e^{-(h-tH)<\xi>^\delta}| \cdot \|\hat{w}(t,\xi)\|$$
$$= C\|w(t)\|,$$

because of the Plancherel formula; therefore, we get

$$w(t,x) + e^{(h-tH)<D>^\delta} \frac{1}{\sqrt{2\pi}} \int_0^t \tilde{R}(t,x,t-t_1)e^{-(h-t_1H)<D>^\delta} w(t_1,x)dt_1$$
$$= e^{(h-tH)<D>^\delta} f(t,x).$$

The final step to show the existence of $W(t,x)$ will be given by the following argument: $(I - R)w = F$ leads to $w = (I + R)^{-1}F = \sum_{n=0}^\infty (-R)^n F$. After an elaborate calculation, we can get the estimate

$$\|\tilde{R}(t,t-t_1)w(t_1)\| \leq M|t-t_1|^{\sigma-1}\|w(t_1)\| \quad (0 < \sigma < 1).$$

This helps us to prove $\sup \|w(t)\|_{L^2\mathbf{R}^l} < +\infty$ for $0 < t < \frac{h}{2H}$. Concerning the uniqueness, we use the lemma obtained by W. Nuij [15] and the property of convergence of pseudo-differential operators of H. Kumano-go [16]. I would like to quote the following (*) by H. Poincaré

"Il n'ya pas de problèmes résolus, et d'autres qui ne le sont pas; Il y a des problèmes plus ou moins résolus."

(*) Centenaire de Jacques Hadamard, mimeographie (1965), see page 8.

References

[1] Y. Ohya, *Le problème de Cauchy pour les équations hyperboliques à caractéristique multiple*, J. Math. Soc. Japan. **16** (1964), 268–286.

[2] J. Leray and Y. Ohya, *Systèmes linéaires, hyperboliques non stricts*, Colloque de Liege CBRM, (1964), 105–144.

[3] P.D. Lax, *Asymptotic solutions of oscillatory initial value problems*, Duke Math. J. **24** (1957), 627–646.

[4] S. Mizohata, *Some remarks on the Cauchy problem*, J. Math. Kyoto Univ. **1** (1961), 107–127.

[5] J. Hadamard, *Les fonctions de classe superieure dans l'équation de Volterra*, Jour. d'Anal. Math. **1** (1951).

[6] A. Lax, *On Cauchy's problem for partial differential equations with multiple characteristics*, Comm. Pure Appl. Math. **9** (1956).

[7] M. Yamaguti, *Le problème de Cauchy et les opérateurs d'intégrale singulière*, Mem. Coll. Sci. Kyoto Univ., Series A. 32 (1) (1959).

[8] E.E. Levi, *Caratteristico multiple e ploblema di Cauchy*, Ann. Math. Pura Appli. **16** (1909), 109–127.

[9] E. DE Giorgi, *Un teorema di unicita per il problema di Cauchy relativo ad equatini differenziali lineari a derivate partiali di tipo parabokico*, Annali di Mat. **40** (1955), 371–377; *Un exempio di non unicita della solutione del problema di Cauchy*, Univ. di Roma, Rendiconti di Math. **14** (1955), 382–387.

[10] L. Nirenberg, *Pseudo-differential operators in global analysis*, Proc. Sympo. Pure Math. **16** (1970), 149–167.

[11] M.D. Bronshtein, *The Cauchy problem for hyperbolic operators with characteristics of variable multiplicity*, Trudy Moskov. Mat. Obsc. **41** (1980), 87–103; *Smoothness of roots of polynomials depending on parameters*, Sibirsk. Mat. Zeit. **20** (1970), 493–501.

[12] Y. Ohya et S. Tarama, *Le problème de Cauchy à caractéristques multiples dans la classe de Gevrey*, Taniguchi Sympo. HEAT, Katata, (1984), 273–306.

[13] S. Wakabayashi, *Remarks on hyperbolic polynomials*, Tukuba J. Math. **10** (1986), 17–28.

[14] S. Tarama, *On the initial value problem for the weakly hyperbolic operators in the Gevrey classes*, Proc. Hyperbolic Equation, Pisa, (1987), 322–339.

[15] W. Nuij, *A note on hyperbolic polynomials*, Math. Scand. **23** (1968), 69–72.

[16] H. Kumano-go, *Pseudo-differential Operators*, M.I.T. Press, Cambridge, 1981.

[17] L. Hörmander, *Linear Partial Differential Operators*, Springer-Verlag, 1963.

Yujiro Ohya Koshien
University Momijigaoka
Takarazuka Hyogo
665-0006 Japan y-ohya@koshien.ac.jp

Absence of Eigenvalues
of Dirac Type Operators

Takashi Ōkaji

ABSTRACT We study the eigenvalue problem for Dirac type operators in $L^2(\mathbf{R}^3)^4$ and show the absence of eigenvalues for a large class of potentials which may diverge at infinity. This result is a generalization of a recent work on the Dirac operator by Kalf, Okaji, O. Yamada [6].

1 Introduction

F. Rellich [10] has shown that if $u \in L^2(U)$ is a solution to the eigenvalue problem $-\Delta u = ku$, $k > 0$, in an exterior domain U of $\mathbf{R}^d$, then u is identically zero. T. Kato [5] extended this result to the Schrödinger equation $-\Delta u + q(x)u = ku$, $x \in U, k > 0$ with $q(x) = o(|x|^{-1})$, $|x| \to \infty$. His result has been generalized by many researchers to a class of second order elliptic equations close to the Laplacian at infinity (Jäger [4], Ikebe–Uchiyama [2], Mochizuki [8])

On the other hand, an analogue to Rellich's theorem for symmetric elliptic systems has been established (P.D. Lax–R.S. Phillips [7] and N. Iwasaki [3]). As for systems with potentials, related results for the Dirac equations or optical systems are extensively studied. In particular, it is shown by Kalf–Okaji–Yamada [6] and Okaji [9] that the eigenvalue problem for systems with potential growing at infinity differs from the one for Schrödinger operators.

If $U \subset \mathbf{R}^3$ is either an exterior domain or the whole space, the eigenvalue problem for the Dirac operator can be formulated as follows.

$$\alpha \cdot pu + m\beta u + Vu + \lambda u = 0, \ u \in L^2(U)^4, \ \lambda \in \mathbf{R}, \ p = -i\nabla, \quad \text{(D)}$$

where $\{\alpha_j\}_{j=0}^3$ is a family of 4×4 matrices satisfying

$$\alpha_j^* = \alpha_j \, , \, \alpha_j\alpha_k + \alpha_k\alpha_j = 2\delta_{jk}, \ \forall j, \ k = 0, \ldots, 3, \ \beta = \alpha_0,$$

$m(x)$ is a scalar function and $V(x)$ is a matrix close to a scalar one at infinity.

The main result of [6] is, roughly speaking, that (D) admits no nontrivial solutions in $L^2(U)^4$ provided that there exists a positive spherically symmetric function q that may diverge at infinity but does not oscillate rapidly such that

$$\tilde{V} = V(x) + \lambda \sim q(|x|), \ m(x) = o(q), \ \text{as } |x| \to \infty.$$

In this paper we shall show a similar result for some Dirac type operators:

$$\left\{ \frac{1}{2}(A \cdot p + p \cdot A) + A_0 + V + \lambda \right\} u = 0,$$

where $\{A_j(x)\}_{j=0,1,2,3}$ is a family of symmetric matrices $(A_j^* = A_j)$ such that

$$A_j A_k + A_k A_j \to 2\delta_{jk} \text{ (Kronecker's Delta) as } |x| \to \infty.$$

The central method of our approach to this kind of problem consists of a series of weighted L^2 estimates based on a local version of the virial theorem. This kind of strategy was first employed in [11] and has been improved in [6] and [9]. We shall give a minor modification to the local version of the virial theorem in order to treat the Dirac type operators. Furthermore, at the final stage of our method, we shall use a new unique continuation theorem which is interesting in itself.

2   Main result

Let $\{A_k\}_{k=1}^3 \subset C^2(U)^{4\times 4}$ be a family of symmetric matrices such that

$$A_j A_k + A_k A_j = 2g^{jk}(x)I, \quad \forall j, \ k = 1, 2, 3, \tag{2.1}$$

where $G = (g^{jk})$ satisfies

$$\Re(G\xi, \xi) \geq \delta |\xi|^2, \quad \forall x \in U, \ \xi \in \mathbf{C}^3 \tag{2.2}$$

and

$$g^{jk}(x) - \delta_{jk} = o(1), \quad r = |x| \to \infty. \tag{2.3}$$

We are interested in the following Dirac type operator $\mathcal{D}$ in U,

$$\mathcal{D} = \sum_{k=1}^3 \frac{1}{2} \{p_k A_k + A_k p_k\}.$$

We emphasize that the principal symbol of $\mathcal{D}^2$ is scalar by the assumption (2.1).

To state our further assumption on the derivatives of A_k, we shall introduce a class of scalar functions. If $I_a = (a, \infty)$ and $0 \leq \sigma \leq \frac{1}{2}$, we define

$$\mathcal{P}_\sigma(I_a) = \{q(r) \in C^2(I_a; \mathbf{R}); \inf_{I_a} q(r) = q_\infty > 0, \ [q']_- = o(r^{-1}q),$$

$$q'(r) = o(r^{-1/2}q^{2-\sigma}), \ q'' = o(r^{-1}q^2)\}.$$

Here,

$$[f(r)]_- = \max(0, -f(r)), \ f' = \frac{d}{dr} f(r), \text{ etc.}.$$

Remark 2.1. $e^r, r^s, (s \geq 0), \log r \in \mathcal{P}(I_a)$.

Thus, we make the following assumptions on the derivatives of $A_k, k = 1, 2, 3$:

$$\nabla_x A_k(x) = o\left(\frac{1}{r}\right), \quad k = 1, 2, 3 \tag{2.4}$$

and for some element q of $\mathcal{P}_\sigma$,

$$\nabla_x^2 A_k(x) = o\left(\frac{q}{r}\right), \quad k = 1, 2, 3. \tag{2.5}$$

In addition, $A_0 \in C^1(U)^{4\times 4}$ denotes a symmetric matrix satisfying

$$A_j A_0 + A_0 A_j = o(r^{-1/2}\sqrt{q}), \quad j = 1, 2, 3 \tag{2.6}$$

and

$$A_0(x) = o(q), \quad \nabla_x A_0(x) = o\left(\frac{q}{r}\right). \tag{2.7}$$

Let a be sufficiently large such that

$$U \supset D_a = \{x \in \mathbf{R}^3; \ |x| > a\}.$$

We shall make the following assumptions on the potential V.

$$V = V_1 + V_2, \quad V_1^* = V_1, \quad V_1 \in C^1(U)^{4\times 4}, \tag{A-1}$$

$$|V_2(x)| \leq K_0/|x|, \tag{A-2}$$

$$V_1(x) - q(|x|)I = o\left(\frac{q^\sigma}{|x|^{1/2}}\right), \tag{A-3}$$

$$\partial_r\{V_1(x) - q(|x|)I\} = o\left(\frac{q}{|x|}\right). \tag{A-4}$$

$$\{\nabla_x - \frac{x}{|x|}\partial_r\}V_1(x) = \mathcal{O}(\frac{q}{|x|}) \text{ as } r \to \infty. \tag{A-5}$$

Theorem 2.2. *Suppose (2.1)–(2.7). If $V(x)$ satisfies (A-1)–(A-5) with $K_0 < 1/2$, then $\mathcal{D}u + A_0 u + V u = 0$ admits no nontrivial solution in $L^2(U)^4$.*

Remark 2.3. It is shown in [6] that the same conclusion as in Theorem 2.2 holds for the Dirac operator (D) if $2K_0 < 1 - b_0$ under the conditions (A-1)–(A-4) and (A-6)–(A-8):

$$m(x) - m_1(|x|) = o\left(\frac{q^\sigma}{|x|^{1/2}}\right), \tag{A-6}$$

$$\partial_r\{m(x) - m_1'(|x|)\} = o\left(\frac{q}{|x|}\right), \tag{A-7}$$

$$|m_1 + rm_1'| \leq b_0 q(r), \quad b_0 < 1. \tag{A-8}$$

3   Proof of Theorem 2.2

3.1   Change of unknown functions

In what follows, $r = |x|$, $\omega = x/|x| \in \mathbf{S}^2$, $\langle u, v \rangle$ denotes the inner product of $\{L^2(\mathbf{S}^2)\}^4$, $\|u\| = \sqrt{\langle u, u \rangle}$ and $T(\mathbf{S}^2)$ stands for the tangent space of $\mathbf{S}^2$.

$$\partial_{x_j} = \omega_j \partial_r + r^{-1}\Omega_j,$$

where $\Omega_j \in T(\mathbf{S}^2)$. Put

$$A_r = \sum_{j=1}^{3} A_j(x)\omega_j, \quad A_\Omega = \frac{1}{2}\sum_{j=1}^{3}\{A_j(x)\Omega_j + \Omega_j A_j(x)\},$$

$$S_\Omega = A_\Omega - A_r, \quad S_\Omega^* = -S_\Omega,$$

$$J = \frac{1}{2}(S_\Omega A_r^{-1} - A_r^{-1}S_\Omega), \quad K = \frac{1}{2}(S_\Omega A_r^{-1} + A_r^{-1}S_\Omega).$$

It turns out that

$$\langle Jf, h \rangle = \langle f, Jh \rangle, \quad \langle Kf, h \rangle = -\langle f, Kh \rangle, \quad \forall f, h \in C^1(\mathbf{S}^2).$$

If $u \in L^2(U)^4$, the integral

$$\int_a^\infty \langle Vru/\sqrt{q}, ru/\sqrt{q} \rangle dr$$

is finite, so that $u/\sqrt{q}$ is more convenient than u itself.

Suppose

$$0 \leq \chi \in C_0^\infty(\mathbf{R}_+), \text{ supp}\chi \subset [s-1, t+1], \ \chi(r) = 1, \ r \in [s, t],$$

$$\varphi \in C^3(\mathbf{R}_+), \ \varphi' \geq 0.$$

Let $u \in L^2(U)^4$ satisfy

$$\mathcal{D}u + (A_0 + V)u = 0, \text{ in } U.$$

Define

$$\zeta = \chi(r)e^\varphi v, \quad v = \frac{ru}{\sqrt{q}}.$$

Then,

$$\{-iA_r\partial_r - i(r^{-1}S_\Omega - A_r\varphi')A_0 + V - iA_r q'/(2q)\}\zeta \tag{3.1}$$
$$= -iA_r\chi'e^\varphi v + ir[A_r, \partial_r]\zeta := f_\chi$$

and

$$[\partial_r - r^{-1}K - (r^{-1}J + \varphi') + i\{(A_0 + V)A_r^{-1} - iq'/(2q)\}]A_r\zeta = if_\chi. \tag{3.2}$$

To describe fundamental relations among K, L and A_r, we introduce a class of matrices of vector fields $L(r)$ on $\mathbf{S}^2$, depending smoothly in r as follows. Let $\tilde{T}(\mathbf{S}^2) = \{L_1(r) + L_0(r); \; L_1 \in T(\mathbf{S}^2), \; L_0 \in L^\infty(\mathbf{S}^2)\}$ and

$$\mathcal{V}_\sigma^1 = \{L \in C(I_a; \; \tilde{T}(\mathbf{S}^2)^{4\times 4}); \; \exists C(r) > 0, \; C > 0, \; \forall u \in C^1(\mathbf{S}^2)^4,$$
$$\|Lu\| \le M_1(r)\|Ju\| + M_2(r)\|u\|,$$
$$M_1(r) = o(r^{-\sigma}), \; M_2(r) = o(q) \text{ as } r \to \infty\}.$$

Similarly,

$$\mathcal{V}_\sigma^0 = \{B \in C^0(\mathbf{S}^2)^{4\times 4}; \; \exists M(r) > 0, \; M(r) = o(r^{-\sigma}),$$
$$\|Bu\| \le M(r)\|u\|, \; \forall u \in C^0(\mathbf{S}^2)^4\}.$$

Lemma 3.1.

$$[K, J] \in \mathcal{V}_0^1, \; K \in \mathcal{V}_0^1.$$

$$(JA_r + A_r J) = [K, A_r] \in \mathcal{V}_0^0.$$

$$[\partial_r, A_r] \in \mathcal{V}_1^0, \; [\partial_r, K], \; [\partial_r, J] \in \mathcal{V}_1^1.$$

In particular if $g^{jk}(x) \equiv \delta_{jk}$, then $K = 0$.

Proof. It is easily verified that

$$2A_r K A_r = S_\Omega A_r + A_r S_\Omega, \tag{3.3}$$

$$\frac{1}{2}[K, J] = -A_r^{-1} S_\Omega^2 A_r^{-1} + S_\Omega A_r^{-2} S_\Omega, \tag{3.4}$$

$$2[K, A_r] = 2(JA_r + A_r J) = A_r^{-1} S_\Omega A_r - A_r S_\Omega A_r^{-1} \tag{3.5}$$

and

$$2A_r[K, A_r]A_r = S_\Omega A_r^2 - A_r^2 S_\Omega. \tag{3.6}$$

Lemma 3.2. $S_\Omega A_r + A_r S_\Omega$ *belongs to* $\mathcal{V}_0^1$. *In particular, this operator has a scalar principal part.*

Proof.

$$2A_r K A_r + 2A_r^2 = S_\Omega A_r + A_r S_\Omega + 2A_r^2$$

$$= \frac{1}{2} \sum_{a,b=1}^{3} (A_a A_b \Omega_a \omega_b + A_b A_a \omega_b \Omega_a + \Omega_a \omega_b A_a A_b + \omega_b \Omega_a A_b A_a$$

$$+ A_a[\Omega_a, A_b]\omega_b + \omega_b[A_b, \Omega_a]A_a)$$

$$= \sum_{a,b=1}^{3} \{g^{ab} \omega_b \Omega_a + \omega_b \Omega_a g^{ab}\}$$

$$+ \frac{1}{2} \sum_{a,b=1}^{3} \{2A_a A_b[\Omega_a, \omega_b] + A_a[\Omega_a, A_b]\omega_b + \omega_b[A_b, \Omega_a]A_a\}$$

$$= \sum_{a,b=1}^{3} \{g^{ab} \omega_b \Omega_a + \omega_b \Omega_a g^{ab}\}$$

$$+ \frac{1}{2} \sum_{a,b=1}^{3} \{2A_a A_b(\delta_{ab} - \omega_a \omega_b) + A_a[\Omega_a, A_b]\omega_b + \omega_b[A_b, \Omega_a]A_a\}$$

$$= \sum_{a,b=1}^{3} \{g^{ab} \omega_b \Omega_a + \omega_b \Omega_a g^{ab}\} + \sum_{a-1}^{3} g^{a,a}(1 - \omega_a^2) - \sum_{a>b} g^{ab} \omega_a \omega_b$$

$$+ \frac{1}{2} \sum_{a,b=1}^{3} \{A_a[\Omega_a, A_b]\omega_b - \omega_b[\Omega_a, A_b]A_a\}.$$

In view of

$$\sum_{a=1}^{3} \omega_a \Omega_a = 0, \quad A_r^2 - I = o(1), \quad \sum_{a=1}^{3} (1 - \omega_a^2) = 2,$$

the assumptions (D-3) and (D-4) imply that $S_\Omega A_r + A_r S_\Omega \in \mathcal{V}_0^1$. Thus the identity

$$2A_r K A_r = A_r^{-1}[S_\Omega A_r + A_r S_\Omega]A_r^{-1}$$

gives the conclusion. □

We continue the proof of Lemma 3.1. Since

$$S_\Omega^2 A_r = -S_\Omega A_r S_\Omega + S_\Omega(A_r S_\Omega + S_\Omega A_r)$$
$$= A_r S_\Omega^2 - (S_\Omega A_r + A_r S_\Omega)S_\Omega + S_\Omega(A_r S_\Omega + S_\Omega A_r) \tag{3.7}$$

we have

$$[S_\Omega^2, A_r] = -(S_\Omega A_r + A_r S_\Omega)S_\Omega + S_\Omega(A_r S_\Omega + S_\Omega A_r).$$

The principal term of $(S_\Omega A_r + A_r S_\Omega)$ is a scalar operator, so that it can be verified that

$$[S_\Omega^2, A_r] \in \mathcal{V}_0^1.$$

Hence, it follows from (3.6) that $[K, A_r] \in \mathcal{V}_0^1$.   □

Let

$$A_0 A_r^{-1} = B_1 + B_2, \quad B_1 = (A_0 A_r^{-1} - A_r^{-1} A_0)/2, \quad B_2 = (A_0 A_r^{-1} + A_r^{-1} A_0)/2.$$

Then,

Lemma 3.3.

$$[K, A_0] \in q\mathcal{V}_0^0, \quad B_1 \in q\mathcal{V}_0^0, \quad B_2 \in \sqrt{\frac{q}{r}}\mathcal{V}_0^0.$$

Proof. The two first properties follows from the hypothesis (2.7) and Lemma 3.2. The last property follows from the hypothesis (2.6).   □

3.2   A local version of the virial theorem

Lemma 3.4. *If* $L_r = \partial_r - r^{-1}K$,

$$\int_{s-1}^{t+1} \langle \partial_r \{r(A_0 + V_1)\}\zeta, \zeta\rangle = 2\mathrm{Re} \int_{s-1}^{t+1} \langle r\{V_2 - iA_r \frac{q'}{2q}\}\zeta, L_r\zeta\rangle dr$$

$$+ 2\mathrm{Re} \int_{s-1}^{t+1} \langle ir A_r \varphi'\zeta, L_r\zeta\rangle dr - 2\mathrm{Re} \int_{s-1}^{t+1} \langle r f_\chi v, L_r\zeta\rangle dr$$

$$- \mathrm{Re} \int_{s-1}^{t+1} \langle [L_r, A_r/i]\zeta, rL_r\zeta\rangle dr$$

$$+ \mathrm{Re} \int_{s-1}^{t+1} \langle \{[K, A_0 + V_1] - i(L_r J A_r + A_r J L_r)\}\zeta, \zeta\rangle dr$$

$$:= I_1 + I_2 + I_3 + I_4 + I_5.$$

Proof. This is a simple consequence of

$$2\mathrm{Re} \int_{s-1}^{t+1} \langle [L_r - (r^{-1}J + \varphi') + i\{(A_0 + V)A_r^{-1} - iq'/(2q)\}]A_r\zeta, riL_r\zeta\rangle dr$$

$$= 2\mathrm{Re} \int_{s-1}^{t+1} \langle ir f_\chi, iL_r\zeta\rangle dr$$

by use of an integration by parts. To see this, it suffices to check

$$\mathrm{Re} \int_{s-1}^{t+1} \langle L_r A_r\zeta, ir L_r\zeta\rangle dr = \mathrm{Re} \int_{s-1}^{t+1} \langle [L_r, -iA_r]\zeta, \zeta\rangle dr,$$

$$-\text{Re}\int_{s-1}^{t+1}\langle JA_r\zeta, iL_r\zeta\rangle dr = -\text{Im}\int_{s-1}^{t+1}\langle (L_r JA_r + A_r JL_r)\zeta, \zeta\rangle dr$$

and

$$\text{Re}\int_{s-1}^{t+1}\langle (A_0 + V_1)\zeta, rL_r\zeta\rangle dr$$

$$= \text{Re}\int_{s-1}^{t+1}[-\langle \partial_r\{r(A_0 + V_1)\}\zeta, \zeta\rangle + \langle [K, A_0 + V_1]\zeta, \zeta\rangle]dr. \quad \square$$

Remark 3.5. If $|x| > a \gg 1$, our assumptions imply

$$(rV)' = q + (V - q) + rq' + r(V - q)' \geq (1 - \varepsilon)q, \ 1 \gg \varepsilon > 0.$$

3.3 L^2-weighted inequality

We shall estimate the integrals $\{I_j\}_{j=1}^5$ from above to obtain

Proposition 3.6. *If $t > s$ is large enough, then*

$$\int_{s-1}^{t+1}[\{(1 - 2K_0 - o(1))q\}\|e^\varphi \chi v\|^2 + r\varphi'\|L_r(e^\varphi A_r\chi v/\sqrt{q})\|^2]dr$$

$$+ \int_{s-1}^{t+1}k_\varphi\|e^\varphi A_r\chi v/\sqrt{q}\|^2 dr$$

$$\leq C\{\int_{s-1}^s + \int_t^{t+1}\}\left[rq + \{|\varphi'| + |\varphi''|\}rq^{-1}\right]\|e^\varphi A_r v\|^2 dr, \quad (3.8)$$

where C is a positive constant independent of choice of φ and

$$k_\varphi \simeq r\varphi'\{(\varphi'' + (r^{-1} - o(r^{-1}))\varphi'\} - \frac{1}{2}(r\varphi'')'$$

$$- o(1)\varphi' - o(1)\{1 + (\varphi')^2 + (r|\varphi''|)^2\}.$$

The proof of Proposition 3.6 is given in the next section.

Once Proposition 3.6 is established, the proof of Theorem 2.2 follows the argument presented in [11] or [6]. We shall give a sketch of the proof.

Lemma 3.7. *Suppose that $v \in L^2(U)$. Let $0 < b < 1$. If s is large enough,*

$$\int_{s+1}^\infty e^{nr^b(\log r)^2}\|\sqrt{q}v\|^2 dr \leq \int_{s-1}^s e^{nr^b(\log r)^2}\|\sqrt{rq}v\|^2 dr.$$

Proof. Taking φ in Lemma 3.6 as $\varphi(r) = n \log \log r$, we see that

$$\int_s^t (\log r)^n \|q^{1/2} v\|^2 dr$$

$$\leq C \left\{ \int_{s-1}^{t+1} o(1)(1 + n^2 (\log r)^{-2})(\log r)^n \|q^{-1/2} v\|^2 dr \right.$$

$$\left. + \left\{ \int_t^{t+1} + \int_{s-1}^s \right\} n (\log r)^{-1} (\log r)^n \|q^{-1/2} v\|^2 dr \right\}. \tag{3.9}$$

The induction hypothesis $(\log r)^{n-1} \sqrt{q} v \in L^2(D_s)^4$ gives

$$\liminf_{t \to \infty} \int_t^{t+1} r \|(\log r)^{n-1} \sqrt{q} v\|^2 dr = 0.$$

Therefore, we obtain

$$\int_s^\infty (\log r)^n \|\sqrt{q} v\|^2 dr < \infty.$$

In view of

$$r^m = \sum_{n=0}^\infty \frac{(m \log r)^n}{n!},$$

we can conclude that

$$\int_s^\infty r^m \|\sqrt{q} v\|^2 dr < \infty.$$

A similar procedure with $\varphi = n \log r$ gives

$$\int_s^\infty \sum_{n=2}^N \frac{1}{n!} (mr^b)^n \|q^{1/2} v\|^2 dr$$

$$\leq C \int_{s-1}^\infty o(1) r^{-2(1-b)} m^2 \sum_{n=2}^N \frac{1}{(n-2)!} (mr^b)^{n-2} \|q^{-1/2} v\|^2 dr$$

$$+ C_m \int_{s-1}^s \|v\|^2 dr \tag{3.10}$$

for all $N = 2, 3, \ldots$. Hence if $0 < b < 1$, it follows from

$$e^{r^b} = \sum_{n=0}^\infty \frac{(r^b)^n}{n!}$$

that

$$\int_{s+1}^\infty e^{nr^b} \|\sqrt{q} v\|^2 dr < +\infty, \quad n = 1, 2, \ldots.$$

Finally if $\varphi = nr^b$, then $k_\varphi > 0$, so that the conclusion follows from Lemma 3.6. □

Letting $n \to \infty$ in the inequality in Lemma 3.7, we have $u = 0$ on $|x| \geq s + 1$. Therefore, the proof of Theorem 2.2 is completed if we show the unique continuation property for $\mathcal{D}$, which will be derived in Section 5.

4   Proof of Proposition 3.6

We begin the proof by an elliptic estimate of the Dirac type operator in the polar coordinates.

Lemma 4.1. *If $k_0 \in C^1(U)^{4\times 4}$ is a symmetric matrix,*

$$\int_{s-1}^{t+1} \{\|L_r k(r) A_r \zeta\|^2 + \frac{1}{2}\|(r^{-1}J + \varphi' - \frac{q'}{2q} + k_0)k A_r \zeta\|^2\}dr$$

$$\leq \int_{s-1}^{t+1} k^2\|i\{f_\chi - (A_0 + V)\zeta\} + (k'k^{-1} - k_0)A_r\zeta\|^2 dr + \frac{1}{2}\int_{s-1}^{t+1} \frac{k^2}{r^2}\|A_r\zeta\|^2 dr$$

$$- \int_{s-1}^{t+1} \left\{r^{-1}k_0 + [L_r, k_0] + \frac{\varphi'}{r} + \varphi'' - (\frac{q'}{2q})' - \frac{q'}{2rq}\right\}\|kA_r\zeta\|^2 dr$$

$$+ \int_{s-1}^{t+1} \{\varphi' + q + o(1)\}\, o(\frac{1}{r})\|kA_r\zeta\|^2 dr. \quad (4.1)$$

Proof. The equation $k\zeta$ should satisfy is

$$\{L_r - (r^{-1}J + \psi(r) + k_0)\}A_r k\zeta = \xi. \quad (4.2)$$

Here

$$\psi(r) = \varphi' - q'/(2q), \quad \xi = (k' - k_0)A_r\zeta - i(V + A_0)k\zeta + if_\chi.$$

Let

$$X = L_r, \quad Y = (r^{-1}J + \psi(r) + k_0).$$

Then

$$\|XA_r k\zeta\|^2 + \|YA_r k\zeta\|^2 + 2\mathrm{Re}\langle X, Y\rangle = \|\xi\|^2$$

and

$$2\mathrm{Re}\int_{s-1}^{t+1} \langle XA_r k\zeta, YA_r k\zeta\rangle dr = \int_{s-1}^{t+1} \langle [Y, X]A_r k\zeta, A_r k\zeta\rangle dr.$$

The ellipticity of $J \in \tilde{T}(S^2)$ and Lemma 3.1 imply

$$\langle -r^{-1}[\partial_r, J] + r^{-2}[J, K]v, v\rangle \leq o(\frac{1}{r})\{\|Yv\|\|v\| + \|\psi v\|\|v\|\}$$

for any $v \in C^\infty(S^2)^4$. In view of

$$[Y, X] = -\psi' - [L_r, k_0] + r^{-2}J - r^{-1}[\partial_r, J] + r^{-2}[J, K]$$

and

$$r^{-2}J = r^{-1}(r^{-1}J + \psi + k_0) - r^{-1}(\psi + k_0),$$

we obtain (4.1).   □

Proposition 3.6 follows from Lemmas 4.2–4.4.

Lemma 4.2. *For any small $\varepsilon > 0$, it holds that*

$$I_1 = 2\text{Re} \int_{s-1}^{t+1} \langle r\{V_2 - iA_r \frac{q'}{2q}\}\zeta, L_r\zeta\rangle dr$$

$$\leq \int_{s-1}^{t+1} \{(2+\varepsilon)K_0 q + r[q']_+ + o(q) - \varphi'' \frac{rq' + (1+\varepsilon)q}{q^2}\}\|\zeta\|^2 dr$$

$$+ C\{\int_{s-1}^{s} + \int_{t}^{t+1}\}\left[rq + \{\varphi' + |\varphi''|\}rq^{-1}\right]\|e^\varphi \tilde{v}\|^2 dr.$$

Proof. The proof of this lemma follows the proof of Lemma 1 in [6], so that we shall only describe its outline.

$$2\text{Re} \int_{s-1}^{t+1} \langle r\{V_2 - iA_r \frac{q'}{2q}\}\zeta, L_r\zeta\rangle dr$$

$$\leq \int_{s-1}^{t+1} K_0\{\|q^{1/2}A_r^{-1}\zeta\|^2 + \|q^{-1/2}A_r L_r\zeta\|^2\} dr$$

$$+ \frac{1}{2}\int_{s-1}^{t+1} \{\|rk^{-1}q'q^{-1}\zeta\|^2 + \|L_r A_r k\zeta\|^2\} dr := E_1 + E_2. \quad (4.3)$$

Note that

$$rk^{-1}q'q^{-1} = \frac{rq'}{\sqrt{rq' + \varepsilon q}}.$$

We shall choose

$$k(r) = \sqrt{rq' + \varepsilon q}/q, \quad \varepsilon > 0,$$

and $k_0 = iB_1 - q'/q$. It follows that

$$k'k^{-1} = -\frac{q'}{q} + \frac{(1+\varepsilon)q' + rq''}{2(rq' + \varepsilon q)}.$$

Define

$$F = k[i\{f_\chi - (B_2 A_r + V)\zeta\} + (k'k^{-1} + q'/q)A_r\zeta].$$

From Lemma 4.1, it follows that

$$E_2 = \frac{1}{2}\int_{s-1}^{t+1} \|L_r A_r k\zeta\|^2 dr \le \frac{1}{2}\int_{s-1}^{t+1}\{(rq' + \varepsilon q)\|\zeta\|^2$$
$$+ \|F\|^2 + (\frac{1}{2r^2} - \frac{\psi + k_0}{r} - \psi' - k_0')\|k\zeta\|^2\}dr. \quad (4.4)$$

A direct calculation gives

$$\|F\|^2 = \|\sqrt{rq' + \varepsilon q}\zeta\|^2 + (o(1) + \varepsilon)\|\sqrt{q}\zeta\|^2 + C_\varepsilon\|kA_r\chi'e^\varphi v\|^2.$$

Indeed, we only need to verify that if

$$F/i = \sqrt{rq' + \varepsilon q}\zeta + H,$$

then

$$\|H\|^2 + 2\mathrm{Re}\langle\sqrt{rq' + \varepsilon q}\zeta, H\rangle = (o(1) + \varepsilon)\|\sqrt{q}\zeta\|^2 + C_\varepsilon\|kA_r\chi'e^\varphi v\|^2.$$

To see this, we use our hypothesis and the identity

$$\mathrm{Re}\langle kif_\chi\zeta, kiV\zeta\rangle = \mathrm{Re}\langle kif_\chi\zeta, ki(V - q)\zeta\rangle.$$

In view of

$$\sup_{|\xi|=1}|A_r^{-1}\xi| = 1 + o(1), \text{ as } r \to \infty,$$

Lemma 4.1 with $k = 1/\sqrt{q}$ and $k_0 = 0$ yields, in the same manner,

$$E_1 = K_0\int_{s-1}^{t+1}\{\|q^{1/2}A_r^{-1}\zeta\|^2 + (1 + \varepsilon/2)\|L_r q^{-1/2}A_r\zeta\|^2$$
$$+ C_\varepsilon\|q^{-1/2}[A_r, L_r]\zeta\|^2 + o(1)\|\sqrt{q}\zeta\|^2\}dr$$
$$\le K_0\int_{s-1}^{t+1}\{(2 + \varepsilon)\|\sqrt{q}\zeta\|^2 - \varphi''\|\zeta/\sqrt{q}\|^2\}dr. \qquad \square$$

Lemma 4.3. *If* $w = \zeta/\sqrt{q}$,

$$I_2 \le \int_{s-1}^{t+1}\{-k_\varphi\|A_r w\|^2 - r\varphi'\|L_r A_r w\|^2 + o(1)\varphi'\|w\|\|\sqrt{q}\zeta\|$$
$$+ o(1)|\varphi''|\|\zeta\|\|A_r w\| + K_0 r^{-1}\|A_r w\|^2 + C\|\chi'e^\varphi v/\sqrt{q}\|^2\}dr.$$

Proof. From

$$2\mathrm{Re}\int_{s-1}^{t+1}\langle r\varphi'(iqw), [A_r, L_r]w\rangle dr = 0,$$

it follows that

$$I_2 = 2\mathrm{Re}\int_{s-1}^{t+1}\langle irA_r\varphi'\zeta, L_r\zeta\rangle dr = 2\mathrm{Re}\int_{s-1}^{t+1}\langle irq\varphi'w, L_r A_r w\rangle dr.$$

Put

$$iqw = -L_r A_r w + (r^{-1}J + \varphi' - q'/q)A_r w - i(A_0 + V - q)w + if_\chi \quad (4.5)$$

in the last expression. It holds that

$$I_2 = -2\int_{s-1}^{t+1} r\varphi' \|L_r A_r w\|^2 dr$$
$$+ 2\mathrm{Re}\int_{s-1}^{t+1} r\varphi' \langle(r^{-1}J + \varphi' - q'/q)A_r w - i(A_0 + V - q)w, L_r A_r w\rangle dr$$
$$+ 2\mathrm{Re}\int_{s-1}^{t+1} r\varphi' \langle if_\chi, L_r A_r w\rangle dr \quad (4.6)$$

and, by virtue of Lemma 3.3,

$$-\mathrm{Re}\int_{s-1}^{t+1} r\varphi' \langle i(A_0 + V - q)w, L_r A_r w\rangle dr$$
$$\leq \int_{s-1}^{t+1} \{r\varphi' \|i(B_2 A_r + V - q)w\| \|L_r A_r w\| + \mathrm{Re}\langle[L_r, r\varphi' i B_1]A_r w, A_r w\rangle\} dr$$
$$\leq \int_{s-1}^{t+1} o(1)r\varphi' \{\|L_r A_r w\|^2 + \|r^{-1/2}\sqrt{q}w\|^2\} dr$$
$$+ \int_{s-1}^{t+1} o(q)\{|(r\varphi')'| + \varphi'\}\|A_r w\|^2\} dr$$
$$\leq \int_{s-1}^{t+1} o(1)[r\varphi' \|L_r A_r w\|^2 + \{(\varphi')^2 + q^2 + (r\varphi'')^2\}\|w\|^2] dr. \quad (4.7)$$

On the other hand,

$$2\mathrm{Re}\int_{s-1}^{t+1} r\varphi' \langle r^{-1}J A_r w, L_r A_r w\rangle dr$$
$$= -\int_{s-1}^{t+1} \{\varphi'' \langle J A_r w, A_r w\rangle + \varphi' \langle[L_r, J]A_r w, A_r w\rangle\} dr. \quad (4.8)$$

Note that $[L_r, J] \in \mathcal{V}_1^1$ implies

$$\varphi'|\langle[L_r, J]A_r w, A_r w\rangle| \leq o(1)\{\|r^{-1}J A_r w\|^2 + \|\varphi' A_r w\|^2\}.$$

Lemma 4.1 with $k_0 = \varphi' - q'/(2q)$ and $k = 1/\sqrt{q}$ gives

$$\|r^{-1}J A_r w\|^2 \leq \|qw\|^2 + \|\varphi'w\|^2 + C\|\chi' e^\varphi v/\sqrt{q}\|^2.$$

To eliminate $\varphi''\langle JA_r w, A_r w\rangle$ in (3.5), we consider

$$\text{Im}\int_{s-1}^{t+1} r\varphi''\langle if_\chi, A_r w\rangle dr = \text{Re}\int_{s-1}^{t+1} \langle L_r A_r w, -r\varphi'' A_r w\rangle dr$$
$$+ \int_{s-1}^{t+1} r\varphi'' \text{Re}\langle r^{-1} J A_r w, A_r w\rangle dr$$
$$+ \int_{s-1}^{t+1} r\varphi'' \text{Im}\langle\{A_0 + V + (\varphi' - q'/q)A_r\}w, A_r w\rangle dr. \quad (4.9)$$

It follows from Lemma 3.3 that

$$r\varphi'' \text{Im}\langle(A_0 + V)w, A_r w\rangle = r\varphi'' \text{Im}\langle(B_2 A_r + V - q)w, A_r w\rangle$$
$$\leq o(1)\{\|r^{1/2}\varphi'' w\|^2 + \|\sqrt{q}w\|^2\}. \quad (4.10)$$

If $\tilde{v} = A_r v$, it follows from (4.6)–(4.9) that

$$I_2 \leq \int_{s-1}^{t+1} [-k_\varphi\|e^\varphi \tilde{v}/\sqrt{q}\|^2 - r\varphi'\|L_r(e^\varphi \tilde{v}/\sqrt{q})\|^2 + o(1)\varphi'\|w\|\|\sqrt{q}\zeta\|$$
$$+ o(1)|\varphi''|\|\zeta\|\|A_r w\| + K_0 r^{-1}\|A_r w\|^2 + C\|\chi' e^\varphi v/\sqrt{q}\|^2]dr. \quad (4.11)$$

$\square$

Lemma 4.4.

$$I_4 + I_5 = \int_{s-1}^{t+1} o(1)[\{q + (\varphi')^2/q\}\|e^\varphi \tilde{v}\|^2 + \|\chi' e^\varphi \tilde{v}\|^2]dr. \quad (4.12)$$

$$I_3 \leq C\left\{\int_{s-1}^{s} + \int_t^{t+1}\right\}\left[rq + \{\varphi' + |\varphi''|\}rq^{-1}\right]\|e^\varphi \tilde{v}\|^2 dr. \quad (4.13)$$

Proof. Observe that

$$A_r J K + K J A_r = A_r[J, K] + [A_r, K]J + K(JA_r + A_r J)$$

and

$$\partial_r J A_r + A_r J \partial_r - (J A_r + A_r L)\partial_r \in \mathcal{V}_0^1.$$

In addition, the conditions (A4) and (A5) give

$$[K, V_1] = [K, V_1 - q] + [K, q] \in q\mathcal{V}_0^0.$$

In view of these observations and Lemma 3.3, combining Lemma 3.1 with Lemma 4.1 with $k = 1$ and $k_0 = -\varphi' + q'q^{-1}$, we can conclude that

$$I_4 + I_5 = \int_{s-1}^{t+1} o(1)\{(q + \varphi')\|e^\varphi \tilde{v}\|^2 + \|\chi' e^\varphi \tilde{v}\|^2\}dr. \quad (4.14)$$

The Schwarz inequality gives

$$\varphi' \le \frac{1}{2}\{q + \frac{(\varphi')^2}{q}\},$$

so that (4.12) follows from (4.14). (4.13) can be easily verified by use of an integration by parts.   □

5   A unique continuation theorem

In this section we shall show that $\mathcal{D}$ has the strong unique continuation property. We say that $u \in L^2_{\mathrm{loc}}(U)$ vanishes of infinite order at $x_0 \in U$ if

$$\int_{|x-x_0|<R} |u|^2 dx = \mathcal{O}(R^n), \quad R \to 0, \ \forall n \in \mathbf{N}.$$

Theorem 5.1. *Suppose (2.1) and (2.2). If $u \in L^2_{\mathrm{loc}}(U)$ satisfies*

$$\mathcal{D}u + Vu = 0, \quad V \in L^\infty_{\mathrm{loc}}(U)^{4\times 4} \tag{5.1}$$

and vanishes of infinite order at $x_0 \in U$, then u is identically zero in U.

Proof. First, we shall reduce $\mathcal{D}$ into the classical Dirac operator at x_0. In fact, there exists an orthogonal transformation $T = (t_{jk})^3_{j,k=1}$ such that $TG(x_0)T^{-1}$ is a diagonal matrix H. Under the transformation $z = T(x - x_0)$ the operator $\mathcal{D}$ has the form

$$\mathcal{D} = -i\sum_{k=1}^3 \frac{1}{2}\{\partial_{z_k}\tilde{A}_k + \tilde{A}_k\partial_{z_k}\},$$

where

$$\tilde{A}_j(x) = \sum_{k=1}^3 t_{jk}A_k(x).$$

Then, it is easily verified that

$$\tilde{A}_j\tilde{A}_k + \tilde{A}_k\tilde{A}_j = \sum_{a,b=1}^3 t_{ka}t_{jb}g_{ab}(x_0 + T^{-1}z)I.$$

The diagonal elements of H are denoted by $g_j > 0$, $j = 1, 2, 3$, and $E = (e_{jk})^3_{j,k=1}$ stands for the matrix

$$e_{jj} = 1/\sqrt{g_j}, \ e_{jk} = 0, \ j \ne k.$$

Under the dilation $y = Ez$, $\mathcal{D}$ has the desired property. Namely,

$$\mathcal{D} = \frac{1}{2}\sum_{j=1}^3\{\hat{A}_j D_{y_j} + D_{y_j}\hat{A}_j\}$$

with

$$\hat{A}_j \hat{A}_k + \hat{A}_k \hat{A}_j = \hat{g}_{jk}(y)I, \ \hat{g}_{jk}(0) = \delta_{jk}.$$

In these new coordinates, it is written

$$\mathcal{D} = \mathcal{D}_0 + \sum_{j=1}^{3} B_j(y)D_{y_j} + C(y), \tag{5.2}$$

where $\mathcal{D}_0 = \sum_{j=1}^{3} \hat{A}_j(0)D_{y_j}$ is the classical Dirac operator,

$$B_j(y) = \mathcal{O}(|y|), \ B_j(y) \in C^2(\tilde{U})^{4 \times 4}, \ C(y) \in C^1(\tilde{U})^{4 \times 4},$$

and $\tilde{U}$ is a domain of $\mathbf{R}^3$ containing the origin. We introduce the polar coordinates

$$y = r\omega, \ r = |y|, \ \omega = y/|y|.$$

In what follows, we use the notation A_j instead of $\hat{A}_j$. Keeping the same notation as in Section 3.1, we have

Lemma 5.2.
$$i\tilde{\mathcal{D}}ru = \{\partial_r - r^{-1}(K + J)\}A_r(ru).$$

Furthermore,

$$\|[K, J]v\| = \mathcal{O}(r\|Jv\| + \|v\|) \text{ as } r \to 0.$$

Proof. This can be verified in the same manner as in Lemma 3.1 because

$$A_j A_k + A_k A_j = 2\delta_{jk} + \mathcal{O}(r) \text{ as } r \to 0. \qquad \square$$

In [1], it has been proved that

$$\frac{1}{4} \int r^{-2n-2}|v|^2 dy \leq \int r^{-2n}\|\mathcal{D}_0 v\|^2 dy, \ v \in C_0^\infty(\tilde{U})^4 \tag{5.3}$$

for any $n \in \mathbf{N}$.

Lemma 5.3. *If $u \in H^1_{\text{loc}}(\tilde{U})^4$ is a solution to (5.2) vanishing of infinite order at the origin, then*

$$\int_{|y|<R} \{|u|^2 + |\nabla_y u|^2\}dy \leq C \exp\{-\delta R^{-1}\}$$

for any small positive R.

Proof. Suppose that $h(r) \in C^\infty([0, \infty)$ satisfies

$$0 \leq h \leq 1, \ h = 0, \text{ on } [2, \infty), \ h = 1 \text{ on } [0, 1].$$

Let M be a large positive number determined later. Applying the inequality (5.3) to $v = h(nM|y|)u(y)$, we obtain

$$\frac{1}{4} \int r^{-2n-2}|h(nMr)u|^2 dy \leq \int r^{-2n}|D_0 h(nM|y|)u|^2 dy. \qquad (5.4)$$

On the other hand, the ellipticity of $\tilde{D}$ gives

$$\int |\nabla_y r^{-n} h(nMr)u|^2 dy \leq C \int \{|\tilde{D}r^{-n}h(nMr)u|^2 + |r^{-n}h(nMr)u|^2\} dy.$$

From the triangle inequality, it follows that

$$\frac{1}{2} \int r^{-2n}|\nabla_y h(nMr)u|^2 dy \leq n^2 \int r^{-2n-2}|h(nMr)u|^2 dy$$

$$+ C \int \{|\tilde{D}r^{-n}h(nMr)u|^2 + \|r^{-n}h(nMr)u\|^2\} dy. \qquad (5.5)$$

From (5.2), (5.4) and (5.5)$\times n^{-2}/4$, it follows that

$$\int \left\{ \frac{1}{8} r^{-2n-2}|h(nMr)u|^2 + \frac{1}{16} n^{-2} r^{-2n} |\nabla_y h(nMr)u|^2 \right\} dy$$

$$\leq 4 \int r^{-2n} h(nM|y|)^2 |(\tilde{D} + Q)u|^2 dy$$

$$+ C_1 \int r^{-2n} h(nM|y|)^2 \{|u|^2 + |y|^2|\nabla_y u|^2\} dy$$

$$+ C_2(nM)^2 \int_{1 \leq nMr \leq 2} |u|^2 dy. \qquad (5.6)$$

Since

$$|y| \leq 2/(nM), \quad \text{on supp}\{h(nM|y|)\},$$

we obtain

$$\frac{1}{16} \int r^{-2n-2}|h(nMr)u|^2 dy \leq C_2(nM)^2 \int_{1 \leq nMr \leq 2} |u|^2 dy \qquad (5.7)$$

if M is large enough. Hence,

$$\int_{|y|<1/2nM} |u|^2 dy \leq C e^{-n \log 2} \int_{1 \leq nMr \leq 2} |u|^2 dy.$$

For any small $R > 0$, one can find n such that $1/(n+1) < R < 1/n$, so that

$$\int_{|y|<R} |u|^2 dy = C' \exp\{-(\log 2)/R\}. \qquad \square$$

For the sake of Lemma 5.3, if $0 < b < 1$, then

$$\int \exp\left\{ nr^{-b} \right\} \{|u|^2 + |\nabla_y u|^2\} dy < \infty.$$

Thus, we can use another Carleman inequality with a stronger weight function.

Lemma 5.4. *If $b > 0$, we have*

$$\frac{b^2 n}{4} \int r^{-b} \exp\left\{nr^{-b}\right\} |A_r u|^2 dy \leq C \int \exp\left\{nr^{-b}\right\} |r(\tilde{D} + Q)u|^2 dy \quad (5.8)$$

for any $u(x) \in C_0^\infty(\Omega\backslash\{0\})^4$ and any large positive number n if Ω is small enough.

Proof. Let $\varphi = nr^{-b}/2$ with $1 > b > 0$. Note that

$$M_r = r\partial_r + \frac{1}{2} - K$$

is skew symmetric. If $v = re^\varphi u$ and $u \in C_0^\infty(\Omega)$, then

$$ir\tilde{D}v = \{M_r - (J + \frac{1}{2} + r\varphi')\}A_r v.$$

Thus,

$$\int_0^\infty \|r(\tilde{D} + Q)v\|^2 dr \geq \frac{1}{2}\int_0^\infty \{\|M_r A_r v\|^2 + \|(J + \frac{1}{2} + r\varphi')A_r v\|^2\}dr$$

$$- \text{Re}\int_0^\infty \langle M_r A_r v, (J + \frac{1}{2} + r\varphi')A_r v\rangle dr - \sup_\Omega |Q| \int_0^\infty \|rv\|^2 dr \quad (5.9)$$

and

$$- 2\text{Re}\int_0^\infty \langle M_r A_r v, (J + \frac{1}{2} + r\varphi')A_r v\rangle dr$$

$$= \int_0^\infty \langle\{r(r\varphi')' + [K, J] + [r\partial_r, J]\}A_r v, A_r v\rangle dr. \quad (5.10)$$

The ellipticity of J implies

$$\langle[K, J]A_r v, A_r v\rangle + \langle[r\partial_r, J])A_r v, A_r v\rangle$$

$$\leq C\{r\|(J + \frac{1}{2} + r\varphi')A_r v\| + r\|(\frac{1}{2} + r\varphi')A_r v\| + \|A_r v\|\}\|A_r v\|. \quad (5.11)$$

If Ω is shrunk sufficiently, it holds that

$$r(r\varphi')' - Cr^2\varphi' \geq \frac{n}{2}b^2 r^{-b} - C\frac{n}{2}br^{-b+1} \geq \frac{n}{4}b^2 r^{-b}.$$

Therefore, (5.9)–(5.11) gives the conclusion (5.8) with aid of the Schwarz inequality.

The strong unique continuation property follows from Lemmas 5.4 and 5.3 by the standard procedure. This achieves the proof of Theorem 5.1. $\qquad\square$

6   Appendix—The global virial theorem

We shall consider general elliptic systems and give a weak analogue to our result. Let $\{A_j\}_{j=0}^d$ be a family of constant symmetric matrices of order N. Consider the eigenvalue problem

$$Au + A_0 u + V(x)u = -\lambda u, \ x \in \mathbf{R}^d, \quad A = \sum_{j=1}^d A_j D_j. \tag{6.1}$$

For any symmetric matrix V, we define the closed interval I_V as the closure of

$$\left\{ (\partial_r \{r(A_0 + V(x))\}\xi, \xi) ; \ x \in \mathbf{R}^d, \ \xi \in S^{d-1} \right\}.$$

By $k_\pm$ we denote

$$k_+ = \max I_V, \ k_- = \min I_V.$$

Theorem 6.1. *Suppose that $V^* = V \in C^1(\mathbf{R}^d)$ and $\sum_{j=1}^d A_j \xi_j$ has no nonzero eigenvalue. If $-\lambda \in \mathbf{R} \backslash I_V$, then (6.1) admits no nontrivial solution in $L^2(\mathbf{R}^d)$.*

Proof. We shall use the following usual virial relation.

Lemma 6.2. *Suppose*

$$F_u(r) = \mathrm{Re}\langle A(\omega)D_r u, u\rangle(r).$$

Let $u \in H^1_{\mathrm{loc}}(U)^N$ be a solution to (6.1) and $Q = A_0 + V + \lambda$. For all $t > s > R$, $\tilde{u}(r) = r^{(d-1)/2}u(r)$ satisfies

$$\int_s^t \langle (rQ)'\tilde{u}, \tilde{u}\rangle dr = s F_{\tilde{u}}(s) - t F_{\tilde{u}}(t). \tag{6.2}$$

From the ellipticity assumption it follows that $u \in H^1(\mathbf{R}^d)$, which implies that

$$F_{\tilde{u}}(r) \in L^1((0, \infty)),$$

so that there exists a sequence $\{t_j\}$ such that $\lim_{j\to\infty} t_j F_{\tilde{u}}(t_j) = 0$. Letting $s \to 0$ and $t = t_j \to \infty$ in the identity (6.2), we conclude that $\int_0^\infty \|\tilde{u}\|^2 dr = 0$. $\square$

References

[1] L. De Carli and T. Ōkaji, *Strong unique continuation property for the Dirac equation*, Publ. RIMS, Kyoto Univ., 35-6 (1999), 825–846.

[2] T. Ikebe and J. Uchiyama, *On the asymptotic behavior of eigenfunctions of second order elliptic operators*, J. Math. Kyoto Univ. **11** (1971), 425–448.

[3] N. Iwasaki, *Local decay of solutions for symmetric hyperbolic systems with dissipative and coercive boundary conditions in exterior domains*, Publ. RIMS, Kyoto Univ. **5** (1969), 193–218.

[4] W. Jäger, *Zur Theorie der Schwingungsgleichung mit variablen Koeffizienten in Aussengebieten*, Math. Z. **102** (1967), 62–88.

[5] T. Kato, *Growth properties of solutions of the reduced wave equation with a variable coefficient*, Comm. Pure Appl. Math. **12** (1959), 403–425.

[6] H. Kalf, T. Okaji and O. Yamada, *Absence of eigenvalues of Dirac operators with potentials diverging at infinity*, preprint.

[7] P.D. Lax and R.S. Phillips, *Scattering Theory*, Academic Press, 1967.

[8] K. Mochizuki, *Growth properties of solutions of second order elliptic differential equations*, J. Math. Kyoto Univ. **16** (1976), 351–373.

[9] T. Ōkaji, *Absence of eigenvalues of the Maxwell operators*, preprint.

[10] F. Rellich, *Über des asymptotische Verhalten der Lösungen von $\Delta u + k^2 u = 0$ in unendlichen Gebieten*, Jber. Deutsch. Math. Ver. **53** (1943), 57–65.

[11] V. Vogelsang, *Absence of embedded eigenvalues of the Dirac equation for long range potentials*, Analysis (1987), 259–274.

Takashi Ōkaji
Department of Mathematics
Kyoto University
Kyoto 606-8502, Japan
okaji@kusm.kyoto-u.ac.jp

The Behaviors of Singular Solutions of Partial Differential Equations in Some Class in the Complex Domain

Sunao Ōuchi

ABSTRACT Let $L(z, \partial_z)$ be a linear partial differential operator with holomorphic coefficients in a neighborhood U of $z = 0$ in $\mathbb{C}^{d+1}$ and K be a nonsingular complex hypersurface. Let $u(z)$ be a solution of the equation $L(z, \partial_z)u(z) = 0$, which has singularities on K. In general there are many singular homogeneous solutions. The purpose of the present paper is to introduce a class of partial differential operators and study of the behaviors of homogeneous solutions of $L(z, \partial_z)$ belonging to this class, by restricting the growth properties of singularities on K.

0 Introduction

In the present paper we consider solutions with singularities of linear partial differential equations with holomorphic coefficients. Studies of solutions with singularities on a surface have been made by many mathematicians. We first mention the pioneer work of Leray [4] concerning the Cauchy problem whose initial surface has characteristic points, and secondly those of Hamada [2], Wagschal [15] and Hamada, Leray and Wagschal [3] concerning Cauchy problems with singular initial data. After these papers there are many papers about the existence of solutions with singularities on a characteristic surface ([6] [9] and [13] etc.). As for the behaviors of solutions near the singularities we refer to Ōuchi [7], [8], [10] [11] and [12]. We show in [7] and [8] that some solutions behave exponentially in some regions but mildly in others, that is, like Stokes phenomenon in the theory of ordinary differential equations. It is shown in [10] and [12] that singular solutions of partial differential equations in some class have asymptotic expansions provided they satisfy some exponential growth estimate near the singularities.

In this paper we consider a class of partial differential operators treated in Ōuchi [11]. Our main purpose is to study in detail the asymptotic behaviors of solutions near the singularities for operators in this class and we find asymptotic terms of solutions (Theorem 3.5).

1 Notation and definitions

We denote the coordinates of $\mathbb{C}^{d+1}$ by $z = (z_0, z_1, \ldots, z_d) = (z_0, z') \in \mathbb{C} \times \mathbb{C}^d$, $|z| = \max\{|z_i|; 0 \leq i \leq d\}$ and $|z'| = \max\{|z_i|; 1 \leq i \leq d\}$. We choose the coordinates so that the hypersurface K is given by $\{z_0 = 0\}$. The differentiation with respect to z_i is denoted by ∂_{z_i} and $\partial_z = (\partial_{z_0}, \partial_{z_1}, \ldots, \partial_d) = (\partial_{z_0}, \partial_{z'})$. For a multi-index $\alpha = (\alpha_0, \alpha_1, \ldots, \alpha_d) = (\alpha_0, \alpha') \in \mathbb{N}^{d+1}$, $\partial_z^\alpha = \prod_{i=0}^d \partial_{z_i}^{\alpha_i}$, $|\alpha| = \alpha_0 + |\alpha'| = \sum_{i=0}^d \alpha_i$. Let $U = U_0 \times U'$ be a polydisk with $U_0 = \{z_0; |z_0| < R_0\}$ and $U' = \{z' \in \mathbb{C}^d; |z'| < R\}$. Set $U_0(\theta) = \{0 < |z_0| < R_0; |\arg z_0| < \theta\}$ and $U(\theta) = U_0(\theta) \times U'$. $\mathcal{O}(U)$ and $(\mathcal{O}(U(\theta)))$ is the set of all holomorphic functions on U $(resp.\ U(\theta))$.

Let $L(z, \partial_z)$ be an m-th order linear partial differential operator with holomorphic coefficients in U. Suppose $L(z, \partial_z)$ is represented as follows:

$$\begin{cases} L(z, \partial_z) = A(z, \partial_{z_0}) + B(z, \partial_z), \\[2mm] A(z, \partial_{z_0}) = \displaystyle\sum_{i=0}^k a_i(z')(z_0\partial_{z_0})^i, \\[2mm] B(z, \partial_z) = \displaystyle\sum_{|\alpha| \leq m} b_\alpha(z)\partial_z^\alpha. \end{cases} \tag{1.1}$$

$A(z, \partial_{z_0})$ is an ordinary differential operator with order k. Let $j_\alpha \in \mathbb{N}$ be the multiplicity of $b_\alpha(z) = 0$ on $\{z_0 = 0\}$. So $b_\alpha(z) = z^{j_\alpha} \tilde{b}_\alpha(z)$ with $\tilde{b}_\alpha(0, z') \neq 0$ on $\{z_0 = 0\}$, provided $b_\alpha(z) \not\equiv 0$. Define

$$\chi(z', \lambda) = \sum_{i=0}^k a_i(z')\lambda^i. \tag{1.2}$$

We give conditions (C_0) and (C_1) on $L(z, \partial_z)$,

$$a_k(0) \neq 0, \tag{C_0}$$

$$j_\alpha - \alpha_0 > 0. \tag{C_1}$$

(C_0) is a condition on $A(z, \partial_{z_0})$ and (C_1) is one on $B(z, \partial_z)$. It follows from (C_0) that there is a neighborhood V' of $z' = 0$ such that $z_0 = 0$ is a regular singularity of $A(z, \partial_{z_0})$ and $\chi(z', \lambda)$ is a polynomial with degree k for $z' \in V'$. In the following presentation we treat $L(z, \partial_{z_0})$ of the form (1.1) satisfying the conditions (C_0) and (C_1).

Let us define a constant γ which is important in this paper.

Definition 1.1.

$$\gamma := \begin{cases} \min\{\dfrac{j_\alpha - \alpha_0}{|\alpha| - k}; \ |\alpha| > k\} & if\ k < m, \\[3mm] +\infty & if\ k = m. \end{cases} \tag{1.3}$$

We represent $B(z, \partial_z)$ in another form to simplify the later calculations. Set $\vartheta = z_0 \partial_{z_0}$. By the relation

$$b_\alpha(z)\partial_z^\alpha = z_0^{j\alpha} \tilde{b}_\alpha(z)\partial_z^\alpha = z_0^{j\alpha - \alpha_0} \tilde{b}_\alpha(z)(z_0^{\alpha_0} \partial_{z_0}^{\alpha_0})\partial_{z'}^{\alpha'}$$
$$= z_0^{j\alpha - \alpha_0} \tilde{b}_\alpha(z)\vartheta(\vartheta - 1)\cdots(\vartheta - \alpha_0 + 1)\partial_{z'}^{\alpha'}$$

we can represent $B(z, \partial_z)$ as,

$$B(z, \partial_z) = \sum_\alpha z_0^{e_\alpha} c_\alpha(z)\vartheta^{\alpha_0}\partial_{z'}^{\alpha'}, \tag{1.4}$$

where $c_\alpha(z) \in \mathcal{O}(U)$ and $e_\alpha \geq 1$ from the condition (C_1). It follows from the definition of γ that

$$\frac{e_\alpha}{\gamma} \geq |\alpha| - k. \tag{1.5}$$

We use the representation (1.4) in the following sections.

At first glance we will have the impression that the class of operators we consider is restricted, however, it contains useful examples. We give them.

(1) Let $P(z, \partial_z) = \partial_{z_0}^k + \sum_{\substack{|\alpha| \leq m \\ \alpha_0 < k}} a_\alpha(z)\partial_z^\alpha$ $(m > k)$. $P(z, \partial_z)$ is a linear partial differential operator with order m and is of the normal form with respect to ∂_{z_0}. By multiplying $P(z, \partial_z)$ by z_0^k, consider $z_0^k P(z, \partial_z)$. Then $z_0^k P(z, \partial_z)$ satisfies (C_0) and (C_1), by setting $A(z_0, \partial_{z_0}) = z_0^k \partial_{z_0}^k$ and $B(z, \partial_z) = \sum_{|\alpha| \leq m, \alpha_0 < k} z_0^k a_\alpha(z)\partial_z^\alpha$.

(2) Let $P(z, \partial_z)$ be an m-th operator of Fuchsian type with weight $(m - h)$ in the sense of Baouendi–Goulaouic [1]. Then $z_0^{m-h} P(z, \partial_z)$ belongs to the class we consider and $\gamma = +\infty$.

(3) We give a concrete example. Let $z = (z_0, z_1) \in \mathbb{C}^2$ and

$$L(z, \partial_z) = z_0 \partial_{z_0} - a(z) + z_0^j c(z)\partial_{z_1}^m, \tag{1.6}$$

where $j \geq 1$ and $c(0, z_1) \not\equiv 0$. Then $\chi(z_1, \lambda) = \lambda - a(0, z_1)$ and $\gamma = j/(m - 1)$ $(m > 1)$, $\gamma = +\infty$ $(m = 1)$.

Let us introduce function spaces on the sectorial region $U(\theta)$.

Definition 1.2. $\mathcal{O}_{(\kappa)}(U(\theta))$ is the set of all $u(z) \in \mathcal{O}(U(\theta))$ such that for any $\varepsilon > 0$ and any θ' with $0 < \theta' < \theta$,

$$|u(z)| \leq M \exp(\varepsilon|z_0|^{-\kappa}) \quad \text{for} \quad z \in U(\theta') \tag{1.7}$$

holds for some constant $M = M(\varepsilon, \theta')$. We put $\mathcal{O}_{(+\infty)}(U(\theta)) = \mathcal{O}(U(\theta))$.

Definition 1.3. $\mathcal{O}_{temp,c}(U(\theta))$ is the set of all $u(z) \in \mathcal{O}(U(\theta))$ such that for any θ' with $0 < \theta' < \theta$,

$$|u(z)| \leq M|z_0|^c \quad \text{for} \quad z \in U(\theta') \tag{1.8}$$

holds for some constant $M = M(\theta')$.

$\mathcal{O}_{temp,c}(U(\theta))$ is the set of all holomorphic functions on $U(\theta)$ having singularities on $z_0 = 0$ with fractional order at most c. Set $\mathcal{O}_{temp}(U(\theta)) = \cup_{c \in \mathbb{R}} \mathcal{O}_{temp,c}(U(\theta))$ and we say that $u(z) \in \mathcal{O}(U(\theta))$ has tempered singularities, or regular singularities, on $\{z_0 = 0\}$ in $U(\theta)$, provided $u(z) \in \mathcal{O}_{temp}(U(\theta))$.

Definition 1.4. $\mathcal{A}sy_{\{\kappa\}}(U(\theta))$ is the set of all $u(z) \in \mathcal{O}(U(\theta))$ such that for any θ' with $0 < \theta' < \theta$ and any $n \in \mathbb{N}$,

$$|u(z) - \sum_{p=0}^{n-1} u_p(z')z_0^p| \le AB^n |z_0|^n \Gamma(\frac{n}{\kappa} + 1) \text{for} \quad z \in U(\theta'), \quad (1.9)$$

where $u_p(z') \in \mathcal{O}(U')$ holds for constants $A = A(\theta')$ and $B = B(\theta')$.

We will apply the results to the operators cited above in Section 5 and $\mathcal{A}sy_{\{\kappa\}}(U(\theta))$ will then appear.

2   Behaviors of singular solutions

Let us consider
$$L(z, \partial_z)u(z) = 0, \quad u(z) \in \mathcal{O}(U(\theta)). \quad (2.1)$$

In general there are many singular solutions. So we restrict solutions of (2.1) by their growth properties. We assume $u(z) \in \mathcal{O}_{(\gamma)}(U(\theta))$, that is, solutions are at most of infra-exponential type with exponent γ defined by (1.3). Further this restriction means that the singularities of solutions become much weaker than the assumption. We give it.

By the condition (C_0), $\chi(0, \lambda) = 0$ has k roots and there exist constants $r' > 0$, a_0, a_1 and b such that for $z' \in \{|z'| \le r'\}$,

$$\{\lambda; \chi(z', \lambda) = 0\} \subset \{\lambda; a_0 \le \Re\lambda \le a_1, |\Im\lambda| \le b\}. \quad (2.2)$$

Then we have

Theorem 2.1. ([11]) Let $u(z) \in \mathcal{O}_{(\gamma)}(U(\theta))$ be a solution of $L(z, \partial_z)u(z) = f(z) \in \mathcal{O}_{temp,c}(U(\theta))$. Then there is a polydisk V centered at $z = 0$ such that $u(z) \in \mathcal{O}_{temp,c'}(V(\theta))$ for any $c' < \min\{c, a_0\}$.

We show Theorem 2.1 by constructing a parametrix and refer the details of the proof to Ōuchi [11]. It follows from Theorem 2.1 that singularities of solutions in $\mathcal{O}_{(\gamma)}(U(\theta))$ are tempered, that is, of fractional order, so we assume $u(z) \in \mathcal{O}_{temp,c}(U(\theta))$ in the following sections.

3   Behaviors of solutions with tempered singularities

In this section we study the behaviors of solutions with tempered singularities and give the main results (Theorems 3.3, 3.5 and 3.6). So let $u(z) \in \mathcal{O}_{temp}(U(\theta))$ be a solution of $L(z, \partial_z)u(z) = 0$ with bound $|u(z)| \le A|z_0|^c$ for $z \in U(\theta)$. For our purposes we consider the Mellin transform of $u(z)$ with respect to z_0,

$$\hat{u}(\lambda, z') = \int_0^T t^{\lambda-1} u(t, z') dt, \tag{3.1}$$

where T is a small positive constant. It follows from the assumption $|u(z)| \le A|z_0|^c$ that $\hat{u}(\lambda, z')$ is holomorphic in λ on $\{\lambda; \Re\lambda > -c\}$. One of the aims of this section is to show that $\hat{u}(\lambda, z')$ has a meromorphic extension with respect to λ to the whole plane and the other is to derive the asymptotic behaviors of $u(z)$ from $\hat{u}(\lambda, z)$. First, we give elementary properties of the Mellin transform. Let $v(t)$ be a smooth function on $(0, T]$ $(T > 0)$ with $|v(t)| \le C|t|^c$. The smoothness of $v(t)$ is often superfluous for the study of the Mellin transform, however, it is enough for our purposes. The Mellin transform $\hat{v}(\lambda)$ of $v(t)$ is defined by

$$\hat{v}(\lambda) = \int_0^T t^{\lambda-1} v(t) dt, \tag{3.2}$$

which is holomorphic on $\{\lambda : \Re\lambda > -c\}$. The inversion of the Mellin transform is

$$v(t) = \lim_{A \to +\infty} \frac{1}{2\pi i} \int_{c'-iA}^{c'+iA} t^{-\lambda} \hat{v}(\lambda) d\lambda \quad \text{for} \quad t \in (0, T), \tag{3.3}$$

where $c' > -c$. We use the notation $\vartheta_t = t\frac{d}{dt}$.

Lemma 3.1. *Suppose that $|(t\frac{d}{dt})^j v(t)| \le C|t|^c$ on $(0, T]$ for $0 \le j \le n$. Then*

$$\int_0^T t^{\lambda-1} \vartheta_t^n v(t) dt = (-\lambda)^n \hat{v}(\lambda) + T^\lambda \Big(\sum_{j=0}^{n-1} (-\lambda)^{n-1-j} \vartheta_t^j v(T)\Big). \tag{3.4}$$

Proof. We obtain (3.4) by integration by parts. □

We give an estimate of holomorphic functions on sectorial regions.

Lemma 3.2. *Let $U_0 = \{|z_0| < R_0\}$, $U' = \{|z'| < R\}$ and $U = U_0 \times U'$. Let $f(z) \in \mathcal{O}_{temp,c}(U(\theta))$ and set $M = \sup\{|z_0|^{-c}|f(z)|; z \in U(\theta)\}$. Let $0 < r_0 < R_0$ and set $V = \{|z_0| < r_0\} \times U'$. Let η be a constant with $0 < \eta < \min\{\theta, \pi/2, (R_0 - r_0)/r_0\}$. Then there is a constant C such that*

$$|\vartheta^s f(z)| \le \frac{MC^s s! |z_0|^c}{(\sin \eta)^s} \quad \text{for } z \in V(\theta - \eta) \tag{3.5}$$

where C is independent of $f(z)$ and η.

Proof. Let $\mathcal{Z}$ be a circle $\zeta - z_0 = |z_0|(\sin \eta)e^{i\varphi}$ $(0 \leq \varphi \leq 2\pi)$. Let $z \in V(\theta - \eta)$. Then $\mathcal{Z} \subset U(\theta)$. By Cauchy's integral formula

$$z_0^s \partial_{z_0}^s f(z) = \frac{z_0^s s!}{2\pi i} \oint_{\mathcal{Z}} \frac{f(\zeta, z')d\zeta}{(\zeta - z_0)^{s+1}}.$$

We have $|\zeta| \leq 2|z_0|$ on $\mathcal{Z}$, so

$$|z_0^s \partial_{z_0}^s f(z)| \leq \frac{Ms!}{2\pi(\sin \eta)^s} \int_0^{2\pi} |\zeta|^c d\varphi \leq \frac{MC^s s!|z_0|^c}{(\sin \eta)^s}.$$

It follows from the relation $z_0^s \partial_{z_0}^s = \vartheta(\vartheta - 1) \cdots (\vartheta - s + 1)$ that (3.5) holds for another C. $\qquad\square$

Now let us calculate $\int_0^T t^{\lambda-1} L(t, z', \partial_t, \partial_{z'}) u(t, z')dt$. Set

$$\Phi(\lambda, z') := \chi(-\lambda, z') = \sum_{h=0}^{k} (-\lambda)^h a_h(z'). \tag{3.6}$$

By Lemma 3.1

$$\int_0^T t^{\lambda-1} A(z', \partial_t) u(t, z')dt = \Phi(\lambda, z')\hat{u}(\lambda, z') + T^\lambda h_A(\lambda, z'),$$

$$h_A(\lambda, z') = \sum_{h=1}^{k} \left(\sum_{i=0}^{h-1} (-\lambda)^{h-1-i} a_h(z')\vartheta_t^i u(T, z') \right). \tag{3.7}$$

By Taylor's expansion of the coefficient of $B(z, \partial_z)$ (see (1.4)), we have $c_\alpha(z) = \sum_{j=0}^{s-1} c_{\alpha,j}(z')z_0^j + c_{\alpha,s}^R(z)z_0^s$ and

$$c_{\alpha,j}(z') \int_0^T t^{\lambda+e_\alpha+j-1} \vartheta_t^{\alpha_0} \partial_{z'}^{\alpha'} u(t, z')dt$$

$$= (-\lambda - e_\alpha - j)^{\alpha_0} c_{\alpha,j}(z') \partial_{z'}^{\alpha'} \hat{u}(\lambda + e_\alpha + j, z') + T^\lambda h_{\alpha,j}(\lambda, z'),$$

where

$$h_{\alpha,j}(\lambda, z') = \sum_{i=0}^{\alpha_0-1} c_{\alpha,j}(z')(-\lambda - e_\alpha - j)^{\alpha_0-1-i} \vartheta_t^i \partial_{z'}^{\alpha'} \hat{u}(T, z').$$

Hence for $N \geq 1$,

$$\int_0^T t^{\lambda-1} B(t, z', \partial_t, \partial_{z'}) u(t, z') dt$$

$$= \sum_{\{\alpha; e_\alpha < N\}} \Big(\sum_{\ell=e_\alpha}^{N-1} (-\lambda - \ell)^{\alpha_0} c_{\alpha, \ell - e_\alpha}(z') \partial_{z'}^{\alpha'} \hat{u}(\lambda + \ell, z')$$

$$+ \int_0^T t^{\lambda+N-1} c_{\alpha, N-e_\alpha}^R (t, z') \vartheta_t^{\alpha_0} \partial_{z'}^{\alpha'} u(t, z') dt + T^\lambda \Big(\sum_{\ell=e_\alpha}^{N-1} h_{\alpha, \ell - e_\alpha}(\lambda, z') \Big) \Big)$$

$$+ \sum_{\{\alpha; e_\alpha \geq N\}} \int_0^T t^{\lambda+e_\alpha-1} c_\alpha(t, z') \vartheta_t^{\alpha_0} \partial_{z'}^{\alpha'} u(t, z') dt.$$

Thus we have partial differential difference equations

$$\int_0^T t^{\lambda-1} L(t, z', \partial_t, \partial_{z'}) u(t, z') dt = \Phi(\lambda, z') \hat{u}(\lambda, z')$$

$$+ \sum_{\ell=1}^{N-1} \mathcal{L}_\ell(\lambda, z', \partial_{z'}) \hat{u}(\lambda + \ell, z') + \hat{u}_N^R(\lambda, z') + T^\lambda H_N(\lambda, z'), \tag{3.8}$$

where

$$\mathcal{L}_\ell(\lambda, z', \partial_{z'}) = \sum_{\{\alpha; \, e_\alpha \leq \ell\}} (-\lambda - \ell)^{\alpha_0} c_{\alpha, \ell - e_\alpha}(z') \partial_{z'}^{\alpha'}, \tag{3.9}$$

$$\hat{u}_N^R(\lambda, z') = \sum_{\{\alpha; \, e_\alpha < N\}} \int_0^T t^{\lambda+N-1} c_{\alpha, N-e_\alpha}^R (t, z') \vartheta_t^{\alpha_0} \partial_{z'}^{\alpha'} u(t, z') dt$$

$$+ \sum_{\{\alpha; e_\alpha \geq N\}} \int_0^T t^{\lambda+e_\alpha-1} c_\alpha(t, z') \vartheta_t^{\alpha_0} \partial_{z'}^{\alpha'} u(t, z') dt \tag{3.10}$$

and

$$H_N(\lambda, z') = h_A(\lambda, z') + \sum_{\{\alpha; \, e_\alpha < N\}} \sum_{\ell=e_\alpha}^{N-1} h_{\alpha, \ell - e_\alpha}(\lambda, z'). \tag{3.11}$$

Theorem 3.3. *Let $u(z) \in \mathcal{O}_{temp,c}(U(\theta))$ be a solution of $L(z, \partial_z) u(z) = 0$. Then $\hat{u}(\lambda, z')$ is meromorphically extensible in λ to the whole plane and its poles are contained in $\cup_{n=0}^{\infty} \{\lambda; \, \Phi(\lambda + n, z') = 0\}$.*

Proof. From (3.8),

$$\hat{u}(\lambda, z') = \frac{\sum_{\ell=1}^{N-1} \mathcal{L}_l(\lambda, z', \partial_{z'}) \hat{u}(\lambda + \ell, z') + \hat{u}_N^R(\lambda, z') + T^\lambda H_N(\lambda, z')}{-\Phi(\lambda, z')} \tag{3.12}$$

and we use this relation to prolong $\hat{u}(\lambda, z')$ meromorphically. Let $V \Subset U$ be a polydisk and $0 < \theta' < \theta$. By Lemma 3.2 $|\vartheta_{z_0}^{\alpha_0} \partial_{z'}^{\alpha'} u(z)| \leq C_0 |z_0|^c$ for $z \in V(\theta')$, so $\hat{u}_N^R(\lambda, z')$ is holomorphic in $\{\lambda; \Re\lambda > -N-c\}$. $H_N(\lambda, z')$ is a polynomial in λ. Hence $\Phi(\lambda, z')^{-1}\big(\hat{u}_N^R(\lambda, z')+T^\lambda H_N(\lambda, z')\big)$ is holomorphic on $\{\lambda; \Re\lambda > -N-c, \Phi(\lambda, z') \neq 0\}$ and we can choose N as large as possible. Let us show that $\hat{u}(\lambda, z')$ is holomorphic on $\{\lambda; \Re\lambda > -n-c, \prod_{p=0}^{n-1}\Phi(\lambda+p, z') \neq 0\}$ for $n = 1, 2, \ldots$. By the assumption $\hat{u}(\lambda, z')$ is holomorphic on $\{\Re\lambda > -c\}$. Let $\Re\lambda > -1-c$. Choose $N = 1$ in (3.12). Then $\hat{u}_1^R(\lambda, z')$ is holomorphic there, so $\hat{u}(\lambda, z')$ is holomorphic on $\{\lambda; \Re\lambda > -1-c, \Phi(\lambda, z') \neq 0\}$ by (3.12). Assume that $n \geq 2$ and $\hat{u}(\lambda, z')$ is holomorphic on $\{\lambda; \Re\lambda > -(n-1)-c, \prod_{p=0}^{n-2}\Phi(\lambda+p, z') \neq 0\}$. Let $\Re\lambda > -n-c$. Then $\Re\lambda+\ell > -(n-\ell)-c$, so $\hat{u}(\lambda+\ell, z')$ $(\ell \geq 1)$ is holomorphic on $\{\lambda; \Re\lambda > -n-c, \prod_{p=0}^{n-\ell-1}\Phi(\lambda+l+p, z') \neq 0\}$. Hence, by taking $N = n$ in (3.12), $\hat{u}(\lambda, z')$ is holomorphic in $\{\lambda; \Re\lambda > -n-c, \prod_{p=0}^{n-1}\Phi(\lambda+p, z') \neq 0\}$. Thus $\hat{u}(\lambda, z')$ is meromorphic on the whole plane and the poles are contained in $\cup_{n=0}^{\infty}\{\lambda; \Phi(\lambda+n, z') = 0\}$.   □

Let us introduce some notation in order to calculate the inverse Mellin transform of $\hat{u}(\lambda, z')$. For sets $X_1, X_2, d(X_1, X_2)$ means the distance between X_1 and X_2. Set

$$Z_n(r') = \cup_{|z'|\leq r'}\{\lambda; \Phi(\lambda+n, z') = 0\}, \quad Z(r') = \cup_{n=0}^{\infty}Z_n(r'), \quad (3.13)$$

$$Z_n(r', \varepsilon_0) = \{\lambda; d(\lambda, Z_n(r')) \leq \varepsilon_0\}, \quad Z(r', \varepsilon_0) = \cup_{n=0}^{\infty}Z_n(r', \varepsilon_0). \quad (3.14)$$

The poles of $\hat{u}(\lambda, z')$ are contained in $Z(r')$ for $|z'| \leq r'$ by Theorem 3.3. The positive constants r' and ε_0 are chosen so small, if necessary. Since $\Phi(\lambda, z') = \chi(-\lambda, z')$, from (2.2),

$$\cup_{|z'|\leq r'}\{\lambda; \Phi(\lambda, z') = 0\} \subset \{\lambda; -a_1 \leq \Re\lambda \leq -a_0, |\Im\lambda| \leq b\}, \quad (3.15)$$

so $Z(r') \subset \{\lambda; \Re\lambda \leq -a_0, |\Im\lambda| \leq b\}$. We also have from (3.12)

Corollary 3.4. *Let* $|z'| \leq r'$, *r' being a constant such that (3.15) holds. If $\hat{u}(\lambda, z')$ is holomorphic in $\{\lambda; \Re\lambda > -a\}$ with $a > a_1$, then $\hat{u}(\lambda, z')$ is an entire function in λ.*

Proof. By the assumption $\hat{u}(\lambda+\ell, z')$ is holomorphic in $\{\lambda; \Re\lambda > -a-\ell\}$. Since $\hat{u}_N^R(\lambda, z')$ is holomorphic in $\{\lambda; \Re\lambda > -N-c\}$ and $H_N(\lambda, z')$ is a polynomial in λ, by choosing large N, $\sum_{\ell=1}^{N-1}\mathcal{L}_l(\lambda, z', \partial_{z'})\hat{u}(\lambda+\ell, z')+\hat{u}_N^R(\lambda, z')+T^\lambda H_N(\lambda, z')$ is holomorphic in $\{\lambda; \Re\lambda > -a-1\}$. Now assume $-a-1 < \Re\lambda \leq -a$. Since $\Phi(\lambda, z') \neq 0$ for $\Re\lambda < -a_1$ and $-a < -a_1$, it follows from (3.12) so it is holomorphic in $\{\lambda; \Re\lambda > -a-1\}$. By repeating this method, $\hat{u}(\lambda, z')$ is holomorphic in $\{\lambda; \Re\lambda > -a-n\}$ for $n = 1, 2, \cdots$ and it is an entire function in λ.   □

Set $\Lambda(n) = \{\lambda \notin Z(r', \varepsilon_0); -n+1/2-c \leq \Re\lambda \leq -n+3/2-c\}$ for $n \geq 0$. Choose $\sigma_n > 0$ $(n \geq 1)$ such that the vertical line $\{\lambda; \Re\lambda = -\sigma_n\} \subset \Lambda(n)$. Let C_n be a Jordan contour in $\{\lambda; \Re\lambda > -\sigma_n\}$ which encloses all the poles located in $\{\lambda; \Re\lambda > -\sigma_n\}$. Define

$$u_n(z) = \frac{1}{2\pi i}\int_{C_n} z_0^{-\lambda}\hat{u}_n(\lambda, z')d\lambda, \quad (3.16)$$

which gives the asymptotic behaviors of $u(z)$, that is, we have

Theorem 3.5. *Let* $u(z) \in \mathcal{O}_{temp}(U(\theta))$ *be a solution of* $L(z, \partial_z)u(z) = 0$ *and* $u_n(z)$ *be the function defined by* (3.16). *Then there is a polydisk* V *centered at* $z = 0$ *such that for any* θ' *with* $0 < \theta' < \theta$ *and any* $n \in \mathbb{N}$,

$$|u(z) - u_n(z)| \leq AB^n|z_0|^{\sigma_n}\Gamma(\frac{n}{\gamma} + 1) \quad in \ V(\theta') \tag{3.17}$$

holds for some constants A *and* B *depending on* θ'.

Theorem 3.6. *Let* $u(z) \in \mathcal{O}_{temp}(U(\theta))$ *be a solution of* $L(z, \partial_z)u(z) = 0$ *satisfying* $|u(z)| \leq A|z_0|^a$ *in* $U(\theta)$ *for some* $a > a_1$. *Then there is a polydisk* V *centered at* $z = 0$ *such that for any* θ' *with* $0 < \theta' < \theta$,

$$|u(z)| \leq C \exp(-c|z_0|^{-\gamma}) \quad in \ V(\theta') \tag{3.18}$$

holds for some positive constants C *and* c.

We give the proofs of Theorems 3.5 and 3.6 in the following sections. We estimate $\hat{u}(\lambda, z')$ in Section 4 and complete the proofs in Section 5.

4   Estimates of $\hat{u}(\lambda, z')$

In order to show Theorems 3.5 and 3.6 we estimate $\hat{u}(\lambda, z')$ outside of poles. We use the method of majorant functions for this purpose. For formal power series of n variables $w = (w_1, \ldots, w_n)$, $A(w) = \sum_\alpha A_\alpha w^\alpha$ and $B(w) = \sum_\alpha B_\alpha w^\alpha$, $A(w) \ll B(w)$ means $|A_\alpha| \leq B_\alpha$ for all $\alpha \in \mathbb{N}^n$. $A(w) \gg 0$ means $A_\alpha \geq 0$ for all $\alpha \in \mathbb{N}^n$. Let us introduce a series of majorant functions $\{\Psi^{(s)}(X)\}_{s \geq 0}$ of one variable X,

$$\Psi^{(s)}(X) := \frac{\Gamma(s + 1)}{(r - X)^{s+1}} \quad (r > 0). \tag{4.1}$$

Obviously $\Psi^{(s)}(X) \gg 0$, $\frac{d\Psi^{(s)}(X)}{dX} = \Psi^{(s+1)}(X)$ and

$$\Psi^{(s)}(X)/\Gamma(s + 1) \ll \Psi^{(s')}(X)/\Gamma(s' + 1),$$

provided $0 \leq s \leq s'$ and $0 < r \leq 1$. We have

Lemma 4.1. (1) *Let* $0 \leq s_1 \leq s_2$. *Suppose* $\Psi^{(s_1)}(X) \ll C\Psi^{(s_2)}(X)$. *Then for any* $s \geq 0$

$$\Psi^{(s+s_1)}(X) \ll C\Psi^{(s+s_2)}(X) \tag{4.2}$$

holds, where C *is the same constant in the assumption.*
(2) *Let* $0 < r < r'$. *Then*

$$(r' - X)^{-1}\Psi^{(s)}(X) \ll \frac{1}{r' - r}\Psi^{(s)}(X). \tag{4.3}$$

(3) Set $X = \sum_{i=1}^{d} z_i$ and $0 < r < r'$. Let $a(z')$ be a holomorphic function in $\{z'; |z'| < r'\}$ with $|a(z')| \le M$ and $v(z') \ll K\Psi^{(s)}(X)$. Then $a(z') \ll Mr'/(r' - X)$ and there is a constant C such that

$$a(z')\partial_{z'}^{\alpha'} v(z') \ll KC\Psi^{(s+|\alpha'|)}(X). \tag{4.4}$$

The proof is not difficult. We refer the proofs of (2) and (3) to [11], [12] or [15]. We give lemmas concerning estimates of $\Phi(\lambda, z')$, $\hat{u}_N^R(\lambda, z')$ and $H_N(\lambda, z')$ (see (3.10) and (3.11)). Let r' be a positive constant such that (3.15) holds and ε_0 be a small constant (see also (3.13) and (3.14)). Let $|z'| \le r'$ and $\lambda \notin Z_0(r', \varepsilon_0)$. Then $\Phi(\lambda, z') \ne 0$, $\Phi(\lambda, z')^{-1}$ is holomorphic and $|\Phi(\lambda, z')| \ge C'(|\lambda| + 1)^k$ holds. Suppose $|z'| \le r'$ and set $X = \sum_{i=1}^{d} z_i$ in the following.

Lemma 4.2. *Let $\lambda \notin Z_0(r', \varepsilon_0)$. Then there is a constant $C > 0$ such that*

$$\Phi(\lambda, z')^{-1} \ll \frac{C}{(|\lambda| + 1)^k}(r' - X)^{-1}. \tag{4.5}$$

Lemma 4.3. *(1) $\hat{u}_N^R(\lambda, z')$ is holomorphic in $\{\lambda; \Re\lambda > -N - c\}$ and there exist constants A and B such that*

$$\hat{u}_N^R(\lambda, z') \ll \frac{AB^N T^{\Re\lambda}}{(\Re\lambda + N + c)}(r' - X)^{-1}. \tag{4.6}$$

(2) $H_N(\lambda, z')$ is a polynomial in λ with degree $\le m - 1$ and there exist constants A and B such that

$$H_N(\lambda, z') \ll AB^N (\sum_{s=0}^{N-1}(|\lambda + s| + 1)^{m-1})(r' - X)^{-1}. \tag{4.7}$$

The proofs of Lemmas 4.2 and 4.3 are easy, so we omit them. Remember $\Lambda(n) = \{\lambda \notin Z(r', \varepsilon_0); -n + 1/2 - c \le \Re\lambda \le -n + 3/2 - c\}$.

Proposition 4.4. *There are constants A and B such that for $\lambda \in \Lambda(n)$, $n = 1, 2, \ldots$,*

$$\hat{u}(\lambda, z') \ll AB^n T^{\Re\lambda} \frac{\prod_{s=1}^{n}(|\lambda + n| + s)^m}{(n!)^{k+m}} \Psi^{(\frac{n}{\gamma}+nk)}(X). \tag{4.8}$$

Proof. $\hat{u}(\lambda, z')$ is holomorphic in $\{\Re\lambda > -c\}$ and

$$|\hat{u}(\lambda, z')| \le MT^{\Re\lambda + c}/(\Re\lambda + c) \le 2MT^{\Re\lambda + c}$$

in $\{\Re\lambda > -c + 1/2\}$. So there is a constant A such that

$$\hat{u}(\lambda, z') \ll AT^{\Re\lambda}\Psi^{(0)}(X) \quad for \ \lambda \in \Lambda(0). \tag{4.9}$$

We show (4.8) by induction on n. Let $\lambda \in \Lambda(1)$ and put $N = 1$ in (3.8). Then

$$\hat{u}(\lambda, z') = -\Phi(\lambda, z')^{-1}(u_1^R(\lambda, z') + T^\lambda H_1(\lambda, z'))$$

and it follows from Lemma 4.1(1) and Lemmas 4.2 and 4.3 that (4.8) holds for $\lambda \in \Lambda(1)$. Suppose $n \geq 2$ and $\lambda \in \Lambda(n)$. Put $N = n$ in (3.8). Then $\lambda + \ell \in \Lambda(n-\ell)$, $1 \leq \ell \leq n - 1$, and by the inductive hypothesis

$$\hat{u}(\lambda + \ell, z') \ll AB^{n-\ell}T^{\Re\lambda+\ell}\frac{\prod_{s=1}^{n-\ell}(|\lambda + n| + s)^m}{(n - \ell)!^{k+m}}\Psi^{(\frac{n-\ell}{\gamma}+(n-\ell)k)}(X),$$

so by Lemma 4.1(2)

$$\mathcal{L}_\ell(\lambda, z', \partial_{z'})\hat{u}(\lambda + \ell, z') \ll AB^{n-\ell}T^{\Re\lambda+\ell}\frac{\prod_{s=1}^{n-\ell}(|\lambda + n| + s)^m}{(n - \ell)!^{k+m}}$$
$$\times C_1^\ell\left(\sum_{\{\alpha;e_\alpha\leq\ell\}}|\lambda + \ell|^{\alpha_0}\Psi^{(\frac{n-\ell}{\gamma}+(n-\ell)k+|\alpha'|)}(X)\right). \tag{4.10}$$

Since

$$\frac{\Psi^{((n-\ell)k)}(X)}{(n - \ell)!^k} \ll \frac{\Psi^{((n-1)k)}(X)}{(n - 1)!^k}$$

and the relations $\frac{e_\alpha}{\gamma} \geq |\alpha| - k$ and $e_\alpha \leq \ell$ hold, we have $\frac{n-\ell}{\gamma} + (n - 1)k + |\alpha'| \leq \frac{n}{\gamma} - \alpha_0 + nk$, and from (4.2)

$$\frac{\Psi^{(\frac{n-\ell}{\gamma}+(n-\ell)k+|\alpha'|)}(X)}{(n - \ell)!^k} \ll \frac{\Psi^{(\frac{n-\ell}{\gamma}+(n-1)k+|\alpha'|)}(X)}{(n - 1)!^k} \ll \frac{\Psi^{(\frac{n}{\gamma}-\alpha_0+nk)}(X)}{(n - 1)!^k}. \tag{4.11}$$

Since $|\lambda + \ell| \leq |\lambda + n| + n - \ell < |\lambda + n| + n$,

$$\frac{(\prod_{s=1}^{n-\ell}(|\lambda + n| + s)^m)|\lambda + \ell|^{\alpha_0}}{(n - \ell)!^m} \leq \frac{(\prod_{s=1}^{n-1}(|\lambda + n| + s)^m)|\lambda + \ell|^{\alpha_0}}{(n - 1)!^m}$$
$$\leq C_3\frac{\prod_{s=1}^{n}(|\lambda + n| + s)^m}{(n - 1)!^m(|\lambda + n| + n)^{m-\alpha_0}} \leq C_3\frac{\prod_{s=1}^{n}(|\lambda + n| + s)^m}{(n - 1)!^m n^{m-\alpha_0}}. \tag{4.12}$$

Thus we have from (4.10), (4.11) and (4.12)

$$\mathcal{L}_\ell(\lambda, z', \partial_{z'})\hat{u}(\lambda + \ell, z')$$
$$\ll AB^{n-\ell}T^{\Re\lambda}C_4^\ell\frac{\prod_{s=1}^{n}(|\lambda + n| + s)^m}{(n - 1)!^{k+m}}\left(\sum_{\{\alpha;e_\alpha\leq\ell\}}n^{-m+\alpha_0}\Psi^{(\frac{n}{\gamma}-\alpha_0+nk)}(X)\right)$$
$$\ll AB^{n-\ell}T^{\Re\lambda}C_5^\ell\frac{\prod_{s=1}^{n}(|\lambda + n| + s)^m}{n!^m(n - 1)!^k}\Psi^{(\frac{n}{\gamma}+nk)}(X).$$

Choose B with $C_5/B \le 1/2$. Then by Lemma 4.2,

$$\Phi(\lambda, z')^{-1} \left(\sum_{\ell=1}^{n-1} \mathcal{L}_\ell(\lambda, z', \partial_{z'}) \hat{u}(\lambda + \ell, z') \right)$$

$$\ll AC'B^{n-1}T^{\Re\lambda} \frac{\prod_{s=1}^n (|\lambda + n| + s)^m}{n!^{m+k}} \Psi^{(\frac{n}{\gamma}+nk)}(X). \tag{4.13}$$

We have by Lemmas 4.2, 4.3 and $|\lambda + s| \le |\lambda + n| + n$ for $0 \le s \le n - 1$,

$$\Phi(\lambda, z')^{-1} \left(u_n^R(\lambda, z') + T^\lambda H_n(\lambda, z') \right)$$

$$\ll C_0 C_1^n T^{\Re\lambda} \left(\sum_{s=0}^{n-1} (|\lambda + s| + 1)^{m-1} \right) \Psi^{(0)}(X) \tag{4.14}$$

$$\ll C' C_2^n T^{\Re\lambda} (|\lambda + n| + n)^{m-1} \Psi^{(0)}(X).$$

So (4.8) holds for $\lambda \in \Lambda(n)$ by (3.8),(4.13) and (4.14). $\square$

Corollary 4.5. *There are constants A, B and a polydisk V' such that for $z' \in V'$ and $\lambda \in \Lambda(n)$,*

$$|\hat{u}(\lambda, z')| \le AB^n T^{\Re\lambda} \frac{\prod_{s=1}^n (|\lambda + n| + s)^m}{n!^m} \Gamma(\frac{n}{\gamma} + 1). \tag{4.15}$$

Proof. In the proof C means various constants. Let $|X| \le r/2$, Then $|\Psi^{(s)}(X)| \le 2^{s+1} s!/r^{s+1}$. Set $V' = \{z'; |z'| < r/2d\}$. Then for $z' \in V$,

$$|\Psi^{(\frac{n}{\gamma}+nk)}(X)| \le C^{n+1} \Gamma(n(\frac{1}{\gamma} + k) + 1) \le C^{n+1} \Gamma(\frac{n}{\gamma} + 1) n!^k \tag{4.16}$$

So it follows from (4.8) and (4.16) that the estimate (4.15) holds. $\square$

5   Proof of Theorems

In this section we give the proofs of Theorems 3.5 and 3.6. Before the proof we first give lemmas. Set

$$\vartheta_t^{-P} v(t) := \int_0^t \frac{dt_{p-1}}{t_{s-1}} \int_0^{t_{p-1}} \frac{dt_{p-2}}{t_{p-2}} \cdots \int_0^{t_1} \frac{v(t_0)}{t_0} dt_0$$

for $v(t)$ being smooth on $(0, T]$ and $|v(t)| \le A|t|^c$ $(c > 0)$.

Lemma 5.1. *Suppose that $|v(t)| \le A|t|^c$ with $c > 0$ on $(0, T]$. Let $\hat{v}(\lambda)$ be the Mellin transform of $v(t)$. Set for $p \in \mathbb{N}$,*

$$v_p(t) = \lim_{A \to +\infty} \frac{1}{2\pi i} \int_{c'-iA}^{c'+iA} t^{-\lambda} \frac{\hat{v}(\lambda)}{(-\lambda)^p} d\lambda \quad (-c < c' < 0). \tag{5.1}$$

Then $v_p(t) = \vartheta_t^{-P} v(t)$.

Proof. Set $w_p(t) := \vartheta_t^{-p} v(t)$. Then $|w_p(t)| \leq A'|t|^c$, so the Mellin transform $\hat{w}_p(\lambda)$ of $w_p(t)$ is holomorphic in $\{\lambda; \Re\lambda > -c\}$ and we have $\lambda\hat{w}_p(\lambda) = w_p(T)T^\lambda - \hat{w}_{p-1}(\lambda)$, hence

$$\hat{w}_p(\lambda) = \left(\sum_{i=0}^{p-1} \frac{(-1)^i w_{p-i}(T)}{\lambda^{i+1}}\right) T^\lambda + \frac{(-1)^p \hat{v}(\lambda)}{\lambda^p}.$$

Therefore we have for $0 < t < T$,

$$w_p(t) = \lim_{A\to+\infty} \frac{1}{2\pi i} \int_{c'-iA}^{c'+iA} t^{-\lambda} \hat{w}_p(\lambda) d\lambda \quad (-c < c' < 0)$$

$$= \lim_{A\to+\infty} \frac{1}{2\pi i} \int_{c'-iA}^{c'+iA} t^{-\lambda} \left((\sum_{i=0}^{p-1} \frac{(-1)^i w_{p-i}(T)}{\lambda^{i+1}})T^\lambda + \frac{(-1)^p \hat{v}(\lambda)}{\lambda^p}\right) d\lambda$$

$$= \lim_{A\to+\infty} \frac{1}{2\pi i} \int_{c'-iA}^{c'+iA} t^{-\lambda} \frac{\hat{v}(\lambda)}{(-\lambda)^p} d\lambda = v_p(t).$$

$\square$

Let $u(z) \in \mathcal{O}_{temp,c}(U(\theta))$ be a solution of $L(z, \partial_z)u(z) = 0$. By considering $z_0^{-c+\varepsilon} u(z)$ ($\varepsilon > 0$), we may assume that $|u(z)| \leq C|z_0|^c$ with $c > 0$ on $U(\theta')$ $(0 < \theta' < \theta)$.

Lemma 5.2. *Let $\lambda \in \Lambda(n)$. Then there exist constants A and B such that*

$$\frac{|\hat{u}(\lambda, z')|}{|\lambda|^{mn+2}} \leq \frac{AB^n T^{\Re\lambda}\Gamma(\frac{n}{\gamma}+1)}{n!^m|\lambda|^2}. \tag{5.2}$$

Proof. Let us return to the estimate (4.15). For $\lambda \in \Lambda(n)$,

$$\frac{\prod_{s=1}^n(|\lambda+n|+s)^m}{|\lambda|^{mn+2}} = \frac{1}{|\lambda|^2}\prod_{s=1}^n\left(\frac{|\lambda+n|+s}{|\lambda|}\right)^m \leq \frac{C^n}{|\lambda|^2},$$

so (5.2) is valid.

$\square$

Set $w_p(t, z') := \vartheta_t^{-p} u(t, z')$. Then

$$w_p(t, z') = \lim_{A\to+\infty} \frac{1}{2\pi i} \int_{c'-iA}^{c'+iA} t^{-\lambda} \frac{\hat{u}(\lambda, z')}{(-\lambda)^p} d\lambda$$

by Lemma 5.1. Putting $p = mn + 2$, by Lemma 5.2 and the deformation of the integration path we have $w_{mn+2}(t, z') = w_{mn+2,C_n}(t, z') + w^*_{mn+2}(t, z')$,

$$w_{mn+2,C_n}(t, z') = \frac{1}{2\pi i} \int_{C_n} \frac{t^{-\lambda}\hat{u}(\lambda, z')}{(-\lambda)^{mn+2}} d\lambda,$$

$$w^*_{mn+2}(t, z') = \frac{1}{2\pi i} \int_{\Re\lambda=-\sigma_n} \frac{t^{-\lambda}\hat{u}(\lambda, z')}{(-\lambda)^{mn+2}} d\lambda. \tag{5.3}$$

Here the constant σ_n and the contour C_n are those defined in Section 3. C_n is in $\{\lambda; -\sigma_n < \Re\lambda < 0\}$ and encloses all the poles of $\hat{u}(\lambda, z')$ in $\{\Re\lambda > -\sigma_n\}$, so $\lambda = 0$ is outside of C_n.

Lemma 5.3. *There exist constants A and B such that, for $t \in (0, T)$,*

$$|w_{mn+2}(t, z') - w_{mn+2,C_n}(t, z')| \leq \frac{AB^n\Gamma(\frac{n}{\gamma} + 1)}{n!^m} t^{\sigma_n}. \tag{5.4}$$

Proof. Let us estimate $w^*_{mn+2}(t, z')$. By Lemma 5.2,

$$|w^*_{mn+2}(t, z')| \leq \frac{AB^n\Gamma(\frac{n}{\gamma} + 1)}{n!^m} \int_{\Re\lambda=-\sigma_n} \frac{(t/T)^{-\Re\lambda}}{|\lambda|^2}|d\lambda| \leq \frac{A_1 B_1^n\Gamma(\frac{n}{\gamma} + 1)}{n!^m} t^{\sigma_n}.$$

$\square$

In the preceding we have discussed the Mellin transform of $u(z_0, z')$ on the real axis. Now let us consider the Mellin transform of $u(z_0, z')$ on $\arg z_0 = \phi$ $(|\phi| < \theta)$. Set

$$w_p(z) = \vartheta_{z_0}^{-p} u(z) := \int_0^{z_0} \frac{d\tau_{p-1}}{\tau_{p-1}} \int_0^{\tau_{p-1}} \frac{d\tau_{p-2}}{\tau_{p-2}} \cdots \int_0^{\tau_1} \frac{v(\tau_0, z')}{\tau_0} d\tau_0.$$

Then $w_p(z) \in \mathcal{O}(U(\theta))$, $|w_p(z)| \leq C|z_0|^c$ on $U(\theta')$. Define

$$\hat{u}^\phi(\lambda, z') = \int_0^T t^{\lambda-1} u(te^{i\phi}, z')dt, \tag{5.5}$$

which is holomorphic in $\{\lambda; \Re\lambda > -c\}$. Obviously, by putting $\phi = 0$, $\hat{u}^0(\lambda, z') = \hat{u}(\lambda, z')$. Set

$$w_p^\phi(te^{i\phi}, z') = \frac{1}{2\pi i} \int_{\Re\lambda=c'} \frac{t^{-\lambda}\hat{u}^\phi(\lambda, z')}{(-\lambda)^p} d\lambda \quad for \ t \in (0, T), \tag{5.6}$$

$$w_{p,C}^\phi(z) = \frac{1}{2\pi i} \int_C \frac{z_0^{-\lambda}e^{i\phi\lambda}\hat{u}^\phi(\lambda, z')}{(-\lambda)^p} d\lambda, \tag{5.7}$$

where $-c < c' < 0$ and C is a closed contour in $\{\lambda; \Re\lambda < 0\}$.

Lemma 5.4. (1) $\hat{u}^0(\lambda, z') - e^{i\phi\lambda}\hat{u}^\phi(\lambda, z')$ *is an entire function in λ.*
(2) $w_{p,C}^\phi(z)$ *does not depend on ϕ, that is, $w_{p,C}^\phi(z) = w_{p,C}^0(z)$.*
(3) *Let $\arg z_0 = \phi$ and set $z_0 = te^{i\phi}$, $t \in (0, T)$. Then $\vartheta_{z_0}^{-p} u(z) = w_p^\phi(te^{i\phi}, z')$.*

Proof. By Cauchy's Theorem,

$$\hat{u}^{\phi}(\lambda, z') = e^{-i\phi\lambda} \int_0^{Te^{i\phi}} z_0^{\lambda-1} u(z_0, z') dz_0$$

$$= e^{-i\phi\lambda} \left(\int_0^T z_0^{\lambda-1} u(z_0, z') dz_0 + \int_T^{Te^{i\phi}} z_0^{\lambda-1} u(z_0, z') dz_0 \right),$$

so the assertion (1) holds. The assertion (2) follows from (1). Let $\arg z_0 = \phi$. Then

$$w_1^{\phi}(te^{i\phi}, z') = \int_0^t \frac{u(t_0 e^{i\phi}, z')}{t_0} dt_0 = \int_0^{te^{i\phi}} \frac{u(\tau_0, z')}{\tau_0} d\tau_0 = \vartheta_{z_0}^{-1} u(z).$$

By repeating this calculation $w_p^{\phi}(te^{i\phi}, z') = \vartheta_{z_0}^{-p} u(z)$ holds for $p \in \mathbb{N}$. $\square$

The proof of Theorem 3.5

Set $w_{p,C_n}(z) = w_{p,C_n}^{\phi}(z)$ which is well defined by Lemma 5.4. Put $p = mn + 2$ and $W = \{|z_0| < T_0\} \times V', 0 < T_0 < T$ and V' being a small polydisk centered at $z' = 0$ in $\mathbb{C}^d$. Then it follows from Lemma 5.3 that there exist constants A and B such that for $z \in W$ with $\arg z_0 = \phi$,

$$|\vartheta_{z_0}^{-mn-2} u(z) - w_{mn+2,C_n}(z)| \leq \frac{AB^n \Gamma(\frac{n}{\gamma} + 1)}{n!^m} |z|^{\sigma_n}. \tag{5.8}$$

Let $0 < \theta' < \theta_0 < \theta$ and put $\phi = \pm\theta_0$. Then the estimate (5.8) holds for $\arg z_0 = \pm\theta_0$, so by the maximal principle (5.8) holds on $z \in W(\theta_0)$. Hence it follows from Lemma 3.2 that for $z \in V(\theta')$, $V = \{|z_0| < T'\} \times V', T' < T_0$,

$$|u(z) - \vartheta_{z_0}^{mn+2} w_{mn+2,C_n}(z)| \leq A_1 B_1^n \Gamma(\frac{n}{\gamma} + 1)|z|^{\sigma_n}, \tag{5.9}$$

and we have Theorem 3.5 by $\vartheta_{z_0}^{mn+2} w_{mn+2,C_n}(z) = u_n(z)$.

The proof of Theorem 3.6

It follows from the assumption on the bound of $u(z)$ and Corollary 3.4 that $\hat{u}(\lambda, z')$ is an entire function in λ. So $u_n(z) \equiv 0$. Hence by Theorem 3.5

$$|u(z)| \leq AB^n \Gamma(\frac{n}{\gamma} + 1)|z|^{\sigma_n} \leq AB^n \Gamma(\frac{n}{\gamma} + 1)|z|^{n-2/3+a}$$

holds in $V(\theta')$ for all $n \in \mathbb{N}$. This implies $|u(z)| \leq C \exp(-c|z_0|^{-\gamma})$.

Remark 5.5. In this paper we only consider homogeneous solutions. But we can treat solutions of $L(z, \partial_z)u(z) = f(z) \in \mathcal{O}_{temp}(U(\theta))$. By using the relation (3.8), we add the information of $\hat{f}(\lambda, z')$ to that of $\hat{u}(\lambda, z')$ and get the behavior of $u(z)$ near the singularities $\{z_0 = 0\}$. In particular the conclusions of Theorems 3.3, 3.5 and 3.6 are valid provided $|f(z)| \leq Ce^{-c|z_0|^{-\gamma}}$ $(c > 0)$ holds.

Let us apply the results to the examples in Section 1.

(1) Let $P(z, \partial_z)$ be a linear partial differential operator of the form

$$P(z, \partial_z) = \partial_{z_0}^k + \sum_{\substack{|\alpha| \leq m \\ \alpha_0 < k}} a_\alpha(z) \partial_z^\alpha, \quad m > k. \tag{5.10}$$

The asymptotic behavior of singular solutions of $P(z, \partial_z)u(z) = f(z)$ are studied in [10]. By Theorem 2.1 and by investigating the Mellin transform of solutions of inhomogeneous equations as remarked above we can give another proof of the following main result in [10].

Theorem 5.6. ([10]) *Let $u(z) \in \mathcal{O}_{(\gamma)}(U(\theta))$ be a solution of $P(z, \partial_z)$ $u(z) = f(z) \in \mathcal{A}sy_{\{\gamma\}}(U(\theta))$. Then $u(z) \in \mathcal{A}sy_{\{\gamma\}}V(\theta)$ for a neighborhood V of $z = 0$.*

We show Theorem 5.6 in [10] by a completely different method from this paper, not using the Mellin transform but estimating derivatives of solutions, and refer to [10] for the details.

(2) Let $P(z, \partial_z)$ be an m-th operator of Fuchsian type in the sense of Baouendi–Goulaouic. The singular solutions of $P(z, \partial_z)u(z) = 0$ are investigated in Mandai [5] and Tahara [14]. They determined the structure of singular homogeneous solutions. In this case $\gamma = +\infty$ and it follows from Theorem 3.5 that $u(z) = \lim_{n \to \infty} u_n(z)$, which is another characterization of singular homogeneous solutions.

(3) Let $L(z, \partial_z) = z_0 \partial_{z_0} - a(z) + z_0^j c(z) \partial_{z_1}^m$, $j \geq 1, m \geq 1$ and $c(0, z_1) \neq 0$. and $u(z) \in \mathcal{O}_{\{\gamma\}} U((\theta))$, $\gamma = j/(m-1)$, be a solution of $L(z, \partial_z)u(z) = 0$. Let V be a small polydisk, c a constant with $c < \Re a(0, z')$ and $0 < \theta' < \theta$. Then $u(z) \in \mathcal{O}_{temp} V((\theta))$ and

$$|u(z) - z_0^{a(0,z')}\left(\sum_{p=0}^{n-1}\left(\sum_{q=0}^{mp} u_{p,q}(z')(\log z_0)^q\right)z_0^p\right)| \leq AB^n \Gamma\left(\frac{n}{\gamma}+1\right)|z_0|^{n+c}$$

holds for $z \in V(\theta')$.

References

[1] Baouendi, M.S. and Goulaouic, C., Cauchy problems with characteristic initial hypersurface, *Comm. Pure Appl. Math.* **26** (1973), 455–475.

[2] Y. Hamada, The singularities of the solutions of the Cauchy problem, *Publ. RIMS, Kyoto Univ.* **5** (1969), 21–40.

[3] Y. Hamada, J. Leray, et C. Wagschal, Système d'équation aux derivées partielles à caractéristique multiples; problème de Cauchy ramifié; hyperbolicité partielle, *J. Math. Pures Appl.* **55** (1976), 297–352.

[4] J. Leray, Uniformisation de la solution du problème linéare analytique de Cauchy près de la variété qui porte les données de Cauchy (Problème de Cauchy I), *Bull. Soc. Math. de France* **85** (1957) 389–429.

[5] T. Mandai, The method of Frobenius to Fuchsian partial differential equations, *J. Math. Soc. Japan* **52** (2000), 645–672.

[6] M. Kashiwara et P. Schapira, Problème de Cauchy pour les systemes microdifferentiels dans le domain complexe, *Inv. Math.* **46** (1978), 17–38.

[7] S. Ōuchi, Asymptotic behaviour of singular solutions of linear partial differential equations in the complex domain, *J. Fac. Sci. Univ. Tokyo* **27** (1980), 1–36.

[8] S. Ōuchi, An integral representation of singular solutions of linear partial differential equations in the complex domain, *J. Fac. Sci. Univ. Tokyo* **27** (1980), 37–85.

[9] S. Ōuchi, Existence of singular solutions and null solutions for linear partial differential operators, *J. Fac. Sci. Univ. Tokyo* **32** (1985), 457–498.

[10] S. Ōuchi, Singular solutions with asymptotic expansion of linear partial differential equations in the complex domain, *Publ. RIMS Kyoto Univ.* **34** (1998), 291–311.

[11] S. Ōuchi, Growth property and slowly increasing behavior of singular solutions of linear partial differential equations in the complex domain, *J. Math. Soc. Japan* **52** (2000), 767–792.

[12] S. Ōuchi, Asymptotic expansion of singular solutions and the characteristic polygon of linear partial differential equations in the complex domain, *Publ. RIMS Kyoto Univ.* **36** (2000), 457–482.

[13] J. Persson, Singular holomorphic solutions of linear partial differential equations with holomorphic coefficients and nonanalytic solutions with analytic coefficients, *Asterisque 89-90, analytic solutions of partial differential equations (Trento 1981)*, Soc. Math. France, 233–247.

[14] H. Tahara, Fuchsian type equations and Fuchsian hyperbolic equations, *Japan. J. Math.* **5** (1979) 245–347.

[15] C. Wagschal, Problème de Cauchy analytique à données méromorphes, *J. Math. Pures Appl.* **51** (1972), 375–397.

Sunao Ōuchi
Department of Mathematics
Sophia University
Kioicho Chiyoda-ku, Tokyo 102-8554, Japan
ouchi@mm.sophia.ac.jp

Systèmes Uniformément Diagonalisables, Dimension Réduite et Symétrie II

Jean Vaillant

À la mémoire de Jean Leray

ABSTRACT Let there be a strong hyperbolic matrix. We state the following result. If the reduced dimension is more than a specified integer, there is a linear basis in which the matrix is symmetric.

Introduction

Nous considérons un système linéaire d'opérateurs différentiels du premier ordre :

$$a(D) = I D_0 + \sum_{k=1}^{k=n} a_k D_k$$

où I est la matrice identité d'ordre m et les a_k sont des matrices carrées réelles d'ordre m.

Soit: $a(\xi) = I\xi_0 + \sum_{k=1}^{k=n} a_k \xi_k$, le symbole principal de $a(D)$. On a défini de façon intrinsèque la dimension réduite de a [5], soit $d(a)$, et ses propriétés ont été établies dans [1] et [6]: $d(a)$ est invariante par les changements linéaires de coordonnées de $\mathbb{R}^{n+1}$ et par les changements linéaires de coordonnées de l'espace $\mathbb{R}^m$ des valeurs.

$d(a) = \text{rang}(a) = $ dimension du sous espace vectoriel de $M(m, \mathbb{R})$ engendré par: $I, a_1, \ldots, a_k, \ldots, a_n$. Si on note $\tilde{a}$ l'application: $\xi \longrightarrow {}^t a(\xi)$, alors $d(a) = d(\tilde{a})$.

Nous avons aussi la définition [6]: a est présymétrique par rapport à N, si et seulement si: il existe une base de $\mathbb{R}^{n+1}$ de premier vecteur N et une base de $\mathbb{R}^m$ telles que dans ces bases les matrices $a_j^i(\xi)$ sont symétriques pour tout ξ. a présymétrique équivaut à $\tilde{a}$ présymétrique.

On dira aussi que la matrice (a_j^i) est présymétrique s'il existe T inversible telle que $T^{-1} a_j^i T$ soit symétrique.

Nous avons un résultat précédent.

Théorème 0 [6]. *Si $a(D)$ est fortement hyperbolique par rapport à N, si $m = 4$ et si $d(a) \geq 4(4+1)/2 - 2 = 8$ alors: $a(\xi)$ est présymétrique.*

Remarques 1°) Si on suppose: $m \geq 3, d(a) \geq m(m+1)/2 - 1$ et a diagonalisable par rapport à N, la présymétrie est obtenue dans [1].

2°) Si $m = 3$, il est montré dans [3] que la condition sur $d(a)$ ne peut être améliorée.

3°) Pour $m = 2$, hyperbolicité forte et présymétrie sont équivalentes [4].

Nous voulons prouver pour $m = 5$, le cas général étant analogue, le suivant.

Théorème. *Si $a(D)$ est fortement hyperbolique par rapport à N, si $m \geq 4$ et si $d(a) \geq m(m+1)/2 - 2$, alors a est présymétrique par rapport à N.*

Nous expliquons le cas $m = 5$, qui est essentiel: les résultats sont annoncés dans [9]. La preuve se décompose en 3 parties. Les deux premières parties ont été étudiées dans [8]. La 3ème partie, la plus difficile est étudiée ici. Les calculs du cas général seront publiés ailleurs.

Le cas des coefficients variables correspondant à la remarque 1 a été considéré dans [2]. Le cas des coefficients variables correspondant au théorème considéré ici est en préparation par les mêmes auteurs.

1   Rappels. Plan de la démonstration. Premiers lemmes

Nous notons aussi Φ^i_j les éléments de la matrice a. Nous savons [5] et [1] que, grâce à la diagonalisabilité de a nous pourrons supposer que

i) pour tout $p < q$, $\Phi^p_q \in$ espace engendré par $\{\Phi^i_j, i > j\} = V$;

ii) pour $1 \leq i \leq n$, $\Phi^i_i(\xi) - (\xi_0 + \psi_i(\xi')) \in V, \xi' = (\xi_1, \ldots, \xi_n)$, ainsi: $d(a) =$ dim. espace engendré par $\{V, \xi_0 + \psi_i\}$; on peut supposer qu'un ψ_i est nul.

On peut supposer que $d(a) = m(m+1)/2 - 2$; les cas $d(a) = m(m+1)/2$ et $d(a) = m(m+1)/2 - 1$ ayant été considérés dans [5] et [1].

Nous avons distingué trois cas:

I $\dim V = \dfrac{m(m-1)}{2} = \dfrac{m(m+1)}{2} - m$.

II $\dim V = \dfrac{m(m-1)}{2} - 1 = \dfrac{m(m+1)}{2} - m - 1$.

III $\dim V = \dfrac{m(m-1)}{2} - 2 = \dfrac{m(m+1)}{2} - m - 2$. Les formes de la diagonale sont linéairement indépendantes; deux formes de V dépendent linéairement des autres formes de V.

C'est le cas étudié ici.

Lemma 1.1. [8] *i) Si la matrice 5×5:*

$$a(\xi') = \begin{pmatrix} & & & & * \\ & \beta(\xi') & & & * \\ & & & & * \\ & & & & * \\ 0\,0\,0\,0 & & & & \psi'(\xi') \end{pmatrix}$$

est uniformément diagonalisable, alors la matrice 4×4: $\beta(\xi')$ est uniformément diagonalisable.

ii) Si la matrice

$$a(\xi') = \begin{pmatrix} & & & * & * \\ & \beta(\xi') & & * & * \\ & & & * & * \\ 0 & 0 & 0 & & \\ 0 & 0 & 0 & \beta'(\xi') \end{pmatrix}$$

est diagonalisable, alors la sous-matrice $\beta(\xi')$ est diagonalisable.

Lemma 1.2. *b est une matrice 4×4, de la forme:*

$$b = \begin{pmatrix} \psi_1 & & \Phi_j^i(\xi') \\ \psi_2 & & i < j \\ \Phi_j^i(\xi') & & \psi_3 \\ i > j & & 0 \end{pmatrix}$$

où $\xi' = (\ldots, \xi_l^k, \ldots)$; $\psi_1, \psi_2, \psi_3, \ldots, \xi_l^k, \ldots$ sont des variables indépendantes et les Φ_j^i sont des formes linéaires en ξ'. On suppose que b est présymétrique: il existe T telle que $T^{-1}bT$ est symétrique. On obtient qu'il existe H diagonale définie positive telle que

$$\tilde{b}H = H\,{}^tb.$$

Preuve. On pose: $TT^t = H$, [1]; H est symétrique définie positive et on a aisément: $bH - H\,{}^tb = 0$; écrivons l'élément de la $4^{\text{ème}}$ ligne, $3^{\text{ème}}$ colonne, on obtient;

$$h_{34}(\psi_4 - \psi_3) + \chi(\xi') = 0$$

où χ est linéaire en ξ'; on a donc:

$$h_{34} = 0.$$

En écrivant les éléments de la $i^{\text{ère}}$ ligne, $j^{\text{ème}}$ colonne, $i > j$, on obtient que H est diagonale; on pose:

$$H = (h_1, h_2, h_3, h_4), \quad h_i > 0. \qquad \square$$

Lemma 1.3. *On considère la matrice:*

$$a(\xi) = \xi_0 I + \begin{pmatrix} 1 & & & & \\ & 0 & & & \mathbf{0} \\ & & 0 & & \\ & & & 0 & \\ \mathbf{0} & & & & 0 \end{pmatrix}\xi_1 + \begin{pmatrix} 0 & \beta_1 & \beta_2 & \beta_3 & \beta_4 \\ \beta_5 & \gamma_1 & & & \\ \beta_6 & & \gamma_2 & & \mathbf{0} \\ \beta_7 & & & \gamma_3 & \\ \beta_8 & & \mathbf{0} & & \gamma_4 \end{pmatrix}\xi_2,$$

les γ_i sont différents 2 à 2.

On suppose que $a(\xi)$ est diagonalisable, alors, on a: $\text{sign } \beta_i = \text{sign } \beta_{i+4}$, *pour* $1 \leq i \leq 4$.

Preuve. C'est une généralisation de [3] et [7].   □

Conséquence 1.4. *Soit:*

$$a(\xi) = \xi_0 I + \begin{pmatrix} 1 & & & \\ & 0 & & \Large 0 \\ & & 0 & \\ & & & 0 \\ \Large 0 & & & 0 \end{pmatrix} \xi_1$$

$$+ \begin{pmatrix} 0 & \beta_1(\xi'') & \beta_2(\xi'') & \beta_3(\xi'') & \beta_4(\xi'') \\ \beta_5(\xi'') & & & & \\ \beta_6(\xi'') & & \Gamma(\xi'') & & \\ \beta_7(\xi'') & & & & \\ \beta_8(\xi'') & & & & \end{pmatrix},$$

$\xi'' = (\dots, \xi_l^k, \dots)$. $\Gamma(\xi'')$ *est une matrice symétrique de formes linéaire en* ξ'', $U(\xi'')$ *est une matrice orthogonale qui diagonalise* $\Gamma(\xi'')$, $\forall \xi''$:

$$^tU(\xi'')\Gamma(\xi'')U(\xi'') = D(\xi''),$$

$D(\xi'')$ *est diagonale; on considère des* ξ'' *telle que les éléments de* $D(\xi'')$ *soient distincts; alors on a:*

$$\text{sign} \sum_i \beta_i U_j^i(\xi'') = \text{sign} \sum_i \beta_{i+4} U_j^i(\xi''), \quad \forall \xi''.$$

Preuve. [6] On remplace ξ'' par $s\xi''$, $s \in \mathbb{R}$ et on se ramène au lemme précédent.   □

Proposition 1.5. *Par un choix convenable d'une base de* $E = \mathbb{R}^{n+1}$, *de premier vecteur* $N = (1, 0, \dots, 0)$, *et d'une base de* $F = \mathbb{R}^m$, *les matrices des application linéaires du cas III ou leurs transposées ont l'une des formes suivantes:*

i)

$$a_j^i(\xi) = \begin{pmatrix} \xi_0 + \psi_1 & & & & \\ \Phi_1^2(\xi'') & \xi_0 + \psi_2 & & \Phi_j^i(\xi''), i < j & \\ \Phi_1^3(\xi'') & \xi_2^3 & \xi_0 + \psi_3 & & \\ \xi_1^4 & \xi_2^4 & \xi_3^4 & \xi_0 + \psi_4 & \\ \xi_1^5 & \xi_2^5 & \xi_3^5 & \xi_4^5 & \xi_0 \end{pmatrix},$$

$\{\xi_j^i, i \leq j < i < 5, (i, j) \neq (2, 1), (i, j) \neq (3, 1)\}$ *et* $\psi_1, \psi_2, \psi_3, \psi_4, \xi_0$ *sont les nouvelles coordonnées:* $\xi'' = (\dots, \xi_j^i, \dots), 1 \leq j < i \leq 5, (i, j) \neq (2, 1), (i, j) \neq (3, 1)$.

ii)

$$a^i_j(\xi) = \begin{pmatrix} \xi_0 + \psi_1 & & & & \\ \Phi^2_1(\xi^5_1, \xi^5_2) & \xi_0 + \psi_2 & & \Phi^i_j(\xi''), i < j & \\ \xi^3_1 & \xi^3_2 & \xi_0 + \psi_3 & & \\ \xi^4_1 & \xi^4_2 & \Phi^4_3(\xi^5_3, \xi^5_4) & \xi_0 + \psi_4 & \\ \xi^5_1 & \xi^5_2 & \xi^5_3 & \xi^5_4 & \xi_0 \end{pmatrix},$$

$\xi'' = (\dots, \xi^i_j, \dots), 1 \le j < i \le 5, (i, j) \ne (2, 1), (i, j) \ne (4, 3).$ *On note:*

$$\Phi^i_j(\xi'') = \sum_{l<k} c^{i\ k}_{j\ l} \xi^l_k \quad ; \quad c^i_j = c^{i\ i}_{j\ j}.$$

Preuve. Dans le cas III, deux formes linéaires en dessous de la diagonale, dépendent linéairement des $m(m + 1)/2 - 2 = 8$ autres.

a) Nous montrons que lorsqu'une des formes dépendantes est sur une même ligne ou une même colonne que l'autre ou la symétrique de l'autre par rapport à la 1ère diagonale, on se ramène au cas i).

Considérons d'abord les matrices où les 2 formes dépendantes sont dans une même colonne.

Si Φ^2_1 et Φ^4_1 sont les formes dépendantes, on échange la 3ème et la 4ème ligne, aussi que la 3ème et 4ème colonne, ce qui revient à faire un changement de base de F, on obtient:

$$\begin{pmatrix} \xi_0 + \psi_1 & \Phi^1_2 & & \dots & \\ \Phi^2_1 & \xi_0 + \psi_2 & & \dots & \\ \Phi^4_1 & \xi^4_2 & \xi_0 + \psi_3 & \xi^4_3 & \\ \xi^3_1 & \xi^3_2 & \Phi^3_4 & \xi_0 + \psi_3 & \\ \xi^5_1 & \xi^5_2 & \xi^5_4 & \xi^5_3 & \xi_0 \end{pmatrix};$$

si $c^3_4 \ne 0$, par changement de coordonnées dans E, on remplace Φ^3_4 par la nouvelle coordonnée $\xi^{4'}_3$ et ξ^4_3 devient $\Phi^{3'}_4$; aux notations près, on est ramené au cas i).

Pour montrer que $c^3_4 \ne 0$, on pose: $\xi^5_1 = \xi^5_2 = \xi^5_3 = \xi^5_4 = 0$; par le lemme 1.1, comme la matrice:

$$b = \begin{pmatrix} \xi_0 + \psi_1 & & \dots & \\ \Phi^2_1 & \xi_0 + \psi_2 & \dots & \\ \xi^3_1 & \xi^3_2 & \xi_0 + \psi_3 & \Phi^3_4 \\ \Phi^4_1 & \xi^4_2 & \xi^4_3 & \xi_0 + \psi_4 \end{pmatrix}$$

est uniformément diagonalisable et de dimension réduite 8, elle est présymétrique; du lemme 1.2 on déduit qu'il existe: $H = (h_1, h_2, h_3, h_4)$ définie positive telle que:

$$bH = H^t b$$

et l'on a:

$$h_3 = c^3_4 h_4,$$

d'où:

$$c_4^3 > 0.$$

Si Φ_1^2, Φ_1^5 sont les formes dépendantes, on échange les $4^{\text{ème}}$ et $5^{\text{ème}}$ lignes et colonnes et on se ramène au cas où Φ_1^2 et Φ_1^4 sont dépendantes, pourvu que: $c_5^4 \neq 0$; on montre que $c_5^4 = 0$ est impossible: si $c_5^4 = 0$, on pose: $\xi_1^3 = \xi_1^4 = \xi_2^3 = \xi_2^4 = \xi_2^5 = \xi_3^4 = \xi_3^5 = 0$; $\xi_4^5 = 1$; $\psi_4 = 0$; on obtient:

$$\begin{pmatrix} \xi_0 + \psi_1 & & & \cdots & \\ c_{15}^{24} & \xi_0 + \psi_2 & & \cdots & \\ 0 & 0 & \xi_0 + \psi_3 & \cdots & \\ 0 & 0 & 0 & \xi_0 & 0 \\ c_{15}^{54} & 0 & 0 & 1 & \xi_0 + \psi_5 \end{pmatrix};$$

on choisit $\psi_5(\psi_1, \psi_2, \psi_3)$ de sorte que 0 soit zéro double du déterminant; alors les mineurs d'ordre 4 sont nuls; celui obtenu en rayant la $4^{\text{ème}}$ ligne et $5^{\text{ème}}$ colonne est identiquement nul ce qui est impossible.

De même, si Φ_1^3, Φ_1^4 sont dépendantes, ou Φ_1^3, Φ_1^5 sont dépendantes, on se ramène à un cas précédent.

On considère la $2^{\text{ème}}$ colonne. Si Φ_2^3, Φ_2^4 sont dépendantes, on se ramène à (Φ_1^3, Φ_1^4) dépendantes par échanger des deux premières lignes et colonne. (Φ_2^3, Φ_2^5) dépendantes, se ramène à (Φ_2^3, Φ_2^4) dépendantes.

De même (Φ_2^4, Φ_2^5) dépendantes se ramène à (Φ_2^3, Φ_2^5) dépendantes et alors (Φ_1^4, Φ_1^5) dépendantes se ramène à (Φ_2^4, Φ_2^5) dépendantes. (Φ_3^4, Φ_3^5) dépendantes se ramène à (Φ_2^4, Φ_2^5) dépendantes. Si Φ_1^2 et Φ_2^3 sont dépendantes, on échange les $1^{\text{ère}}$ et $2^{\text{ème}}$ lignes et colonne et on se ramène au cas où Φ_1^2 et Φ_1^3 sont dépendantes. On procède de même dans les cas où (Φ_1^2, Φ_2^4) ou (Φ_1^2, Φ_2^5) ou (Φ_2^3, Φ_3^4) ou (Φ_2^3, Φ_3^5) ou (Φ_3^4, Φ_4^5) sont dépendantes. Si (Φ_1^3, Φ_2^3) sont dépendantes, on se ramène à (Φ_2^3, Φ_3^3) dépendantes; si (Φ_1^3, Φ_3^3) sont dépendantes, on se ramène au précédent; si (Φ_2^4, Φ_3^4) sont dépendantes, on se ramène à (Φ_3^4, Φ_4^4) dépendantes, et si (Φ_1^4, Φ_4^5) sont dépendantes on se ramène à un cas précédent.

Les couples de 2 formes dépendantes symétriques d'un couple précédent par rapport à la $2^{\text{ème}}$ diagonale se ramènent aux précédents, en considérant les matrices transposées et en inversant l'ordre des vecteurs de base F.

b) Les autres cas se ramènent à ii). Si (Φ_1^2, Φ_3^3) sont dépendantes, on se ramène à (Φ_1^2, Φ_3^4) dépendantes, par échange de lignes et de colonnes; si (Φ_1^2, Φ_4^5) sont dépendantes, on se ramène encore au cas précédent; (Φ_1^3, Φ_2^4) dépendantes se ramène à (Φ_1^2, Φ_3^4) dépendantes; (Φ_1^3, Φ_2^5) dépendantes se ramène au cas précédent; (Φ_1^4, Φ_2^3) dépendantes se ramène à (Φ_1^3, Φ_2^4) dépendantes; (Φ_1^4, Φ_2^5) dépendantes se ramène à (Φ_2^5, Φ_4^5) dépendantes, si $c_{15}^{44} \neq 0$ et à (Φ_1^3, Φ_2^5) dépendantes sinon; (Φ_1^5, Φ_2^3) dépendantes se ramène à (Φ_1^4, Φ_2^3) dépendantes; (Φ_1^5, Φ_2^4) dépendantes se ramène à (Φ_2^4, Φ_3^4) dépendantes si $c_{14}^{53} \neq 0$ et à (Φ_2^4, Φ_1^5) dépendantes sinon.

Les cas des couples symétriques par rapport à le $2^{\text{ème}}$ diagonale se ramènent aux précédents.

Dans le cas ii), on peut supposer que Φ_1^2 ne dépend que de ξ_1^5 et ξ_2^5 sinon on se ramène à un cas du i); de même on peut supposer que Φ_3^4 ne dépend que de ξ_3^5 et ξ_4^5. $\qquad\square$

2   Preuve du théorème; cas III i)

On suppose en fait (Φ_2^3, Φ_2^4) dependantes.

Lemma 2.1. $c_{42}^{31} = c_{43}^{31} = c_{44}^{31} = 0$; $c_{42}^{11} = c_{43}^{11} = c_{44}^{13} = 0$; $c_{32}^{11} = c_{34}^{11} = c_{34}^{13} = 0$; $c_{23}^{11} = c_{24}^{11} = c_{24}^{13} = 0$; $c_{31}^{2k} = k_3^2 c_{21}^{3k}$, pour $1 \leq k < l \leq 4$; $(k, l) \neq (2, 3)$, $(k, l) \neq (2, 4)$, $c_{41}^{2k} = k_4^2 c_{21}^{4k}$, pour $1 \leq k < l \leq 4$; $(k, l) \neq (2, 3)$, $(k, l) \neq (2, 4)$, $c_4^1 = c_2^1 k_4^2$; $k_4^2 = k_3^2 c_4^3$; $c_3^1 = c_2^1 k_3^2$, $c_j^i > 0$, $k_j^i > 0$.

Preuve. On pose $\xi_1^5 = \xi_2^5 = \xi_3^5 = \xi_4^5 = 0$; la matrice b, 4×4 obtenue en rayant la $5^{\text{ème}}$ ligne et la $5^{\text{ème}}$ colonne de (a_j^i) est uniformément diagonalisable (Lemme 1.1); sa dimension réduite est 8; elle est donc présymétrique; il existe donc (Lemme 1.2) une matrice diagonale définie positive H telle que:

$$bH = H^t \bar{b}. \qquad (1)$$

On explicite les éléments des ($i^{\text{ème}}$ ligne, $j^{\text{ème}}$ colonne), $i > j$ de (1); on obtient:

$$i = 4, j = 3 : c_{42}^{31} = c_{43}^{31} = c_{44}^{31} = 0; \ c_4^3 = \frac{h_3}{h_4} > 0;$$

$$i = 4, j = 2 : c_{4l}^{2k} = k_4^2 c_{2l}^{4k}, 1 \leq k < l \leq 4, (k, l) \neq (2, 3), (k, l) \neq (2, 4),$$

$$c_4^2 = \frac{h_2}{h_4};$$

$$i = 4, j = 1 : c_{42}^{11} = c_{33}^{11} = c_{44}^{13} = 0; \ c_4^1 = \frac{h_1}{h_4} > 0;$$

$$i = 3, j = 2 : c_{3l}^{2k} = k_3^2 c_{2l}^{3k}, i \leq k < l \leq 4, (k, l) \neq (2, 3), (k, l) \neq (2, 4);$$

$$i = 3, j = 1 : c_{32}^{11} = c_{34}^{11} = c_{34}^{13} = 0; \ c_3^1 = \frac{h_1}{h_3} > 0;$$

$$i = 2, j = 1 : c_{23}^{11} = c_{24}^{11} = c_{24}^{13} = 0; \ c_2^1 = \frac{h_1}{h_2} > 0,$$

d'où le résultat du lemme. $\qquad\square$

Lemma 2.2. $c_{55}^{42} = c_{55}^{43} = c_{54}^{43} = 0$; $c_{45}^{32} = c_{45}^{33} = c_{45}^{34} = 0$; $c_{55}^{23} = c_{54}^{23} = c_{55}^{24} = 0$; $c_{55}^{32} = c_{54}^{33} = c_{55}^{34} = 0$; $c_{31}^{2k} = \tilde{k}_3^2 c_{21}^{3k}$, pour $2 \leq k < l \leq 5$; $(k, l) \neq (2, 3)$, $(k, l) \neq (2, 4)$, $c_{41}^{2k} = \tilde{k}_4^2 c_{21}^{4k}$, pour $2 \leq k < l \leq 5$; $(k, l) \neq (2, 3)$, $(k, l) \neq (2, 4)$, $c_5^3 = c_4^3 c_5^4$; $c_5^2 = \tilde{k}_4^2 c_5^4$; $\tilde{k}_4^2 = \tilde{k}_3^2 c_4^3$, $c_j^i > 0$, $\tilde{k}_j^i > 0$.

De plus: $\tilde{k}_3^2 k_4^2 = k_3^2 \tilde{k}_4^2$ et, si c_{24}^{33} ou $c_{24}^{43} \neq 0$: $k_3^2 = \tilde{k}_3^2$, $k_4^2 = \tilde{k}_4^2$.

Preuve. On pose $\xi_1^2 = \xi_1^3 = \xi_1^4 = \xi_1^5$; la matrice b_1 obtenue en rayant la $1^{\text{ère}}$ ligne et la $1^{\text{ère}}$ colonne de (a_j^i) est, comme précédemment présymétrique; il existe $\tilde{H}$ diagonale définie positive telle que:

$$b_1 \tilde{H} = \tilde{H}^t b_1. \tag{2}$$

En explicitant les éléments des ($i^{\text{ème}}$ ligne, $j^{\text{ème}}$ colonne), $i > j$ de (2), on obtient le résultat.   □

Lemma 2.3. $c_{5l}^{1k} = c_{5l}^{2k} = c_{5l}^{3k} = c_{5l}^{4k} = 0;\ 1 \le k < l \le 4,\ (k,l) \ne (2,3)$, $(k,l) \ne (2,4),\ c_{j5}^{1k} = 0,\ j > 1\ et\ k = 2\ ou\ 4$.

Preuve. On choisit d'abord $(k,l) = (1,2)$; on pose alors dans $(a_j^i(\xi))$: $\xi_j^j = 0$ pour: $1 \le i < j \le 5, (i,j) \ne (1,2), (i,j) \ne (2,3), (i,j) \ne (2,4)$. On considère:

$$\text{dét } a_j^i(\xi) = \text{dét}(\xi_0 I + a_j^i(\xi_1^5)):$$

comme: $\xi_j^5 = 0,\ 1 \le j \le 4,\ \xi_0 = -\psi_5$ est un zéro du déterminant, pour tout ψ_5;

$$\text{dét } a_j^i(\xi) = (\xi_0 + \psi_5) P(\xi_0, \psi_1, \psi_3, \psi_4),$$

(on peut poser $\psi_2 = 0$). $P(-\psi_5, \psi_1, \psi_3, \psi_4)$ est un polynôme de degré 1 en ψ_3 (le coefficient de ψ_3 est de la forme: $(\psi_5)^3$+des termes de degré inférieur en ψ_5 et n'est pas identiquement nul). On choisit ψ_3 fonction de ψ_1, ψ_4, ψ_5 de sorte que $-\psi_5$ soit zéro double de dét $a_j^i(\xi)$. Comme $a_j^i(\xi)$ est diagonalisable, pour ce choix de ψ_3 tous les mineurs d'ordre 4 sont nuls, en particulier celui obtenu en rayant la $5^{\text{ème}}$ ligne et la $3^{\text{ème}}$ colonne. On obtient un polynôme en ψ_1, ψ_5, ψ_4 de degré 3 en ψ_5. Le coefficient de $(\psi_5)^3$ est c_{52}^{31}, on a donc $c_{52}^{31} = 0$; en considérant les autres coefficients, on obtient: $c_{52}^{21} c_{22}^{31} = 0,\ c_{22}^{31} c_{22}^{41} c_{52}^{41} = 0,\ c_{52}^{11} c_{22}^{34} = 0$.

Si $c_{22}^{31} \ne 0$, on obtient d'abord

$$c_{52}^{21} = c_{52}^{11} = 0.$$

Si de plus $c_{22}^{41} \ne 0$, on a: $c_{52}^{41} = 0$, d'où le résultat du lemme; si $c_{22}^{41} = 0$, on remarque que, si on pose: $\psi_4 = \psi_5, \xi_0 = -\psi_5$ est zéro double de dét $a_j^i(\xi)$, le mineur obtenu en rayant la $5^{\text{ème}}$ et la $4^{\text{ème}}$ colonne est identiquement nul et on obtient encore:

$$c_{52}^{41} = 0.$$

Si $c_{22}^{31} = 0$, on remarque que si on pose, $\psi_3 = \psi_5, \xi_0 = -\psi_5$ est zéro double de dét $a_j^i(\xi)$.

$$\text{dét } a_j^i(\xi) = (\xi_0 + \psi_5)^2 Q(\xi_0, \psi_1, \psi_4).$$

$Q(-\psi_5, \psi_1, \psi_4)$ est un polynôme de degré 1 en ψ_1; on choisit ψ_1 fonction de ψ_5 et ψ_4 de sorte que $-\psi_5$ soit zéro triple; les mineurs d'ordre 3 correspondants sont alors identiquement nuls; celui obtenu en rayant les $3^{\text{ème}}$ et $5^{\text{ème}}$ lignes, $1^{\text{ère}}$ et $3^{\text{ème}}$ colonnes donne:

$$c_{52}^{11} = c_{52}^{21} = 0, \qquad c_{22}^{41} \cdot c_{52}^{41} = 0.$$

Si $c_{22}^{41} \neq 0$, on a le résultat du lemme; si $c_{22}^{41} = 0$, on pose $\psi_3 = \psi_4 = \psi_5, \xi_0 = -\psi_5$ est zéro triple on obtient encore

$$c_{52}^{41} = 0.$$

Les autres égalités se démontrent de façon analogue.   □

Lemma 2.4. *On considère les éléments* $c_{\beta 5}^{\alpha 1}, \begin{pmatrix} \alpha \\ \beta \end{pmatrix} \in \mathcal{J};$

$$\mathcal{J} = \begin{pmatrix} 1 & 1 & 1 & 2 & 3 & 4 \\ 2 & , & 3 & , & 4 & , & 5 & , & 5 & , & 5 & , \end{pmatrix}.$$

On a: $c_{\beta 5}^{\alpha 1} = 0.$

Preuve. On pose: $\xi_1^2 = \xi_1^3 = \xi_1^4 = 0; c_{55}^2 \xi_2^5 + c_{55}^{21} \xi_1^5 = 0, c_{55}^3 \xi_3^5 + c_{55}^{31} \xi_1^5 = 0,$
$c_{55}^4 \xi_4^5 + c_{55}^{41} \xi_1^5 = 0; \psi_1 = 0.$

$$\text{dét } a(\xi) = [(\xi_0 + \psi_5)\xi_0 - c_5^1(\xi_1^5)^2] \text{ dét} \begin{pmatrix} \xi_0 + \psi_2 & * & * \\ * & \xi_0 + \psi_3 & * \\ * & * & \xi_0 + \psi_4 \end{pmatrix}$$

$$\underline{\xi_0} = \frac{-\psi_5 - \sqrt{(\psi_5)^2 + 4c_5^1(\xi_1^5)^2}}{2}$$

est un zéro de dét $a(\xi)$. $\underline{\xi_0}$ est zéro double, si on choisit convenablement ψ_2 (en fonction de $\psi_3, \psi_4, \psi_1, \xi_1^5, \xi_3^4$), resp. ψ_3 (en fonction de $\psi_2, \psi_4, \psi_1, \xi_1^5, \xi_3^4$), resp. ψ_4 (en fonction de $\psi_2, \psi_3, \psi_1, \xi_1^5, \xi_3^4$). Le mineur obtenu en rayant alors la 5ème colonne et le 2ème ligne de $a(\xi)$ (resp. la 5ème colonne et le 3ème ligne) (resp. la 5ème colonne et la 4ème ligne) est nul. Dans ce mineur le coefficient de $\underline{\xi_0}(\underline{\xi_0} + \psi_3)(\underline{\xi_0} + \psi_4)$ (resp. $\underline{\xi_0}(\underline{\xi_0} + \psi_2)(\underline{\xi_0} + \psi_4))$ (resp. $\underline{\xi_0}(\underline{\xi_0} + \psi_2)(\underline{\xi_0} + \psi_3)$) est nul; soit:

$$c_{25}^{11} = c_{35}^{11} = c_{45}^{11} = 0.$$

Par symétrie par rapport à la 2ème diagonale, on obtient:

$$c_{55}^{41} = c_{55}^{31} = c_{55}^{21} = 0.$$

□

Proposition 2.5. *On note* $\mathcal{J}_1$ *l'ensemble de couples d'indices;*

$$\mathcal{J}_1 = \left\{ \begin{array}{cccccc} 1 & 1 & 2 & 4 & 1 \\ 2 & , & 4 & , & 5 & , & 5 & , & 5 & , \end{array} \right\}.$$

Si il existe $\begin{pmatrix} \alpha \\ \beta \end{pmatrix} \in \mathcal{J}_1, \begin{pmatrix} \alpha' \\ \beta' \end{pmatrix} \in \mathcal{J}_1, \begin{pmatrix} \alpha' \\ \beta' \end{pmatrix} \neq \begin{pmatrix} \alpha \\ \beta \end{pmatrix}$ *tels que* $c_{2\beta}^{3\alpha} \neq 0$ *et* $c_{2\beta'}^{3\alpha'} \neq 0$ *alors* $a(\xi')$ *est présymétrique.*

Preuve. On pose dans $(a(\xi'))$:

$$\xi_1^3 = \xi_3^4 = \xi_3^5 = 0; \quad c_{22}^{31}\xi_1^2 + c_{24}^{31}\xi_1^4 + c_{25}^{32}\xi_2^5 + c_{25}^{34}\xi_4^5 + c_{25}^{31}\xi_1^5 = 0. \quad (3)$$

La matrice b, 4×4, obtenue en rayant la 3ème ligne et la 3ème colonne de $(a_j^i(\xi'))$ est (lemmes 1.1 et 1.2) présymétrique: il existe une matrice diagonale que nous noterons encore H telle que:

$$bH = H'b.$$

En explicitant les (4ème ligne, 3ème colonne), (4ème ligne, 2ème colonne), (4ème ligne, 1ère colonne), (3ème ligne, 2ème colonne), (2ème ligne, 1ère colonne), on obtient:

$$c_5^1 = c_2^1 c_5^2 = c_4^1 c_5^4;$$

on déduit alors des lemmes 2.2 et 2.1 que:

$$c_5^1 = c_2^1 \tilde{k}_3^2 c_4^3 c_5^4 = c_2^1 k_3^2 c_4^3 c_5^4,$$

d'où,

$$k_3^2 = \tilde{k}_3^2, \quad k_4^2 = \tilde{k}_4^2;$$

on a aussi

$$c_{45}^{21} = k_4^2 c_{25}^{41}.$$

On a encore: (3); on pose $\psi_3 = 0$; 0 est un zéro de dét $a(\xi)$; en choisissant convenablement ψ_2, c'est un zéro double, le mineur obtenu en rayant la 3ème ligne, 2ème colonne est identiquement nul; dans ce mineur on considère le coefficient de ψ_1, ψ_4, ψ_5 et on obtient:

$$c_{35}^{21} - k_3^2 c_{25}^{31} = 0.$$

Comme on a: $c_3^1 = c_2^1 k_3^2$, $c_4^1 = c_2^1 k_3^2 k_4^3$, $c_5^1 = c_2^1 k_3^2 k_4^3 c_5^4$, $c_5^2 = k_3^2 k_4^3 c_5^4$, $c_5^3 = c_4^3 c_5^4$, $c_5^1 = c_2^1 k_3^2 k_4^3 c_5^4$, la matrice diagonale:

$$D = \left\{ 1, \frac{1}{\sqrt{c_2^1}}, \frac{1}{\sqrt{c_3^1}}, \frac{1}{\sqrt{c_4^1}}, \frac{1}{\sqrt{c_5^1}}, \right\}$$

est telle que:

$$D^{-1}\big(a_j^i(\xi)\big)D$$

soit symétrique.   □

Proposition 2.6. *On note*

$$\mathcal{J}_2 = \left\{ \frac{1}{2}, \frac{1}{3}, \frac{2}{5}, \frac{3}{5}, \frac{1}{5} \right\}.$$

S'il existe $\begin{pmatrix} \alpha \\ \beta \end{pmatrix} \in \mathcal{J}_2$ *et* $\begin{pmatrix} \alpha' \\ \beta' \end{pmatrix} \in \mathcal{J}_2$, $\begin{pmatrix} \alpha \\ \beta \end{pmatrix} \neq \begin{pmatrix} \alpha' \\ \beta' \end{pmatrix}$, *tels que* $c_{2\beta}^{4\alpha} \neq 0$ *et* $c_{2\beta'}^{4\alpha'} \neq 0$, *alors* $a_j^i(\xi)$ *est présymétrique.*

Preuve. En échangeant les $3^{\text{ème}}$ et $4^{\text{ème}}$ lignes, $3^{\text{ème}}$ et $4^{\text{ème}}$ colonnes, on est ramené à la proposition 2.5, aux notations près.   □

Proposition 2.7. *S'il existe* $\begin{pmatrix} \alpha \\ \beta \end{pmatrix} \in \mathcal{J}_1 \begin{pmatrix} \alpha \\ \beta \end{pmatrix} \neq \begin{pmatrix} 1 \\ 5 \end{pmatrix}$, *tels que* $c_{2\beta}^{3\alpha} \neq 0$, *alors* $a_j^i(\xi)$ *est présymétrique.*

Compte tenu de la proposition 2.5, on peut supposer que $\forall \begin{pmatrix} \alpha' \\ \beta' \end{pmatrix} \neq \begin{pmatrix} \alpha \\ \beta \end{pmatrix}$, $\begin{pmatrix} \alpha' \\ \beta' \end{pmatrix} \in \mathcal{J}_1$, on a $c_{2\beta'}^{3\alpha'} = 0$.

La preuve résulte des lemme suivants.

Lemma 2.8. *S'il existe* $\alpha \in \{2, 4\}$ *tel que* $c_{2\alpha}^{31} \neq 0$ *[resp.* $c_{25}^{3\alpha} \neq 0$*], on a* $\Phi_3^2 = k_3^2 \Phi_2^3$, *[resp.* $\Phi_3^2 = \tilde{k}_3^2 \Phi_2^3$*].*
Si de plus $c_{25}^{33} \neq 0$ *[resp.* $c_{23}^{31} \neq 0$*], on a* $k_3^2 = \tilde{k}_3^2$, $k_4^2 = \tilde{k}_4^2$.

Preuve. On obtient aisément que $c_{35}^{21} = 0$.

Supposons que $c_{2\alpha}^{31} \neq 0$. On pose $\xi_1^3 = \xi_3^4 = \xi_2^5 = \xi_4^5 = \xi_1^5 = 0$; $\psi_3 = 0$
$\sum_{\beta \in \{2, 4\}} c_{2\beta}^{31} \xi_1^\beta + c_{25}^{33} \xi_3^5 = 0$; on a

$$\det\left(a_j^i(\xi)\right) = \left[\xi_0(\xi_0 + \psi_5) - c_5^3(\xi_3^5)^2\right] \det \begin{pmatrix} \xi_0 + \psi_1 & * & * \\ * & \xi_0 + \psi_2 & * \\ * & * & \xi_0 + \psi_4 \end{pmatrix}.$$

On a le zéro:

$$\xi_0 = \frac{-\psi_5 - \sqrt{(\psi_5)^2 + 4c_5^3(\xi_3^5)^2}}{2};$$

le zéro est double si le déterminant est nul pour ξ_0; on choisit ψ_2 de sorte qu'il en soit aussi; alors le mineur obtenu, pour ces valeurs, en rayant la $2^{\text{ème}}$ colonne et la $2^{\text{ème}}$ ligne est nul; le coefficient de $\psi_1 \psi_4 \psi_5$ dans le mineur est nul, d'où: $(k_3^2 - \tilde{k}_3^2) c_{25}^{33} = 0$.

Si $c_{25}^{33} \neq 0$, on a $k_3^2 = \tilde{k}_3^2$, si $c_{25}^{33} = 0$, on a $\Phi_3^2(\xi) \equiv \Phi_3^2(\xi_1^2, \xi_1^3, \xi_1^4, \xi_3^4) = k_3^2 \Phi_3^2(\xi_1^2, \xi_1^3, \xi_1^4, \xi_3^4)$.

Si $c_{25}^{3\alpha} \neq 0$, on procède exactement de même.   □

Lemma 2.9. *S'il existe* $\alpha \in \{2, 3\}$ *tel que:* $c_{2\alpha}^{41} \neq 0$ *[resp.* $c_{25}^{4\alpha} \neq 0$*], on a:* $\Phi_4^2 = k_4^2 \Phi_2^4$, *[resp.* $\Phi_4^2 = \tilde{k}_4^2 \Phi_2^4$*].*
Si de plus $c_{25}^{44} \neq 0$ *[resp.* $c_{24}^{41} \neq 0$*], on a:* $k_3^2 = \tilde{k}_3^2$, $k_4^2 = \tilde{k}_4^2$.

Preuve. On échange les $3^{\text{ème}}$ et $4^{\text{ème}}$ lignes et les $3^{\text{ème}}$ et $4^{\text{ème}}$ colonnes et on adapte les notations.   □

Lemma 2.10. *i) Si* $c_{23}^{31} \neq 0$ *[resp.* $c_{25}^{33} \neq 0$*], on a:* $c_{35}^{21} = k_3^2 c_{25}^{31}$ *[resp.* $\tilde{k}_3^2$*]*
ii) Si $c_{23}^{31} \neq 0$ *et* $c_{25}^{33} \neq 0$, *on a:* $k_3^2 = \tilde{k}_3^2$, $k_4^2 = \tilde{k}_4^2$, $\Phi_3^2 = k_3^2 \Phi_2^3$.

Preuve. On pose $\xi_1^2 = \xi_1^4 = \xi_2^5 = \xi_4^5 = \xi_3^4 = 0; \psi_2 = 0; c_{23}^{31}\xi_1^3 + c_{25}^{33}\xi_3^5 + c_{25}^{31}\xi_1^5 = 0$.

On a

$$\text{dét}\left(a_j^i(\xi)\right) = \left[\xi_0(\xi_0 + \psi_4) - k_4^2(c_{23}^{41}\xi_1^3 + c_{25}^{43}\xi_3^5 + c_{25}^{41}\xi_1^5)^2\right]$$

$$\times \text{dét}\begin{pmatrix} \xi_0 + \psi_1 & * & * \\ * & \xi_0 + \psi_3 & * \\ * & * & \xi_0 + \psi_5 \end{pmatrix}.$$

On a le zéro:

$$\xi_0 = \frac{-\psi_4 - \sqrt{(\psi_4)^2 + 4k_4^2(\dots)^2}}{2};$$

en choisissant convenablement ψ_3, c'est un zéro double, le mineur obtenu en rayant la 2ème colonne et la 3ème ligne est alors nul; le coefficient de $\psi_1\psi_4\psi_5$ dans ce mineur est $k_3^2 c_{23}^{31}\xi_1^3 + \tilde{k}_3^2 c_{25}^{33}\xi_3^5 + c_{35}^{21}\xi_1^5 = 0$; si $c_{23}^{31} \neq 0$, on a:

$$(\tilde{k}_3^2 - k_3^2)c_{25}^{33}\xi_3^5 + (c_{35}^{21} - k_3^2 c_{25}^{31})\xi_1^5 = 0,$$

d'où $c_{35}^{21} = k_3^2 c_{25}^{31}$;

si $c_{25}^{33} \neq 0$, on a: $c_{35}^{21} = \tilde{k}_3^2 c_{25}^{31}$;

si $c_{23}^{31} \neq 0$ et $c_{25}^{33} \neq 0$, on a: $k_3^2 = \tilde{k}_3^2$, d'où (2.2): $k_4^2 = \tilde{k}_4^2$.   □

Lemma 2.11. *i) Si* $c_{24}^{41} \neq 0$ *[resp.* $c_{25}^{44} \neq 0$*], on a:* $c_{45}^{21} = k_4^2 c_{25}^{41}$ *[resp.* $\tilde{k}_4^2.$*]*
ii) Si $c_{24}^{41} \neq 0$ *et* $c_{25}^{44} \neq 0$*, on a:* $k_4^2 = \tilde{k}_4^2$*;* $k_3^2 = \tilde{k}_3^2$*;* $\Phi_4^2 = k_4^2 \Phi_2^4$*.*

Lemma 2.12. $\alpha \in \{2, 4\}$, $\beta \in \{2, 4\}$, $\beta \neq \alpha$.
i) Si on a: $c_{2\alpha}^{31} \neq 0$ *[resp.* $c_{25}^{3\alpha} \neq 0$*], on a:* $c_{35}^{21} = 0$, $c_{45}^{21} = \tilde{k}_4^2 c_{25}^{41}$ *[resp.* $\tilde{k}_4^2$*],* $c_5^1 = c_\beta^1 c_5^\beta$*.*

ii) Si on a: $c_{2\alpha}^{31} \neq 0$ *et* $c_{2\beta}^{41} \neq 0$ *[resp.* $c_{25}^{3\alpha} \neq 0$ *et* $c_{25}^{4\beta} \neq 0$*], on a: de plus:* $k_3^2 = \tilde{k}_3^2,$, $k_4^2 = \tilde{k}_4^2.$ $a_j^i(\xi)$ *est présymétrique.*

Preuve. On pose $\xi_1^3 = \xi_3^4 = \xi_3^\beta = 0$; $\sum c_{2\delta}^{3\gamma} = 0$, pour $\begin{pmatrix} \gamma \\ \delta \end{pmatrix} \in \mathcal{J}_1$.

La matrice b, 4×4 obtenue en rayant la 3ème ligne et la 3ème colonne de $a_j^i(\xi)$ est (1.1) et (1.2) présymétrique; il existe H diagonale telle que $bH = H^t b$. En explicitant comme précédemment, on obtient:

si $c_{2\alpha}^{31} \neq 0$, [d'où $\xi_1^\alpha = 0$]:

$$(\tilde{k}_4^2 - k_4^2)(c_{22}^{41}\xi_1^2 + c_{24}^{41}\xi_1^4) + (\tilde{k}_4^2 c_{2\beta}^{41} - c_{45}^{21})\xi_1^5 = 0;$$

d'où $c_{45}^{21} = \tilde{k}_4^2 c_{25}^{41}$; on a: aussi $c_5^1 = c_\beta^1 c_5^\beta$.

Si de plus $c_{2\beta}^{41} \neq 0$, comme ξ_1^β est libre, $k_4^2 = \tilde{k}_4^2$, $k_3^2 = \tilde{k}_3^2$; d'où $c_5^1 = c_2^1 k_3^2 c_4^3 c_5^4$.

Comme $c_{25}^{31} = 0$, on a: aisément $c_{35}^{21} = 0$ (en annulant tous les ξ sauf ξ_1^5 et en construisant un zéro double). La présymetrie s'obtient comme dans la proposition 2.5.

Si $c_{25}^{3\alpha} \neq 0$, on procède de même.   □

Lemma 2.13. $\alpha \in \{2, 3\}$, $\beta \in \{2, 3\}$, $\beta \neq \alpha$.

i) Si on a: $c_{2\alpha}^{41} \neq 0$ *[resp.* $c_{25}^{4\alpha} \neq 0$*], on a:* $c_{45}^{21} = 0$, $c_{35}^{21} = k_3^2 c_{25}^{31}$ *[resp.* $\tilde{k}_3^2$*],* $c_5^1 = c_\beta^1 c_5^\beta$.

ii) Si on a: $c_{2\alpha}^{41} \neq 0$ *et* $c_{2\beta}^{31} \neq 0$ *[resp.* $c_{25}^{4\alpha} \neq 0$ *et* $c_{25}^{3\beta} \neq 0$*], on a: de plus:* $k_3^2 = \tilde{k}_3^2$, $k_4^2 = \tilde{k}_4^2$. $a_j^i(\xi)$ *est présymétrique.*

Preuve. On échange le $3^{\text{ème}}$ et $4^{\text{ème}}$ lignes et colonnes et on se ramène au lemme précédent au notations près.   □

Preuve de la proposition 2.7. Comme il existe $\alpha \in \{2, 4\}$ tel que $c_{2\alpha}^{31} \neq 0$ [ou $c_{25}^{3\alpha} \neq 0$]. on a: donc par le lemme 2.8: $\Phi_3^2 = k_3^2 \Phi_2^3$ [ou $\tilde{k}_3^2$].

a) Si on a, de plus; il existe $\beta \in \{2, 3\}$ tel que $c_{2\beta}^{41} \neq 0$ [resp. $c_{25}^{4\beta} \neq 0$], on a: par le lemme 2.9; $\Phi_4^2 = k_4^2 \Phi_2^4$ [resp. $\tilde{k}_4^2$].

Si $\left(c_{22}^{31} \neq 0 \text{ ou } c_{25}^{32} \neq 0\right)$ et $\left(c_{23}^{41} \neq 0 \text{ ou } c_{25}^{43} \neq 0\right)$, on a: (2.12) et (2.13): $c_5^1 = c_4^1 c_5^4 = c_2^1 c_5^2$, d'où [2.1 et 2.2] $k_3^2 = \tilde{k}_3^2$, $k_4^2 = \tilde{k}_4^2$. On obtient la présymétrie comme à la fin de la proposition 2.5.

Il en est de même, si: $\left(c_{24}^{31} \neq 0 \text{ ou } c_{25}^{34} \neq 0\right)$ et $\left(c_{22}^{41} \neq 0 \text{ ou } c_{25}^{42} \neq 0\right)$. Il reste à examiner les cas

$\left(c_{22}^{31} \neq 0 \text{ ou } c_{25}^{32} \neq 0\right)$ et $\left(c_{22}^{41} \neq 0 \text{ ou } c_{25}^{42} \neq 0\right)$
$\left(c_{24}^{31} \neq 0 \text{ ou } c_{25}^{34} \neq 0\right)$ et $\left(c_{23}^{41} \neq 0 \text{ ou } c_{25}^{43} \neq 0\right)$.

a_1) On suppose $c_{22}^{31} \neq 0$ et $c_{22}^{41} \neq 0$. Il résulte des hypothèses de la proposition 2.7 que: $c_{24}^{31} = c_{25}^{32} = c_{25}^{34} = c_{25}^{31} = 0$ $c_{23}^{41} = c_{25}^{42} = c_{25}^{43} = c_{25}^{41} = 0$.

On a aussi, d'après le lemme 2.13 ii) si $c_{23}^{31} \neq 0$, on a: $k_3^2 = \tilde{k}_3^2$, $k_4^2 = \tilde{k}_4^2$, d'où la présymétrie.

Il en est de même, si $c_{24}^{43} \neq 0$ ou $c_{24}^{33} \neq 0$.

On suppose donc maintenant $c_{23}^{31} = c_{24}^{43} = c_{24}^{33} = 0$.

De même grâce au lemme 2.12 ii) si $c_{24}^{41} \neq 0$, on a aussi la présymétrie; on suppose donc $c_{24}^{41} = 0$ grâce au lemme 2.8, si $c_{25}^{33} \neq 0$, on a: la présymétrie; on suppose donc $c_{25}^{33} = 0$. Enfin grâce au lemme 2.9, si $c_{25}^{44} \neq 0$, on a: la présymétrie; on suppose donc $c_{25}^{44} = 0$. On a donc

$$(a_j^i(\xi)) = \begin{pmatrix} \xi_0 + \psi_1 & c_2^1 \xi_1^2 & c_3^1 \xi_1^3 & c_4^1 \xi_1^4 & c_5^1 \xi_4^5 \\ \xi_1^2 & \xi_0 + \psi_2 & k_3^2 c_{22}^{31} \xi_1^2 & k_4^2 c_{22}^{41} \xi_1^2 & c_5^2 \xi_2^5 \\ \xi_1^3 & c_{22}^{31} \xi_1^2 & \xi_0 + \psi_3 & c_4^3 \xi_3^4 & c_5^3 \xi_3^5 \\ \xi_1^4 & c_{22}^{41} \xi_1^2 & \xi_3^4 & \xi_0 + \psi_4 & c_5^4 \xi_4^5 \\ \xi_1^5 & \xi_2^5 & \xi_3^5 & \xi_4^5 & \xi_0 + \psi_5 \end{pmatrix}$$

$c_3^1 = c_2^1 k_3^2$; $c_4^1 = c_2^1 k_3^2 c_4^3$; $c_5^1 = c_2^1 k_3^2 c_4^3 c_5^4$; $c_5^2 = \tilde{k}_3^2 c_5^3$; $c_5^3 = c_4^3 c_5^4$. On définit la matrice diagonale

$$D = \left(1, \frac{1}{\sqrt{c_2^1}}, \frac{1}{\sqrt{c_3^1}}, \frac{1}{\sqrt{c_4^1}}, \frac{1}{\sqrt{c_5^1}}\right)$$

et on calcule $D^{-1}a(\xi)D$. On obtient la nouvelle matrice $a^i_j(\xi')$

$$
\begin{pmatrix}
\psi_1 & \sqrt{c^1_2\xi^2_1} & \sqrt{c^1_3\xi^3_1} & \sqrt{c^1_4\xi^4_1} & \sqrt{c^1_5\xi^5_1} \\
\sqrt{c^1_2\xi^2_1} & \psi_2 & \sqrt{k^2_3c^{31}_{22}\xi^2_1} & \sqrt{k^2_4c^{41}_{22}\xi^2_1} & c^2_5\sqrt{\dfrac{c^1_2}{c^1_5}}\xi^5_2 \\
\sqrt{c^1_3\xi^3_1} & \sqrt{k^2_3c^{31}_{22}\xi^2_1} & \psi_3 & \sqrt{c^3_4\xi^4_3} & c^3_5\sqrt{\dfrac{c^1_3}{c^1_5}}\xi^5_3 \\
\sqrt{c^1_4\xi^4_1} & \sqrt{k^2_4c^{41}_{22}\xi^2_1} & \sqrt{c^3_4\xi^4_3} & \psi_4 & c^4_5\sqrt{\dfrac{c^1_4}{c^1_5}}\xi^5_4 \\
\sqrt{c^1_5\xi^5_1} & \sqrt{\dfrac{c^1_5}{c^1_2}}\xi^5_2 & \sqrt{\dfrac{c^1_5}{c^1_3}}\xi^5_3 & \sqrt{\dfrac{c^1_5}{c^1_4}}\xi^5_4 & \psi_5
\end{pmatrix}
$$

on pose $\sqrt{c^1_\alpha}\xi^\alpha_1 = \xi^{\alpha\prime}_1$, $\alpha \in \{2, 3, 4, 5\}$; $\sqrt{\dfrac{c^1_5}{c^1_\beta}}\xi^5_\beta = \xi^{5\prime}_\beta$, $\beta \in \{2, 3, 4\}$; $\rho = \dfrac{\tilde{k}^2_3}{k^2_3}$,

$\lambda \neq 0$; on obtient, en ne changeant pas les notations pour les ξ^α_1, ξ^5_β:

$$
\begin{pmatrix}
\psi_1 & \xi^2_1 & \xi^3_1 & \xi^4_1 & \xi^5_1 \\
\xi^2_1 & \psi_2 & \lambda\xi^2_1 & \mu\xi^4_1 & \rho\xi^5_2 \\
\xi^3_1 & \lambda\xi^2_1 & \psi_3 & \xi^4_3 & \xi^5_3 \\
\xi^4_1 & \mu\xi^2_1 & \xi^4_3 & \psi_4 & \xi^5_4 \\
\xi^5_1 & \xi^5_2 & \xi^5_3 & \xi^5_4 & \psi_5
\end{pmatrix}
$$

ou bien

$$
\begin{pmatrix}
\psi_5 & \xi^5_4 & \xi^5_3 & \rho\xi^5_2 & \xi^5_1 \\
\xi^5_4 & \psi_4 & \xi^4_3 & \mu\xi^2_1 & \xi^4_1 \\
\xi^5_3 & \xi^4_3 & \psi_3 & \lambda\xi^2_1 & \xi^3_1 \\
\xi^5_2 & \mu\xi^2_1 & \lambda\xi^2_1 & \psi_2 & \xi^2_1 \\
\xi^5_1 & \xi^4_1 & \xi^3_1 & \xi^2_1 & \psi_1
\end{pmatrix} ;
$$

on pose $\xi^4_3 = \xi^4_1 = \xi^3_1 = 0$, $\xi^5_4 = \xi^5_3 = 0$; la matrice

$$
b = \begin{pmatrix}
\psi_4 & 0 & \mu\xi^2_1 & 0 \\
0 & \psi_3 & \lambda\xi^2_1 & 0 \\
\mu\xi^2_1 & \lambda\xi^2_1 & \psi_2 & \xi^2_1 \\
0 & 0 & \xi^2_1 & \psi_1
\end{pmatrix}
$$

est symétrique; ses valeurs propres sont distinctes pour ξ^2_1 et ψ_i généraux; on obtient un diagonaliseur Δ en cherchant les vecteurs propres: soit (u_1, u_2, u_3, u_4) l'un d'eux; on a: pour la valeur propre $\underline{\xi}_0$:

$$
\begin{cases}
(\underline{\xi}_0 + \psi_4)u_1 + \mu\xi^2_1 u_3 = 0 \\
(\underline{\xi}_0 + \psi_3)u_2 + \lambda\xi^2_1 u_3 = 0 \\
\xi^2_1 u_3 + (\underline{\xi}_0 + \psi_1)u_4 = 0
\end{cases}
$$

Comme on peut choisir $\underline{\xi}_0 \neq -\psi_i$, et $u_3 \neq 0$, on a: $u_4 \neq 0$. On transforme $a(\xi)$ par la matrice $\begin{pmatrix} 1 & 0 \\ 0 & \Delta \end{pmatrix}$; on obtient

$$\begin{pmatrix} \psi_5 & \rho u_3 \xi_2^5 + u_4 \xi_1^5 & * & * & * \\ u_3 \xi_2^5 + u_4 \xi_1^5 & * & & & \\ * & & * & & 0 \\ * & & & * & \\ * & & 0 & & * \end{pmatrix}$$

Les u ne dépendent pas de ξ_2^5 et ξ_1^5; par la conséquence 1.4, on doit avoir: $\rho = 1$, soit $k_3^2 = \tilde{k}_3^2$, d'où la symétrie.

$a_2)$ $c_{22}^{31} \neq 0$ et $c_{25}^{42} \neq 0$. On est ramène à considérer le cas: $c_{24}^{31} = c_{25}^{32} = c_{25}^{34} = c_{25}^{31} = c_{25}^{33} = c_{24}^{33} = 0$; $c_{22}^{41} = c_{23}^{41} = c_{25}^{43} = c_{25}^{41} = c_{24}^{41} = c_{24}^{43} = 0$.

En calculant $D^{-1}a(\xi)D$, puis par un changement de variable, comme au cas précédent, on se ramène à:

$$a_j^i(\xi) = \begin{pmatrix} \psi_1 & \xi_1^2 & \xi_1^3 & \xi_1^4 & \xi_1^5 \\ \xi_1^2 & \psi_2 & \Phi_2^3(\xi_1^2, \xi_1^3) & \rho\Phi_2^4(\xi_2^5, \xi_4^5) & \rho\xi_2^5 \\ \xi_1^3 & \Phi_2^3(\xi_1^2, \xi_1^3) & \psi_3 & \xi_3^4 & \xi_3^5 \\ \xi_1^4 & \Phi_2^4(\xi_2^5, \xi_4^5) & \xi_3^2 & \psi_4 & \xi_4^5 \\ \xi_1^5 & \xi_2^5 & \xi_3^5 & \xi_4^5 & \psi_5 \end{pmatrix} \qquad \rho = \dfrac{\tilde{k}_3^2}{k_3^2}$$

ou par échange de lignes et colonnes

$$\begin{pmatrix} \psi_2 & \xi_1^2 & \Phi_2^3(\xi_1^2, \xi_1^3) & \rho\Phi_2^4(\xi_2^5, \xi_4^5) & \rho\xi_2^5 \\ \xi_1^2 & \psi_1 & \xi_1^3 & \xi_1^4 & \xi_1^5 \\ \Phi_2^3(\xi_1^2, \xi_1^3) & \xi_1^3 & \psi_3 & \xi_3^4 & \xi_2^5 \\ \Phi_2^4(\xi_2^5, \xi_4^5) & \xi_1^4 & \xi_3^4 & \psi_4 & \xi_4^5 \\ \xi_2^5 & \xi_1^5 & \xi_3^5 & \xi_4^5 & \psi_5 \end{pmatrix}.$$

On pose $\xi_1^3 = \xi_1^5 = \xi_3^4 = \xi_4^5 = 0$, on diagonalise la matrice symétrique.

$$\begin{pmatrix} \psi_1 & \xi_1^4 \\ \xi_1^4 & \psi_4 \end{pmatrix} \quad \text{et} \quad \begin{pmatrix} \psi_3 & \xi_3^5 \\ \xi_3^5 & \psi_5 \end{pmatrix}$$

par les matrices orthogonales

$$\begin{pmatrix} s_2^2 & s_4^2 \\ s_2^4 & s_4^4 \end{pmatrix} \quad \text{et} \quad \begin{pmatrix} s_3^3 & s_5^3 \\ s_3^5 & s_5^5 \end{pmatrix}.$$

Par la conséquence 1.4, on déduit que $\xi_1^2 s_2^2 + \rho c_{25}^{42} s_2^4 \xi_2^5$ et $\xi_1^2 s_2^2 + c_{25}^{42} s_4^4 \xi_2^5$ sont proportionnels, d'où: $\rho = 1$, $k_3^2 = \tilde{k}_3^2$ et la présymétrie.

$a_3)$ $c_{25}^{32} \neq 0$ et $c_{25}^{42} \neq 0$. On est ramène à considérer le cas: $c_{22}^{31} = c_{24}^{31} = c_{25}^{34} = c_{25}^{31} = c_{23}^{31} = c_{25}^{33} = c_{24}^{33} = 0$ $c_{22}^{41} = c_{23}^{41} = c_{25}^{43} = c_{25}^{41} = c_{24}^{41} = c_{25}^{44} = c_{24}^{43} = 0$.

On se ramène à

$$
\begin{pmatrix}
\psi_2 & \xi_1^2 & \rho\lambda\xi_2^5 & \rho\mu\xi_2^5 & \rho\xi_2^5 \\
\xi_1^2 & \psi_1 & \xi_1^3 & \xi_1^4 & \xi_1^5 \\
\lambda\xi_2^5 & \xi_1^3 & \psi_3 & \xi_3^4 & \xi_3^5 \\
\mu\xi_2^5 & \xi_1^4 & \xi_3^4 & \psi_4 & \xi_4^5 \\
\xi_2^5 & \xi_1^5 & \xi_3^5 & \xi_4^5 & \psi_5
\end{pmatrix}
\qquad \lambda' \neq 0,\ \mu \neq 0;\ \rho = \frac{\tilde{k}_3^2}{k_3^2}.
$$

On pose: $\xi_1^4 = \xi_3^4 = \xi_1^5 = \xi_3^5 = \xi_3^5 = 0$; en diagonalisant la sous-matrice 4×4 en bas à droite, on obtient: $s_2^2\xi_1^2 + \rho\lambda s_2^3\xi_2^5$ et $s_2^2\xi_1^2 + \lambda s_2^3\xi_2^5$ sont proportionnels, d'où $\rho = 1$, $k_3^2 = \tilde{k}_3^2$ et la présymétrie.

a$_4$) $c_{24}^{31} \neq 0$ et $c_{23}^{41} \neq 0$. On a à considère le cas: $c_{22}^{31} = c_{25}^{32} = c_{25}^{34} = c_{25}^{31} = c_{25}^{33} = c_{24}^{33} = 0$ $c_{22}^{41} = c_{25}^{42} = c_{25}^{43} = c_{25}^{41} = c_{25}^{44} = c_{24}^{43} = 0$.

On se ramène à

$$
\begin{pmatrix}
\psi_5 & \rho\xi_4^5 & \rho\xi_3^5 & \xi_2^5 & \xi_1^5 \\
\xi_4^5 & \psi_4 & \xi_3^4 & \mu\xi_1^3 + \mu'\xi_1^4 & \xi_1^4 \\
\xi_3^5 & \xi_3^4 & \psi_3 & \lambda\xi_1^3 + \lambda'\xi_1^4 & \xi_1^3 \\
\xi_2^5 & \mu\xi_1^3 + \mu'\xi_1^4 & \lambda\xi_1^3 + \lambda'\xi_1^4 & \psi_2 & \xi_1^2 \\
\xi_1^5 & \xi_1^4 & \xi_1^3 & \xi_1^2 & \psi_1
\end{pmatrix}
\qquad \mu \neq 0;\ \lambda' \neq 0,\ \rho = \frac{k_3^2}{\tilde{k}_3^2}.
$$

On pose $\xi_3^4 = \xi_1^2 = \xi_1^4 = 0$, $\xi_2^5 = \xi_4^5 = 0$, on diagonalise la sous-matrice 4×4 en bas à droite, comme précédemment une colonne s'écrit (u_1, u_2, u_3, u_4) avec $u_2 \neq 0$ et $u_4 \neq 0$ on en déduit encore $\rho = 1$ $k_3^2 = \tilde{k}_3^2$ et la présymétrie.

a$_5$) $c_{24}^{31} \neq 0$ et $c_{25}^{43} \neq 0$. On a à considèrer le cas: $c_{22}^{31} = c_{25}^{32} = c_{25}^{34} = c_{25}^{31} = c_{25}^{33} = c_{24}^{33} = 0$; $c_{22}^{41} = c_{23}^{41} = c_{25}^{42} = c_{25}^{41} = c_{24}^{41} = c_{24}^{43} = 0$.

On se ramène à

$$
\begin{pmatrix}
\psi_1 & \xi_1^2 & \xi_1^3 & \xi_1^4 & \xi_1^5 \\
\xi_1^2 & \psi_2 & \Phi_2^3(\xi_1^3, \xi_1^4) & \rho\Phi_2^4(\xi_3^5, \xi_4^5) & \xi_2^5 \\
\xi_1^3 & \Phi_2^3(\xi_1^3, \xi_1^4) & \psi_3 & \xi_3^4 & \dfrac{1}{\rho}\xi_3^5 \\
\xi_1^4 & \Phi_2^4(\xi_3^5, \xi_4^5) & \xi_3^4 & \psi_4 & \dfrac{1}{\rho}\xi_4^5 \\
\xi_1^5 & \xi_2^5 & \xi_3^5 & \xi_4^5 & \psi_5
\end{pmatrix}
\qquad \rho = \frac{\tilde{k}_3^2}{k_3^2}.
$$

On pose $\xi_1^2 = \xi_2^5 = 0$, $c_{24}^{31}\xi_1^4 + c_{23}^{31}\xi_1^3 = 0$; $c_{25}^{43}\delta_3^5 + c_{25}^{44}\delta_4^5 = 0$; on exprime ξ_1^4 et ξ_3^5 en fonction de ξ_1^3 et ξ_4^5; la matrice 4×4 notée b obtenue en rayant la $2^{\text{ème}}$ ligne et $2^{\text{ème}}$ colonne de $a(\xi)$ est présymétrique; il existe H telle que $bH = H^t b$. On en déduit que $\rho = 1$, d'où $k_3^2 = \tilde{k}_3^2$ et la présymétrie.

a$_6$) c_{25}^{34} et $c_{25}^{43} \neq 0$, on a à considérer le cas: $c_{22}^{31} = c_{24}^{31} = c_{25}^{32} = c_{25}^{31} = c_{23}^{31} = c_{24}^{33} = 0$; $c_{22}^{41} = c_{23}^{41} = c_{25}^{42} = c_{25}^{41} = c_{24}^{41} = c_{24}^{43} = 0$.

On se ramène à

$$
\begin{pmatrix}
\psi_1 & \xi_1^2 & \xi_1^3 & \xi_1^4 & \xi_1^5 \\
\xi_1^2 & \psi_2 & \Phi_2^3(\xi_3^5,\xi_4^5) & \Phi_2^4(\xi_3^5,\xi_4^5) & \xi_2^5 \\
\xi_1^3 & \Phi_2^3(\xi_3^5,\xi_4^5) & \psi_3 & \xi_3^4 & \dfrac{1}{\rho}\xi_3^5 \\
\xi_1^4 & \Phi_2^4(\xi_3^5,\xi_4^5) & \xi_3^4 & \psi_4 & \dfrac{1}{\rho}\xi_4^5 \\
\xi_1^5 & \xi_2^5 & \xi_3^5 & \xi_4^5 & \psi_5
\end{pmatrix}
\qquad
\rho = \dfrac{\tilde{k}_3^2}{k_3^2}.
$$

On définit la matrice diagonale $D = \left(1, 1, \dfrac{1}{\sqrt{\rho}}, \dfrac{1}{\sqrt{\rho}}, 1\right)$, on calcule $D^{-1}aD$; on

pose $\sqrt{\rho}\,\xi_1^3 = \xi_1^{3\prime}$, $\sqrt{\rho}\,\xi_1^4 = \xi_1^{4\prime}$, $\dfrac{\xi_1^5}{\sqrt{\rho}} = \xi_1^{5\prime}$, $\dfrac{\xi_4^5}{\sqrt{\rho}} = \xi_4^{5\prime}$, $\Phi_2^{3\prime} = \rho\Phi_2^3$, $\Phi_2^{4\prime} = \rho\Phi_2^4$;

on obtient, en omettant les primes:

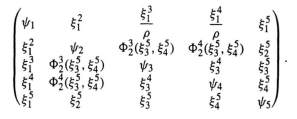

On pose: $\xi_2^5 = \xi_3^4 = \xi_3^5 = 0$ $\xi_1^2 = \xi_1^3 = 0$; on diagonalise la sous-matrice 4×4 en bas à droite; on écrit la conséquence 1.4, on obtient: $\dfrac{\xi_1^4}{\rho}u_3 + \xi_1^5 u_4$ et $\xi_1^4 u_3 + \xi_1^5 u_4$ sont proportionnels, avec $u_3 \neq 0$, $u_4 \neq 0$, d'où $\rho = 1$ et la symétrie.

a_7) a_8) Le cas $\left(c_{25}^{32} \neq 0 \text{ et } c_{22}^{41} \neq 0\right)$ [resp. $\left(c_{23}^{41} \neq 0 \text{ et } c_{25}^{34} \neq 0\right)$] se ramène au cas a_2) [resp. a_5)] par échange des $3^{\text{ème}}$ et $4^{\text{ème}}$ lignes et colonnes.

b) On suppose que $\forall\beta \in \{2, 3\}$, on a: $c_{2\beta}^{41} = c_{25}^{4\beta} = 0$.

b_1) Si $c_{24}^{41} \neq 0$ et $c_{25}^{44} \neq 0$, le lemmes 2.11. et 2.12 s'appliquent, et on a: $k_3^2 = \tilde{k}_3^2$, $k_4^2 = \tilde{k}_4^2$, $\Phi_3^2 = k_3^2\Phi_2^3$, $\Phi_4^2 = k_4^2\Phi_2^4$, $c_5^1 = c_2^1 k_3^2 c_4^3 c_5^4$; on obtient la présymétrie.

b_2) Si $c_{24}^{41} \neq 0$, $c_{25}^{44} = 0$, on distingue:

b_2') Si on a: $c_{2\alpha}^{31} \neq 0$, $\alpha \in \{2, 4\}$ et si $c_{25}^{33} \neq 0$, on a: [lemme 2.8] $k_3^2 = \tilde{k}_3^2$, $k_4^2 = \tilde{k}_4^2$ et [lemme 2.11] $c_{45}^{21} = k_4^2 c_{25}^{41}$; par le lemme 2.12, on a: $c_5^1 = c_2^1 k_3^2 c_4^3 c_5^4$. On obtient la présymétrie.

Si on a: $c_{2\alpha}^{31} \neq 0$, $\alpha \in \{2, 4\}$, $c_{25}^{33} = 0$, on a: [lemme 2.11 et 2.12]: $\Phi_4^2(\xi_1^4, \xi_1^5, \xi_3^4) = k_4^2\Phi_2^4(\xi_1^4, \xi_1^5, \xi_3^4)$ et $\Phi_3^2 \equiv \Phi_3^2(\xi_1^\alpha, \xi_1^3, \xi_3^4) = k_3^2\Phi_2^4(\xi_1^\alpha, \xi_1^3, \xi_3^4)$, $c_5^1 = c_\beta^1 c_5^\beta$, $\beta \neq \alpha$, $\beta \in \{2, 4\}$.

Si $c_{24}^{33} \neq 0$ ou $c_{24}^{43} \neq 0$, on a: $k_3^2 = \tilde{k}_3^2$, d'où la présymétrie.

Supposons $c_{24}^{33} = c_{24}^{43} = 0$.

Le lemme 2.12 implique $c_{45}^{21} = \tilde{k}_4^2 c_{25}^{41}$ et le lemme 2.11 implique $c_{45}^{21} = k_4^2 c_{25}^{41}$.

Si $c_{25}^{41} \neq 0$, on a: $\tilde{k}_4^2 = \tilde{k}_5^2$, $\tilde{k}_3^2 = \tilde{k}_3^2$, puis la présymétrie.

Si $c_{25}^{41} = 0$, on ramène $a_j^i(\xi)$ à la forme:

si $\alpha = 2$:

$$\begin{pmatrix} \psi_5 & \xi_4^5 & \xi_3^5 & \rho\xi_2^5 & \xi_3^5 \\ \xi_4^5 & \psi_4 & \xi_3^4 & \mu\xi_1^4 & \xi_1^4 \\ \xi_3^5 & \xi_3^4 & \psi_3 & \lambda\xi_1^3 + \lambda'\xi_1^2 & \xi_1^3 \\ \xi_2^5 & \mu\xi_1^4 & \lambda\xi_1^3 + \lambda'\xi_1^2 & \psi_2 & \xi_1^2 \\ \xi_1^5 & \xi_1^4 & \xi_1^3 & \xi_1^2 & \psi_1 \end{pmatrix} \quad \lambda' \neq 0, \mu \neq 0;$$

on pose $\xi_3^4 = \xi_1^4 = \xi_1^3 = 0$; $\xi_4^5 = \xi_1^5 = 0$ on diagonalise la sous-matrice en bas à droite comme précédemment on peut choisir la 1ère colonne de Δ avec $u_3 \neq 0$, $u_2 \neq 0$: on obtient que: $u_2\xi_3^5 + \rho u_3\xi_2^5$ doit être proportionnel à $u_2\xi_3^5 + u_3\xi_2^5$ d'où $\rho = 1$ et la symétrie.

si $\alpha = 4$,

$$\begin{pmatrix} \psi_5 & \rho\xi_4^5 & \rho\xi_3^5 & \xi_2^5 & \xi_1^5 \\ \xi_4^5 & \psi_4 & \xi_3^4 & \mu\xi_1^4 & \xi_1^4 \\ \xi_3^5 & \xi_3^4 & \psi_3 & \lambda\xi_1^3 + \lambda'\xi_1^4 & \xi_1^3 \\ \xi_2^5 & \mu\xi_1^4 & \lambda\xi_1^3 + \lambda'\xi_1^4 & \psi_2 & \xi_1^2 \\ \xi_1^5 & \xi_1^4 & \xi_1^3 & \xi_1^2 & \psi_1 \end{pmatrix} \quad \lambda' \neq 0, \mu \neq 0;$$

on pose $\xi_3^4 = \xi_1^3 = \xi_1^2 = 0$ $\xi_4^5 = \xi_1^5 = 0$ et on obtient encore la présymétrie.

b_2'') Si $c_{25}^{3\alpha} \neq 0$, $\alpha \in \{2, 4\}$ et $c_{23}^{31} \neq 0$ ou $c_{24}^{33} \neq 0$ ou $c_{24}^{43} \neq 0$, on obtient la présymétrie comme dans le cas b_2').

Si $c_{23}^{31} = c_{24}^{33} = c_{24}^{43} = 0$, on a de même $\Phi_4^2(\xi_1^4, \xi_1^5) = k_4^2\Phi_2^4(\xi_1^4, \xi_1^5)$ et $\Phi_3^2(\xi_\alpha^5, \xi_3^5) = k_3^2\Phi_2^3(\xi_\alpha^5, \xi_3^5)$, $c_5^1 = c_\beta^1 c_5^\beta$, $\beta \neq \alpha$.

Si $c_{25}^{41} \neq 0$, on a aussi $k_4^2 = \tilde{k}_4^2$ $k_3^2 = \tilde{k}_3^2$ et la présymétrie.

Si $c_{25}^{41} = 0$ on ramène $a_j^i(\xi)$ à la forme

si $\alpha = 2$

$$\begin{pmatrix} \psi_2 & \xi_1^2 & \rho(\lambda\xi_3^5 + \lambda'\xi_2^5) & \mu\xi_1^4 & \rho\xi_2^5 \\ \xi_1^2 & \psi_1 & \xi_1^3 & \xi_1^4 & \xi_1^5 \\ \lambda\xi_3^5 + \lambda'\xi_2^5 & \xi_1^3 & \psi_3 & \xi_3^4 & \xi_3^5 \\ \mu\xi_1^4 & \xi_1^4 & \xi_3^4 & \psi_4 & \xi_4^5 \\ \xi_2^5 & \xi_1^5 & \xi_3^5 & \xi_4^5 & \psi_1 \end{pmatrix} \quad \lambda' \neq 0, \mu \neq 0, \rho = \dfrac{\tilde{k}_3^2}{k_3^2}$$

si $\alpha = 4$

$$\begin{pmatrix} \psi_5 & \dfrac{\xi_4^5}{\rho} & \dfrac{\xi_3^5}{\rho} & \xi_2^5 & \xi_1^5 \\ \xi_4^5 & \psi_4 & \xi_3^4 & \mu\xi_1^4 & \xi_1^4 \\ \xi_3^5 & \xi_3^4 & \psi_3 & \rho(\lambda\xi_3^5 + \lambda'\xi_2^5) & \xi_1^3 \\ \xi_2^5 & \mu\xi_1^4 & \lambda\xi_3^5 + \lambda'\xi_2^5 & \psi_2 & \xi_1^2 \\ \xi_1^5 & \xi_1^4 & \xi_1^3 & \xi_1^2 & \psi_1 \end{pmatrix} \quad \lambda' \neq 0, \mu \neq 0, \rho = \dfrac{\tilde{k}_3^2}{k_3^2}.$$

Dans le premier cas, on pose $\xi_1^4 = \xi_1^5 = \xi_3^4 = \xi_3^5 = 0$, on obtient $\xi_1^2 u_1 + \rho\lambda'\xi_2^5$ proportionnels à $\xi_1^2 u_1 + \lambda'\xi_2^5$ avec $u_1 \neq 0$, $u_2 \neq 0$ d'où $\rho = 1$ et la symétrie.

Dans le second cas, on pose $\xi_4^5 = \xi_3^4 = \xi_1^4 = 0$; on supprime la $2^{\text{ème}}$ ligne et la $2^{\text{ème}}$ colonne et on écrit que la matrice 4×4 obtenue est présymétrique.

b_3) Si $c_{24}^{41} = 0$, $c_{25}^{44} \neq 0$, on distingue

b_3') $c_{2\alpha}^{31} \neq 0$, $\alpha \in \{2, 4\}$.

Si $c_{25}^{33} \neq 0$ ou $c_{24}^{33} \neq 0$ ou $c_{24}^{43} \neq 0$, on a la présymétrie comme dans le début du cas b_2'.

Si $c_{25}^{33} = c_{24}^{33} = c_{24}^{43} = 0$, on a:

$$\Phi_3^2 \equiv \Phi_3^2(\xi_1^\alpha, \xi_1^3) = k_3^2 \Phi_2^3(\xi_1^\alpha, \xi_1^3)$$

et

$$\Phi_4^2 \equiv \Phi_3^2(\xi_4^5, \xi_1^5) = \bar{k}_4^2 \Phi_2^4(\xi_4^5, \xi_1^5)$$

$c_5^1 = c_\beta^1 c_5^\beta$, $\beta \neq \alpha$.

Si $\alpha = 2$, on se ramène a

$$\begin{pmatrix} \psi_2 & \xi_1^2 & \Phi_2^3(\xi_1^2, \xi_1^3) & \rho\Phi_2^4(\xi_4^5, \xi_1^5) & \rho\xi_2^5 \\ \xi_1^2 & \psi_1 & \xi_1^3 & \xi_1^4 & \xi_1^5 \\ \Phi_2^3(\xi_1^2, \xi_1^3) & \xi_1^3 & \psi_3 & \xi_3^4 & \xi_3^5 \\ \Phi_2^4(\xi_4^5, \xi_1^5) & \xi_1^4 & \xi_3^4 & \psi_4 & \xi_4^5 \\ \xi_2^5 & \xi_1^5 & \xi_3^5 & \xi_4^5 & \psi_5 \end{pmatrix}$$

et si $\alpha = 4$ à

$$\begin{pmatrix} \psi_1 & \xi_1^2 & \xi_1^3 & \xi_1^4 & \xi_1^5 \\ \xi_1^2 & \psi_2 & \Phi_2^3(\xi_1^4, \xi_1^3) & \rho\Phi_2^4(\xi_1^5, \xi_1^5) & \xi_2^5 \\ \xi_1^3 & \Phi_2^3(\xi_1^4, \xi_1^3) & \psi_3 & \xi_3^4 & \dfrac{1}{\rho}\xi_3^5 \\ \xi_1^4 & \Phi_2^4(\xi_4^5, \xi_1^5) & \xi_3^4 & \psi_4 & \dfrac{1}{\rho}\xi_4^5 \\ \xi_1^5 & \xi_2^5 & \xi_3^5 & \xi_4^5 & \psi_5 \end{pmatrix}$$

On obtient la symétrie dans le premier cas en posant $\xi_1^3 = \xi_1^5 = \xi_3^4 = \xi_4^5 = 0$ et on construit Δ de la forme

$$\begin{pmatrix} 0 & 0 & * & * \\ u_2 & * & 0 & 0 \\ 0 & 0 & * & * \\ u_4 & * & 0 & 0 \end{pmatrix}$$

avec $u_2 \neq 0$ et $u_4 \neq 0$ et en écrivant que $c_{22}^{31}\xi_1^2 u_2 + \rho\xi_2^5 u_4$ est proportionnel à $c_{22}^{31}\xi_1^2 u_2 + \xi_2^5 u_4$ on a la présymétrie.

Dans le $2^{\text{ème}}$ cas, on pose $\xi_1^4 = \xi_3^4 = \xi_4^5 = \xi_1^5 = 0$ et on symétrise la sous-matrice 4×4 obtenue en rayant la $4^{\text{ème}}$ ligne et la $4^{\text{ème}}$ colonne de $\left(a_j^i(\xi')\right)$.

b_3'') on a: $c_{25}^{3\alpha} \neq 0$, $\alpha \in \{2, 4\}$.

Si $c_{23}^{31} \neq 0$ ou $c_{24}^{33} \neq 0$ ou $c_{24}^{43} \neq 0$ on a la presymétrie comme on début de b_2').

Si $c_{23}^{31} = c_{24}^{33} = c_{24}^{43} = 0$ on a: $\Phi_4^2(\xi_4^5, \xi_1^5) = \tilde{k}_4^2 \Phi_2^4(\xi_4^5, \xi_1^5)$ et $\Phi_3^2 \equiv \Phi_3^2(\xi_\alpha^5, \xi_3^5) = \tilde{k}_4^2 \Phi_2^3(\xi_\alpha^5, \xi_3^5)$, $c_5^1 = c_\beta^1 c_5^\beta$.

Si $\alpha = 2$, on se ramène à:

$$\begin{pmatrix} \psi_2 & \xi_1^2 & \rho\Phi_2^3(\xi_2^5, \xi_3^5) & \rho\Phi_2^4(\xi_4^5, \xi_1^5) & \rho\xi_2^5 \\ \xi_1^2 & \psi_1 & \xi_1^3 & \xi_1^4 & \xi_1^5 \\ \Phi_2^3(\xi_2^5, \xi_3^5) & \xi_1^3 & \psi_3 & \xi_3^4 & \xi_3^5 \\ \Phi_2^4(\xi_4^5, \xi_1^5) & \xi_1^4 & \xi_3^4 & \psi_4 & \xi_4^5 \\ \xi_2^5 & \xi_1^5 & \xi_3^5 & \xi_4^5 & \psi_5 \end{pmatrix} \qquad \rho = \frac{\tilde{k}_3^2}{k_3^2}$$

Si $\alpha = 4$, on se ramène à:

$$\begin{pmatrix} \psi_5 & \dfrac{1}{\rho}\xi_4^5 & \dfrac{1}{\rho}\xi_3^5 & \xi_2^5 & \xi_1^5 \\ \xi_4^5 & \psi_4 & \xi_3^4 & \rho\Phi_2^4(\xi_4^5, \xi_1^5) & \xi_1^4 \\ \xi_3^5 & \xi_3^4 & \psi_3 & \rho\Phi_2^3(\xi_3^5, \xi_4^5) & \xi_1^3 \\ \xi_2^5 & \Phi_2^4(\xi_4^5, \xi_1^5) & \Phi_2^3(\xi_3^5, \xi_4^5) & \psi_2 & \xi_1^2 \\ \xi_1^5 & \xi_1^4 & \xi_1^3 & \xi_1^2 & \psi_1 \end{pmatrix} \qquad \rho = \frac{\tilde{k}_3^2}{k_3^2}.$$

Dans le premier cas, on pose $\xi_1^4 = \xi_1^5 = \xi_3^5 = \xi_3^5 = 0$ et on construit un diagonaliseur Δ où la $1^{\text{ère}}$ colonne est telle que $u_1 \neq 0$, $u_2 \neq 0$, $u_3 = 0$, $u_4 = 0$, on obtient $\xi_1^2 u_1 + \rho c_{25}^{32} \xi_2^5 u_2$ proportionnel à $\xi_1^2 u_1 + c_{25}^{32} \xi_2^5 u_2$, d'où la symétrie.

Dans le deuxième cas on pose $\xi_4^5 = \xi_3^4 = \xi_1^4 = \xi_1^5 = 0$ et on symétrise la matrice 4×4 obtenue en rayant les $4^{\text{ème}}$ ligne et colonne.

b_4) On a $c_{24}^{41} = c_{25}^{44} = 0$

b_4') $c_{2\alpha}^{31} \neq 0$, $\alpha \in \{2, 4\}$. Si $c_{25}^{33} \neq 0$ ou $c_{24}^{33} \neq 0$ ou $c_{24}^{43} \neq 0$ on a la présymétrie comme au cas b_2').

Si $c_{25}^{33} = c_{24}^{33} = c_{24}^{43} = 0$, on a: $\Phi_3^2 \equiv \Phi_3^2(\xi_1^\alpha, \xi_1^3) = k_3^2 \Phi_3^2(\xi_1^\alpha, \xi_1^3)$ et $\Phi_4^2 \equiv \Phi_4^2(\xi_1^5) = \tilde{k}_4^2 \Phi_2^4(\xi_1^5)$, $c_5^1 = c_\beta^1 c_5^\beta$.

On se ramène si $\alpha = 2$ à

$$\begin{pmatrix} \psi_2 & \xi_1^2 & \Phi_2^3(\xi_1^2, \xi_1^3) & \rho\lambda\xi_1^5 & \rho\xi_2^5 \\ \xi_1^2 & \psi_1 & \xi_1^3 & \xi_1^4 & \xi_1^5 \\ \Phi_2^3(\xi_1^2, \xi_1^3) & \xi_1^3 & \psi_5 & \xi_3^4 & \xi_3^5 \\ \lambda\xi_1^5 & \xi_1^4 & \xi_3^4 & \psi_4 & \xi_4^5 \\ \xi_2^5 & \xi_1^5 & \xi_3^5 & \xi_4^5 & \psi_r \end{pmatrix} \qquad \rho = \frac{\tilde{k}_3^2}{k_3^2};$$

si $\alpha = 4$

$$\begin{pmatrix} \psi_5 & \dfrac{\xi_4^5}{\rho} & \dfrac{\xi_3^5}{\rho} & \xi_1^5 & \xi_2^5 \\ \xi_4^5 & \psi_4 & \xi_3^4 & \xi_1^4 & \rho\lambda\xi_1^5 \\ \xi_2^5 & \xi_3^4 & \psi_3 & \xi_1^3 & \mu\xi_1^4 + \mu'\xi_1^3 \\ \xi_1^5 & \xi_1^4 & \xi_1^3 & \psi_1 & \xi_1^2 \\ \xi_2^5 & \lambda\xi_1^5 & \mu\xi_1^4 + \mu'\xi_1^3 & \xi_1^2 & \psi_2 \end{pmatrix} \qquad \rho = \frac{\tilde{k}_3^2}{k_3^2}; \quad \mu \neq 0$$

Dans le premier cas, on pose $\xi_1^3 = \xi_1^4 = \xi_3^4 = \xi_3^5 = \xi_4^5 = 0$, on construit le diagonaliseur Δ:

$$\Delta = \begin{pmatrix} u_1 & 0 & 0 & * \\ 0 & 1 & 0 & 0 \\ 0 & 0 & 1 & 0 \\ u_4 & 0 & 0 & v_4 \end{pmatrix} \quad \text{avec } u_1 \neq 0,\, u_4 \neq 0;$$

on doit avoir $\xi_1^2 u_1 + \rho \xi_2^5 u_4$ proportionnel à $\xi_1^2 u_1 + \xi_2^5 u_4$, d'où le résultat.

Dans le deuxième cas, on pose $\xi_3^4 = \xi_1^5 = \xi_1^3 = \xi_1^2 = 0$ et

$$\Delta = \begin{pmatrix} u_1 & * & 0 & 0 \\ 0 & 0 & v_2 & *0 \\ u_3 & * & 0 & 0 \\ 0 & 0 & v_4 & * \end{pmatrix} \quad \text{avec } v_2 \neq 0,\, v_4 \neq 0;$$

on a: $\dfrac{\xi_3^5}{\rho} v_2 + \xi_2^5 v_4$ proportionnel à $\xi_3^5 v_2 + \xi_2^5 v_4$ d'où le résultat.

b$_4''$) $c_{25}^{3\alpha} \neq 0$, $\alpha \in \{2, 4\}$. Si $c_{23}^{31} \neq 0$ ou $c_{24}^{33} \neq 0$ ou $c_{24}^{43} \neq 0$, on obtient la présymétrie.

Si $c_{23}^{31} = c_{24}^{33} = c_{24}^{43} = 0$, on a:

$$\Phi_3^2 \equiv \Phi_3^2(\xi_5^\alpha, \xi_5^3) = \tilde{k}_3^2 \Phi_3^2(\xi_5^\alpha, \xi_5^3)$$

et

$$\Phi_4^2 \equiv \Phi_4^2(\xi_1^5) = \tilde{k}_4^2 \Phi_4^2(\xi_1^5).$$

On se ramène à
si $\alpha = 2$

$$\begin{pmatrix} \xi_2 & \xi_1^2 & \rho(\mu\xi_2^5 + \mu'\xi_3^5) & \lambda\xi_1^5 & \rho\xi_2^5 \\ \xi_1^2 & \psi_1 & \xi_1^3 & \xi_1^4 & \xi_1^5 \\ \mu\xi_2^5 + \mu'\xi_3^5 & \xi_1^3 & \psi_3 & \xi_3^4 & \xi_3^5 \\ \lambda\xi_1^5 & \xi_1^4 & \xi_3^4 & \psi_4 & \xi_4^5 \\ \xi_2^5 & \xi_1^5 & \xi_3^5 & \xi_4^5 & \psi_5 \end{pmatrix} \quad \mu \neq 0, \quad \rho = \dfrac{\tilde{k}_3^2}{k_3^2}$$

si $\alpha = 4$

$$\begin{pmatrix} \xi_5 & \dfrac{1}{\rho}\xi_4^5 & \dfrac{1}{\rho}\xi_3^5 & \xi_2^5 & \xi_1^5 \\ \xi_4^5 & \psi_4 & \xi_3^4 & \lambda\xi_1^5 & \xi_1^4 \\ \xi_3^5 & \xi_3^4 & \psi_3 & \rho(\mu\xi_4^5 + \mu'\xi_3^5) & \xi_1^3 \\ \xi_2^5 & \lambda\xi_1^5 & \mu\xi_4^5 + \mu'\xi_3^5 & \psi_2 & \xi_1^2 \\ \xi_1^5 & \xi_1^4 & \xi_1^3 & \xi_1^2 & \psi_1 \end{pmatrix} \quad \mu \neq 0, \quad \rho = \dfrac{\tilde{k}_3^2}{k_3^2}$$

dans le premier cas, on pose $\xi_1^4 = \xi_3^4 = \xi_1^5 = \xi_3^5 = \xi_4^5 = 0$ et par un calcul analogue aux précédents, on a la présymétrie.

Dans le 2ème cas, on pose $\xi_4^5 = \xi_3^4 = \xi_1^5 = \xi_1^4 = 0$ et on symétrise la matrice 4×4 obtenue. $\qquad\qquad\square$

Proposition 2.14. *Si* $\exists \begin{pmatrix} \alpha \\ \beta \end{pmatrix} \in \mathcal{J}_2$, $\begin{pmatrix} \alpha \\ \beta \end{pmatrix} \neq \begin{pmatrix} 1 \\ 5 \end{pmatrix}$ *tel que* $c_{2\beta}^{4\alpha} \neq 0$ *alors* $a(\xi)$ *est presymétrique.*

Preuve. Elle se ramène à celle de la proposition précédent par échange des $3^{\text{ème}}$ et $4^{\text{ème}}$ lignes et colonnes et changement de notations. $\square$

Proposition 2.15. *Si* $\forall \begin{pmatrix} \alpha \\ \beta \end{pmatrix} \in \mathcal{J}_1$ $\begin{pmatrix} \alpha \\ \beta \end{pmatrix} \neq \begin{pmatrix} 1 \\ 5 \end{pmatrix}$ *on a:* $c_{2\beta}^{3\alpha} = 0$ *et si* $\forall \begin{pmatrix} \gamma \\ \delta \end{pmatrix} \in \mathcal{J}_2$, $\begin{pmatrix} \gamma \\ \delta \end{pmatrix} \neq \begin{pmatrix} 1 \\ 5 \end{pmatrix}$ *on a:* $c_{2\delta}^{4\gamma} = 0$ *alors* $a(\xi)$ *est presymétrie.*

Preuve. Si $\left(c_{23}^{31} \neq 0 \text{ et } c_{25}^{33} \neq 0 \right)$, ou $\left(c_{24}^{41} \neq 0 \text{ et } c_{25}^{44} \neq 0 \right)$, les lemmes 2.10 et 2.11 impliquent: $k_3^2 = \tilde{k}_3^2, k_4^2 = \tilde{k}_4^2, \Phi_3^2 = k_3^2\Phi_2^3, \Phi_4^2 = k_4^2\Phi_2^4$. Il reste à étudier c_5^1. Supposons $c_{23}^{31} \neq 0$ et $c_{25}^{33} \neq 0$; l'autre cas se réduit à celui là par échange des $3^{\text{ème}}$ et $4^{\text{ème}}$ lignes et colonnes.

On pose $\xi_1^2 = \xi_2^5 = 0$; $c_{23}^{31}\xi_1^3 + c_{25}^{33}\xi_3^5 + c_{24}^{33}\xi_3^4 + c_{25}^{31}\xi_1^5 = 0$; on exprimera ξ_1^3 à l'aide des autres variables, ensuite, on suppose d'abord $c_{24}^{41} \neq 0$ ou $c_{25}^{44} \neq 0$ et on pose: $c_{24}^{41}\xi_1^4 + c_{25}^{44}\xi_4^5 + c_{24}^{43}\xi_3^4 + c_{25}^{41}\xi_1^5 = 0$; on exprime ξ_1^4 ou ξ_4^5 à l'aide des autres variables.

On considère la matrice b 4×4 obtenue en rayant la $2^{\text{ème}}$ ligne et la $2^{\text{ème}}$ colonne de a; elle est présymétrique (1.1) et (1.2). Il existe H telle que:

$$bH = H^t b.$$

on explicite et on obtient:

$$c_5^1 = \frac{h_1}{h_4} = \frac{h_1}{h_2}\frac{h_2}{h_4} = c_3^1 c_5^3;$$

d'où la présymétrie de a.

Si $c_{24}^{41} = c_{25}^{44} = 0$ et $c_{24}^{43} \neq 0$ on procède de même et on a la presymétrie.

Si $c_{24}^{41} = c_{25}^{44} = c_{24}^{43} = 0$, on réduit comme précédemment $\left(a_j^i(\xi') \right)$ à la forme

$$\begin{pmatrix} \psi_5 & \rho\xi_4^5 & \rho\xi_3^5 & \rho\xi_2^5 & \xi_1^5 \\ \xi_4^5 & \psi_4 & \xi_3^4 & \lambda\xi_1^5 & \xi_1^4 \\ \xi_3^5 & \xi_3^4 & \psi_3 & \Phi_2^3(\xi_1^3, \xi_3^5, \xi_3^4, \xi_1^5) & \xi_1^3 \\ \xi_2^5 & \lambda\xi_1^5 & \Phi_2^3(\xi_1^3, \xi_3^5, \xi_3^4, \xi_1^5) & \psi_2 & \xi_1^2 \\ \xi_1^5 & \xi_1^4 & \xi_1^3 & \xi_1^2 & \psi_1 \end{pmatrix} \qquad \rho = \frac{c_2^3 k_3^2 c_4^3 c_5^4}{c_5^1};$$

Supposons $\lambda \neq 0$; on pose $\xi_3^4 = \xi_1^3 = \xi_3^4 = 0$; $c_{25}^{33}\xi_3^5 + c_{25}^{31}\xi_1^5 = 0$; $\xi_4^5 = 0$; la sous-matrice 4×4, en bas à droite, est diagonalise par Δ; on peut choisir Δ:

$$\Delta = \begin{pmatrix} u_1 & * & * & 0 \\ 0 & 0 & 0 & 1 \\ u_3 & * & * & 0 \\ u_4 & * & * & 0 \end{pmatrix},$$

avec $u_1 \neq 0$, $u_3(\xi_1^4, \xi_1^5, \psi_4, \psi_2, \psi_1) \neq 0$, $u_4 \neq 0$.

On utilise la conséquence 1.4, on obtient $\rho u_2 \xi_2^5 + u_4 \xi_1^5$ et $u_2 \xi_2^5 + u_4 \xi_1^5$ sont de même signe, d'où $\rho = 1$ et la symétrie de a.

Si $\lambda = 0$, on pose $\xi_2^5 = \xi_1^5 \neq 0$; $c_{25}^{33}\xi_3^5 + c_{23}^{33}\xi_1^3 + c_{25}^{31}\xi_1^5 + c_{24}^{33}\xi_3^4 = 0$; on exprime ξ_3^5 à l'aide des autres variables et on symétrise **la matrice** 4×4 obtenue en barrant la 4ème ligne et colonne; on obtient encore $\rho = 1$.

b) On suppose que $\left(c_{23}^{31} = 0 \text{ ou } c_{25}^{33} = 0\right)$ et $\left(c_{24}^{41} = 0 \text{ ou } c_{25}^{44} = 0\right)$.

On distingue des sous-cas et on procède **comme au a)**.

b$_1$) si $c_{23}^{31} \neq 0$ et $c_{24}^{41} \neq 0$, on pose $\xi_1^2 = \xi_2^5 = 0$; $c_{23}^{31}\xi_1^3 + c_{24}^{33}\xi_3^4 + c_{25}^{31}\xi_1^5 = 0$; $c_{24}^{41}\xi_1^3 + c_{24}^{43}\xi_3^4 + c_{25}^{41}\xi_1^5 = 0$; on exprime ξ_1^3 et ξ_1^4 à l'aide de ξ_3^4 et ξ_1^5; on symétrise la matrice 4×4 obtenue en rayant la 2ème ligne et la 2ème colonne; si $c_{24}^{33} \neq 0$ ou $c_{25}^{31} \neq 0$, on obtient $c_5^1 = c_3^1 c_5^3$ en posant $\xi_1^3 = \xi_3^4 = \xi_1^5 = \xi_1^5 = 0$ et on symétrise la matrice obtenue en rayant la 3ème ligne et la 3ème colonne; on a: $c_5^2 = k_4^2 c_5^4$ d'où la présymétrie; si $c_{24}^{33} = c_{25}^{31} = 0$, on pose: $\xi_1^3 = \xi_3^4 = \xi_3^5 = 0$, on symétrise la matrice 4×4 obtenue en rayant la 3ème ligne et la 3ème colonne et on obtient:

$$c_5^1 = c_2^1 c_5^2, \qquad c_5^2 = k_4^2 c_5^4,$$

d'où la présymétrie;

b$_2$) Si $c_{23}^{31} \neq 0$ et $c_{25}^{44} \neq 0$ on a (2.10) et (2.11)

$$\Phi_3^2 \equiv \Phi_3^2(\xi_1^3, \xi_3^4, \xi_1^5) = k_3^2 \Phi_2^3$$
$$\Phi_4^2 \equiv \Phi_4^2(\xi_4^5, \xi_3^4, \xi_1^5) = \tilde{k}_4^2 \Phi_2^4.$$

On montre d'abord que $\tilde{k}_4^2 = k_4^2$: on pose $\xi_1^3 = \xi_3^4 = \xi_3^5 = \xi_1^5 = 0$ et on diagonalise la matrice 4×4 obtenue en rayant les 3ème ligne et colonne.

Ensuite on procède comme au b$_1$) et on obtient la presymétrie.

b$_3$) Si $c_{23}^{31} \neq 0$, $c_{24}^{41} = c_{25}^{44} = 0$, on a:

$$\Phi_3^2 \equiv \Phi_3^2(\xi_1^3, \xi_3^4, \xi_1^5) = k_3^2 \Phi_2^3.$$

Si $c_{24}^{43} \neq 0$,

$$\Phi_4^2 \equiv \Phi_4^2(\xi_3^4, \xi_1^5) = k_4^2 \Phi_2^4.$$

Si $c_{24}^{43} = 0$,

$$\Phi_4^2 \equiv c_{45}^{21}\xi_1^5 \quad \text{et} \quad \Phi_2^4 \equiv c_{25}^{41}\xi_1^5$$

si $c_{24}^{33} \neq 0$ ou $c_{25}^{31} \neq 0$, on montre que $c_{45}^{21} = k_4^2 c_{25}^{41}$ en posant $\xi_1^2 = \xi_1^3 = \xi_1^4 = \xi_2^5 = \xi_3^5 = \xi_1^5 = 0$ et en symétrisant la matrice 3×3 obtenue, si $c_{24}^{33} = c_{25}^{31} = 0$, en posant, si $c_{25}^{41} \neq 0$ $\xi_1^3 = \xi_3^4 = \xi_3^5 = 0$, on obtient le résultat, si $c_{25}^{41} = 0$ on a $c_{45}^{21} = 0$. Il nous reste à exprimer c_5^2 et c_5^1

En posant $\xi_1^3 = \xi_3^4 = \xi_3^5 = \xi_1^5 = 0$ et en symétrisant on obtient $c_5^2 = k_4^2 c_5^4$, $k_4^2 = \tilde{k}_4^2$.

Ensuite on ramène comme précédemment a à la forme

$$\begin{pmatrix} \psi_5 & \rho\xi_4^5 & \rho\xi_3^5 & \rho\xi_2^5 & \xi_1^5 \\ \xi_4^5 & \psi_4 & \xi_3^4 & \lambda\xi_1^5 + \lambda'\xi_3^4 & \xi_1^4 \\ \xi_3^5 & \xi_3^4 & \psi_3 & \Phi_2^3(\xi_1^3, \xi_3^4, \xi_1^5) & \xi_1^3 \\ \xi_2^5 & \lambda\xi_1^5 + \lambda'\xi_3^4 & \Phi_2^3(\xi_1^3, \xi_3^4, \xi_1^5) & \psi_2 & \xi_1^2 \\ \xi_1^5 & \xi_1^4 & \xi_1^3 & \xi_1^2 & \psi_1 \end{pmatrix} \qquad \rho = \frac{c_2^1 k_3^2 c_4^3 c_5^4}{c_5^2};$$

si $c_{25}^{31} \neq 0$, on pose $\xi_1^2 = \xi_1^4 = \xi_3^4 = 0$; $c_{23}^{31}\xi_1^3 + c_{25}^{31}\xi_1^5 = 0$; on exprime ξ_1^3 en fonction de ξ_1^5; on diagonalise le sous-matrice 4×4 en bas à droite, on obtient Δ sous la forme

$$\Delta = \begin{pmatrix} 0 & 0 & * & * \\ u_3 & * & 0 & 0 \\ 0 & 0 & * & * \\ u_4 & * & 0 & 0 \end{pmatrix} \qquad u_2(\xi_1^5) \neq 0; \ u_4(\xi_1^5) \neq 0;$$

par la conséquence 1.4, on obtient $\rho\xi_3^5 u_2 + \xi_1^5 u_4$ est du même signe que $\xi_3^5 u_2 + \xi_1^5 u_4$ on en déduit $\rho = 1$ et la symétrie.

si $c_{25}^{31} = 0$ on pose $\xi_1^3 = \xi_1^4 = \xi_3^4 = 0$, on symétrise la sous-matrice 4×4 en rayant la $3^{\text{ème}}$ ligne et la $3^{\text{ème}}$ colonne et on obtient $\rho = 1$ et la symétrie.

b_4) $c_{25}^{33} \neq 0$ et $c_{25}^{44} \neq 0$ on procède comme dans le cas b_1).

b_5) $c_{25}^{33} \neq 0$ et $\left(c_{24}^{41} = c_{25}^{44} = 0\right)$ on procède comme dans le cas b_3).

Les cas b_6) $c_{25}^{33} \neq 0$ et $c_{24}^{41} \neq 0$, b_7) $c_{24}^{41} \neq 0$ et $\left(c_{23}^{31} = c_{21}^{33} = 0\right)$ b_8) $c_{25}^{44} \neq 0$ et $\left(c_{23}^{31} = c_{25}^{33} = 0\right)$ se ramènent aux cas précédents en échangeant les $3^{\text{ème}}$ ligne et colonne.

b_9) Il reste à étudier le cas $c_{23}^{31} = c_{25}^{33} = c_{24}^{41} = c_{25}^{44} = 0$

i) Si c_{24}^{33} ou $c_{24}^{43} \neq 0$; on a: $k_3^2 = \bar{k}_3^2$; on pose: $\xi_1^2 = \xi_1^3 = \xi_1^4 = \xi_2^5 = \xi_3^5 = \xi_4^5 = 0$; en rayant les $1^{\text{ère}}$ et $5^{\text{ème}}$ lignes et colonnes, on obtient une matrice 3×3 présymétrique; on en déduit que: $c_{35}^{21} = k_3^2 c_{25}^{21}$ et $c_{45}^{21} = k_4^2 c_{25}^{41}$.

Ensuite, on ramène $a(\xi')$ à la forme

$$\begin{pmatrix} \psi_5 & \rho\xi_4^5 & \rho\xi_3^5 & \rho\xi_2^5 & \xi_1^5 \\ \xi_4^5 & \psi_4 & \xi_3^4 & \lambda\xi_3^4 + \lambda'\xi_1^5 & \xi_1^4 \\ \xi_3^5 & \xi_3^4 & \psi_3 & \mu\xi_3^4 + \mu'\xi_1^5 & \xi_1^3 \\ \xi_2^5 & \lambda\xi_3^4 + \lambda'\xi_1^5 & \mu\xi_3^4 + \mu'\xi_1^5 & \psi_2 & \xi_1^2 \\ \xi_1^5 & \xi_1^4 & \xi_1^3 & \xi_1^2 & \psi_1 \end{pmatrix},$$

$$\lambda' = c_{25}^{41}, \ \mu' = c_{25}^{31}, \ \rho = \frac{c_2^1 k_3^2 c_4^3 c_5^4}{c_5^1},$$

on pose $\xi_1^2 = \xi_1^3 = \xi_3^4 = 0$; $\xi_3^5 = \xi_2^5 = 0$; on construit Δ tel que

$$\Delta = \begin{pmatrix} u_1 & \cdots & \cdots & \cdots \\ u_2 & \cdots & \cdots & \cdots \\ u_3 & \cdots & \cdots & \cdots \\ u_4 & \cdots & \cdots & \cdots \end{pmatrix} \qquad u_1(\xi_1^4, \xi_1^3, \psi) \neq 0, u_4 \neq 0,$$

on obtient que $\rho \xi_4^5 u_1 + \xi_1^5 u_4$ est de même signe que $\xi_4^5 u_1 + \xi_1^5 u_4$, d'où $\rho = 1$ et la symétrie.

ii) $c_{24}^{33} = c_{24}^{43} = 0$. On pose $\xi_1^2 = \xi_1^3 = \xi_1^4 = \xi_2^5 = \xi_3^5 = \xi_4^5 = 0$ on symétrise la matrice 3×3 obtenue comme en i), on obtient $c_{35}^{21} = kc_{25}^{31}$, $c_{45}^{21} = kc_4^3 c_{25}^{41}$, $k \neq 0$; on calcule $D^{-1}a(\xi')D$, avec la matrice diagonale

$$\left(1, 1, \frac{1}{\sqrt{k}}, \frac{1}{\sqrt{k}}, 1\right);$$

on se ramène , comme précédemment a(ξ') à la forme

$$\begin{pmatrix} \psi_1 & c_2^1 \xi_1^2 & \dfrac{c_3^1 \xi_1^3}{k} & \dfrac{c_4^1 \xi_1^4}{kc_4^3} & c_5^1 \xi_1^5 \\ \xi_1^2 & \psi_2 & \sqrt{k}\mu' \xi_1^5 & \sqrt{kc_4^3}\lambda' \xi_1^5 & c_5^2 \xi_2^5 \\ \xi_1^3 & \sqrt{k}\mu' \xi_1^5 & \psi_3 & \xi_3^4 & kc_5^3 \xi_3^5 \\ \xi_1^4 & \sqrt{kc_4^3}\lambda' \xi_1^5 & \xi_5^4 & \psi_4 & kc_4^3 \xi_4^5 \\ \xi_1^5 & \xi_2^5 & \xi_3^5 & \xi_4^5 & \psi_5 \end{pmatrix}.$$

Si $\mu' \neq 0$ (resp. $\lambda' \neq 0$), on pose $\xi_2^5 = \xi_3^5 = \xi_4^5 = 0$; $\xi_1^4 = 0$ (resp. $\xi_1^3 = 0$) on symétrise la sous-matrice 4×4 en bas à droite, par Δ:

$$\Delta = \begin{pmatrix} u_1 & \cdots & \cdots & \cdots \\ u_2 & \cdots & \cdots & \cdots \\ u_3 & \cdots & \cdots & \cdots \\ u_4 & \cdots & \cdots & \cdots \end{pmatrix} \quad u_1 \neq 0, u_2 \neq 0, \text{(rsp. } u_3 \neq 0\text{),}$$

on doit avoir $c_2^1 u_1 \xi_1^2 + \dfrac{1}{k} c_3^1 \xi_1^3 u_2 \left[\text{resp. } c_2^1 u_1 \xi_1^2 + \dfrac{c_4^1 \xi_1^4 u_3}{kc_4^3}\right]$ et $\xi_1^2 u_1 + \xi_1^3 u_2 \left[\text{resp. } \xi_1^2 u_1 + \xi_1^4 u_3\right]$ de même signe; on en déduit $k = k_3^2$;

Si μ' [resp. $\lambda' \neq 0$] on obtient par symétrie par rapport à la $2^{\text{ème}}$ diagonale, de même $k = \bar{k}_3^2$ et a(ξ') s'écrit: en revenant à la première forme:

$$\begin{pmatrix} \psi_1 & c_2^1 \xi_1^2 & c_3^1 \xi_1^3 & c_5^1 \psi_4 1 & c_5^1 \xi_1^5 \\ \xi_1^2 & \psi_2 & k_3^2 \mu' \xi_1^5 & k_4^2 \lambda' \xi_1^5 & c_5^2 \xi_2^5 \\ \xi_1^3 & \mu' \xi_1^5 & \psi_3 & c_4^3 \xi_3^4 & c_5^3 \xi_3^5 \\ \xi_1^4 & \lambda' \xi_1^5 & \xi_3^4 & \psi_4 & c_5^4 \xi_4^5 \\ \xi_1^5 & \xi_2^5 & \xi_3^5 & \xi_4^5 & \psi_1 \end{pmatrix} \quad \lambda' \neq 0, \text{ ou } \mu' \neq 0,$$

et se ramène, en transformant par la matrice

$$\left(1 \quad \frac{1}{\sqrt{c_2^1}} \quad \frac{1}{\sqrt{c_3^1}} \quad \frac{1}{\sqrt{c_4^1}} \quad \frac{1}{\sqrt{c_5^1}}\right)$$

à

$$\begin{pmatrix} \psi_5 & \rho\xi_4^5 & \rho\xi_3^5 & \rho\xi_2^5 & \xi_1^5 \\ \xi_4^5 & \psi_4 & \xi_3^4 & \mu'\xi_1^5 & \xi_1^4 \\ \xi_3^5 & \xi_3^4 & \psi_3 & \mu\xi_1^5 & \xi_1^3 \\ \xi_2^5 & \mu'\xi_1^5 & \mu\xi_1^5 & \psi_2 & \xi_3^2 \\ \xi_2^5 & \xi_1^4 & \xi_1^3 & \xi_1^2 & \psi_1 \end{pmatrix} \qquad \lambda' \neq 0, \text{ ou } \mu' \neq 0, \ \rho = \frac{c_2^1 k_3^2 c_4^3 c_5^4}{c_5^1}.$$

Comme au i), on obtient la symétrie.

Si $\lambda' = \mu' = 0$, on pose $\xi_1^3 = \xi_3^4 = \xi_3^5 = 0$, on symétrise la matrice 4×4 obtenue en rayant la $3^{\text{ème}}$ ligne et la $3^{\text{ème}}$ colonne et on obtient le résultat. $\square$

3   Preuve du théorème; cas III ii)

Lemma 3.1. $c_{43}^{31} = c_{44}^{31} = c_{43}^{32} = c_{44}^{32} = 0;$ $c_{43}^{21} = c_{44}^{21} = c_{43}^{22} = 0;$ $c_{43}^{11} = c_{43}^{12} = c_{44}^{12} = 0;$ $c_{33}^{21} = c_{34}^{21} = c_{34}^{22} = 0;$ $c_{34}^{11} = c_{33}^{12} = c_{34}^{12} = 0;$ $c_{23}^{11} = c_{24}^{11} = c_{23}^{12} = c_{24}^{12} = 0;$ $c_3^1 = k_2^1 c_3^2;$ $c_4^1 = k_2^1 c_4^2;$ $k_2^1 > 0;$ $c_3^2 > 0;$ $c_4^2 > 0.$

Preuve. On pose $\xi_1^5 = \xi_2^5 = \xi_3^5 = \xi_4^5 = 0$; on obtient comme précédemment une matrice 4×4 présymétrique d'ou le résultat par les lemmes 1.1 et 1.2. $\square$

Lemma 3.2. $c_{5\beta}^{\gamma\alpha} = 0,$ *pour* $\begin{pmatrix} \alpha \\ \beta \end{pmatrix} \in \left\{ \begin{pmatrix} 2 \\ 4 \end{pmatrix}, \begin{pmatrix} 2 \\ 3 \end{pmatrix}, \begin{pmatrix} 1 \\ 4 \end{pmatrix}, \begin{pmatrix} 1 \\ 3 \end{pmatrix} \right\},$ $\gamma \in \{1, 2, 3, 4\}.$

Preuve. Si $\begin{pmatrix} \alpha \\ \beta \end{pmatrix} = \begin{pmatrix} 2 \\ 4 \end{pmatrix}$; on annule tous les ξ_j^i sauf ξ_2^4 que l'on pose égal à 1; on pose $\psi_1 = \psi_5 = \psi_3 = 0$, $\xi_0 = 0$ est racine triple; d'où par annulation d'un mineur d'ordre 3 correspondants, on obtient $c_{54}^{32} = 0$; on choisit ψ_2 ou ψ_4 tel que $\psi_2\psi_4 - c_4^2(\xi_2^4)^2 = 0$; $\xi = 0$ est racine quadruple, d'où $c_{54}^{12} = 0$, $c_{54}^{22} = 0$, $c_{54}^{42} = 0$; les autres résultats s'obtiennent de même. $\square$

Lemma 3.3. $c_{25}^{1\alpha} = c_{35}^{2\alpha} = c_{35}^{1\alpha} = c_{45}^{2\alpha} = c_{55}^{2\alpha} = c_{55}^{1\alpha} = 0,$ *pour* $\alpha \in \{\xi_3^5, \xi_4^5\}.$

Preuve. On annule tous les ξ_j^i sauf ξ_3^5 (resp. ξ_4^5). $\square$

Lemma 3.4. $c_{25}^{13} = c_{25}^{14} = 0;$ $c_{55}^{43} = c_{55}^{34} = 0;$ $c_5^3 = k_4^3 c_5^4$ *(définition);* $k_4^3 > 0;$ $c_5^4 > 0,$ $c_{45}^{33} = k_4^3 c_{35}^{43},$ $c_{45}^{34} = k_4^3 c_{35}^{44}.$

Preuve. On pose $\xi_1^5 = \xi_2^5 = \xi_1^4 = \xi_2^4 = \xi_1^3 = \xi_2^3 = 0$; il apparaît une sous-matrice 2×2 et une sous-matrice 3×3; alors on applique le lemme 1.1. $\square$

Lemma 3.5. $c_{55}^{41} = c_{55}^{42} = c_{55}^{21} = c_{55}^{12} = c_{45}^{21} = c_{45}^{22} = c_{45}^{11} = c_{45}^{12} = 0;$ $c_{25}^{11} = k_2^1 c_{15}^{21},$ $c_{25}^{12} = k_2^1 c_{15}^{22},$ $c_5^1 = k_2^1 c_5^2.$

Preuve. On pose $\xi_3^5 = \xi_4^5 = 0$; $c_{3}^1\xi_1^3 + c_{35}^{11}\xi_1^5 + c_{35}^{12}\xi_2^5 = 0$; $c_3^2\xi_2^3 + c_{35}^{21}\xi_1^5 + c_{35}^{22}\xi_2^5 = 0$; on exprime ξ_1^3 et ξ_2^3 à l'aide de ξ_1^5 et ξ_2^5; en rayant la $3^{\text{ème}}$ colonne et la $3^{\text{ème}}$ colonne de (a_j^i), on obtient une matrice 4×4 présymétrique par le lemme 1.1; on lui applique le lemme 1.2 et en détaillant les calculs on obtient les résultat. $\square$

Lemma 3.6. $c_{55}^{31} = c_{55}^{32} = c_{35}^{21} = c_{35}^{22} = c_{35}^{11} = c_{35}^{12} = 0.$

Preuve. On pose $\xi_1^4 = \xi_2^4 = \xi_3^5 = \xi_4^5 = 0$; il apparaît une matrice 4×4 en rayant les $4^{\text{ème}}$ lignes et colonnes, à laquelle on applique les lemmes 1.1 et 1.2. $\square$

Lemma 3.7. $c_{45}^{32} = c_{45}^{31} = 0.$

Preuve. Pour démontrer que $c_{45}^{32} = 0$ on annule tous les ξ_j^i excepté ξ_2^5 et par construction d'un point double, on a le résultat; on procède de même pour l'autre égalité. $\square$

Proposition 3.8. *Si $c_{15}^{22} \neq 0$, a est presymétrie.*

Preuve. On pose d'abord $\xi_1^5 = \xi_2^5 = \xi_1^3 = \xi_1^4 = 0$; la matrice 4×4 obtenue en rayant la $1^{\text{ère}}$ ligne et la $1^{\text{ère}}$ colonne est présymétrique, d'où: $c_4^2 = c_3^2 k_4^3$.

On transforme la matrice $\left(a_j^i(\xi')\right)$ par la matrice diagonale:

$$\left(1, \frac{1}{\sqrt{c_2^1}}, \frac{1}{\sqrt{c_3^1}}, \frac{1}{\sqrt{c_4^1}}, \frac{1}{\sqrt{c_5^1}}\right);$$

après un changement de variable dans $\mathbb{R}^{n+1}$ et dans $\mathbb{R}^m$, et en transposant, on obtient a(ξ') sous la forme

$$\begin{pmatrix} \psi_5 & \rho\xi_4^5 & \rho\xi_3^5 & \xi_2^5 & \xi_1^5 \\ \xi_4^5 & \psi_4 & \Phi_3^4(\xi_3^5, \xi_4^5) & \xi_2^4 & \xi_1^4 \\ \xi_3^5 & \Phi_3^4(\xi_3^5, \xi_4^5) & \psi_3 & \xi_2^3 & \xi_1^3 \\ \xi_2^5 & \xi_2^4 & \xi_2^3 & \psi_2 & \Phi_1^2(\xi_1^5, \xi_2^5) \\ \xi_1^5 & \xi_1^4 & \xi_1^3 & \Phi_1^2(\xi_1^5, \xi_2^5) & \psi_1 \end{pmatrix}$$

avec $\rho = \dfrac{c_3^2 k_4^3 c_5^4}{c_5^2}$.

i) Si $c_{35}^{44} \neq 0$, on pose $\xi_1^3 = \xi_1^5 = \xi_1^4 = \xi_2^4 = 0$; $c_{35}^{44}\xi_4^5 + c_{35}^{43}\xi_3^5 = 0$; on diagonalise le sous-matrice 4×4 en bas à droite le diagonaliseur est de la forme

$$\begin{pmatrix} 1 & 0 & * & * \\ 0 & u_2 & * & * \\ 0 & u_3 & * & * \\ 0 & u_4 & * & * \end{pmatrix}$$

avec $u_2 \neq 0$, $u_3 \neq 0$ u_2 et u_3 ne dépendent que de ψ et de ξ_2^5; on applique la conséquence 1.4: $\rho\xi_3^5 u_2 + \xi_2^5 u_3$ et $\xi_3^5 u_2 + \xi_2^5 u_3$ doivent être de même signe pour tout ξ_3^5, d'où $\rho = 1$ et la symétrie.

ii) Si $c^{44}_{35} = 0$, on pose $\xi^5_3 = \xi^3_2 = \xi^3_1 = \xi^4_1 = \xi^5_1 = 0$, on obtient aisément que $\rho\xi^5_4 u_1 + \xi^5_2 u_3$ et $\xi^5_4 u_1 + \xi^5_2 u_3$ doivent être de même signe avec $u_1 \neq 0$, $u_3 \neq 0$, ξ^5_4 est libre, d'où le résultat.   □

Proposition 3.9. *Si $c^{22}_{15} = 0$, $a^i_j(\xi)$ est présymétrique.*

Preuve. On pose $\xi^3_1 = \xi^4_1 = \xi^5_1 = 0$; la matrice 4×4 obtenue en rayant la 1ère ligne et la 1ère colonne de $\left(a^i_j(\xi')\right)$ est présymétrique (lemme 1.1); on applique le lemme 1.2 et on obtient $c^2_4 = c^2_3 k^3_4$, $c^2_5 = c^2_4 c^4_5$; on en déduit aisément la présymétrie.   □

References

[1] Tatsuo Nishitani, *Symmetrization of hyperbolic systems with real constant coefficients*, Ann. Scuola Norm. Sup. Pisa Cl. Sci. (4) **21** (1994), no. 1, 97–130.

[2] Tatsuo Nishitani and Jean Vaillant, *Smoothly symmetrisable systems and the reduced dimension*, Tsukuba J. Math. **25** (2001), no. 1, 165–177.

[3] Yorimasa Oshime, *Canonical forms of 3 × 3 strongly hyperbolic systems with real constant coefficients*, J. Math. Kyoto Univ. **31** (1991), no. 4, 937–982.

[4] Gilbert Strang, *On strong hyperbolicity*, J. Math. Kyoto Univ. **6** (1967), 397–417.

[5] Jean Vaillant, *Symétrisabilité des matrices localisées d'une matrice fortement hyperbolique en un point multiple*, Ann. Scuola Norm. Sup. Pisa Cl. Sci. (4) **5** (1978), no. 2, 405–427.

[6] _____, *Systèmes fortement hyperboliques 4 × 4, dimension réduite et symétrie*, Ann. Scuola Norm. Sup. Pisa Cl. Sci. (4) **29** (2000), no. 4, 839–890.

[7] _____, *Symétrie des opérateurs fortement hyperboliques 4 × 4 ayant un point triple caractéristique dans $\mathbb{R}^3$*, Ann. Univ. Ferrara Sez. VII (N.S.) **45** (1999), no. suppl., 339–363 (2000), Workshop on Partial Differential Equations (Ferrara, 1999).

[8] _____, *Systèmes uniformément diagonalisables, dimension réduite et symétrie I*, Bull. Soc. Roy. Sci. Liège, (à la mémoire de P. Laubin), **70** (2001), no. 4–6, 407–433.

[9] ———, *Uniformly diagonalizable real systems, reduced dimension and symetry*, International Congress of the International Society for Analysis, its Applications and Computation (ISAAC), held at the University of Berlin, Germany, 2001, (*à paraître*).

Jean Vaillant
Mathématiques – B.C. 172
Université Paris VI
4, place Jussieu
F-75252 Paris Cedex 05, France
vint@ccr.jussieu.fr

On Hypoellipticity of the Operator $\exp[-|x_1|^{-\sigma}]D_1^2 + x_1^4 D_2^2 + 1$

Seiichiro Wakabayashi and Nobuo Nakazawa

ABSTRACT Let $L(x, D) = f_\sigma(x_1)D_1^2 + x_1^4 D_2^2 + 1$, where $x = (x_1, x_2) \in \mathbb{R}^2$, $\sigma > 0$ and $f_\sigma(t) = \exp[-|t|^{-\sigma}]$ if $t \neq 0$ and $f_\sigma(0) = 0$. We shall prove that $L(x, D)$ is hypoelliptic at $x = (0, 0)$ if and only if $\sigma < 2$.

1   Introduction

Let $x = (x_1, x_2) \in \mathbb{R}^2$ and

$$P(x, D) = a(x)D_1^2 + b(x)D_2^2 + ic(x)D_1 + id(x)D_2 + 1,$$

where $D = (D_1, D_2) = -i(\partial_1, \partial_2) = -i(\partial/\partial x_1, \partial/\partial x_2)$, $a(x), b(x), c(x), d(x)$ $\in C^\infty(\mathbb{R}^2)$, $a(x) \geq 0$, $b(x) \geq 0$ and $c(x)$ and $d(x)$ are real-valued. Assume that

(A) $\partial^\alpha a(x) = \partial^\alpha b(x) = \partial^\beta c(x) = \partial^\beta d(x) = 0$ if $a(x)b(x) = 0, \alpha, \beta \in (\mathbb{Z}_+)^2$, $|\alpha| = 2$ and $|\beta| = 1$.

Here $\mathbb{Z}_+ = \mathbb{N} \cup \{0\}$, $|\alpha| = \alpha_1 + \alpha_2$ and $\partial^\alpha = \partial_1^{\alpha_1} \partial_2^{\alpha_2}$ for $\alpha = (\alpha_1, \alpha_2) \in (\mathbb{Z}_+)^2$.
 Under the condition (A) we have the following.

Theorem 1.1. *$P(x, D)$ is locally solvable at every $x^0 \in \mathbb{R}^2$, i.e., there is a neighborhood ω of x^0 such that for any $f \in \mathcal{E}'$ there is $u \in \mathcal{D}'$ satisfying $P(x, D)u = f$ in ω, where $\mathcal{E}' = \{u \in \mathcal{D}'; \text{supp } u \text{ is compact}\}$.*

It is well known that $^t P(x, D)$ is locally solvable if $P(x, D)$ is hypoelliptic, where $^t P(x, D)$ denotes the transposed operator of $P(x, D)$ (see [6], [7]). So, taking this fact into account we assume throughout this article that the condition (A) is satisfied.

Definition 1.2. Let $x^0 \in \mathbb{R}^2$. We say that P is hypoelliptic at x^0 if there is a neighborhood ω of x^0 such that

$$\omega \cap \text{sing supp } Pu = \omega \cap \text{sing supp } u \quad \text{for } u \in \mathcal{E}',$$

where sing supp u denotes the singular support of u.

It follows from Nakazawa's results in [2] that $P(x, D)$ is hypoelliptic at $x = 0$ if there is a neighborhood ω of the origin such that for any $\nu > 0$ there is $C_\nu > 0$ satisfying

(H) $$\sum_{|\alpha|\neq 0,\, |\alpha|+|\beta|=2} (\log\langle\xi\rangle)^{|\alpha|}|P_{2(\beta)}^{(\alpha)}(x,\xi)|\langle\xi\rangle^{-|\beta|}$$

$$+ \log\langle\xi\rangle(|\partial_1 a(x)+c(x)| + |\partial_2 b(x)+d(x)|)$$

$$\leq \nu(P_2(x,\xi)+1) + C_\nu\langle\xi\rangle^{-1} \quad \text{for } x \in \omega \text{ and } |\xi| \geq 1,$$

where $P_2(x,\xi) = a(x)\xi_1^2 + b(x)\xi_2^2$. For example, if there is a neighborhood ω of the origin such that for any $\varepsilon > 0$ there is $C_\varepsilon > 0$ satisfying

$$\exp[-\varepsilon/m(x)] \leq C_\varepsilon \min\{a(x), b(x)\} \quad \text{if } x \in \omega \text{ and } m(x) \neq 0,$$

where $m(x) = \sqrt{a(x)} + \sqrt{b(x)} + |c(x)| + |d(x)|$, then $P(x, D)$ is hypoelliptic at $x = 0$. In particular, the operator $L(x, D) \equiv f_\sigma(x_1)D_1^2 + x_1^4 D_2^2 + 1$ is hypoelliptic at $x = 0$ if $0 \leq \sigma < 2$, where

$$f_\sigma(t) = \begin{cases} \exp[-|t|^{-\sigma}] & (t \neq 0), \\ 0 & (t = 0) \end{cases}$$

for $\sigma > 0$ and $f_0(t) = 1/e$. The above result can be improved if the condition

(B) $a(x)b(x) \neq 0$ if $x_1 \neq 0$

is satisfied. In fact, Nakazawa proved in [2] that $P(x, D)$ is hypoelliptic at $x = 0$ if the condition (B) is satisfied and

(H)′ there is a neighborhood ω of the origin such that for any $\nu > 0$ there is $C_\nu > 0$ satisfying

$$\sum_{\substack{|\alpha|\neq 0,\, \alpha_1 = 0 \\ |\alpha|+|\beta|=2}} (\log\langle\xi\rangle)^{|\alpha|}|P_{2(\beta)}^{(\alpha)}(x,\xi)|\langle\xi\rangle^{-|\beta|} + \log\langle\xi\rangle(|\partial_2 b(x)+d(x)|)$$

$$\leq \nu(P_2(x,\xi)+1) + C_\nu\langle\xi\rangle^{-1} \quad \text{for } x \in \omega.$$

Therefore, $P(x, D)$ is hypoelliptic at $x = 0$ if the condition (B) is satisfied and there is a neighborhood ω of the origin such that for any $\varepsilon > 0$ there is $C_\varepsilon > 0$ satisfying

$$\exp[-\varepsilon/(\sqrt{b(x)} + |d(x)|)] \leq C_\varepsilon \min\{a(x), b(x)\}$$

if $x \in \omega$ and $\sqrt{b(x)} + |d(x)| \neq 0$. In particular, the operator $Q(x, D) \equiv x_1^4 D_1^2 + f_\sigma(x_1)D_2^2 + 1$ is hypoelliptic at $x = 0$ if $\sigma > 0$. Moreover, $Q(x, D)$ is not hypoelliptic if $\sigma = 0$. Indeed, $u(x) = x_1 \exp[ix_1^{-1} + \sqrt{2}ex_2]$ ($x_1 \neq 0$) is a nonsmooth null solution of $Q(x, D)$ if $\sigma = 0$ (see, also, [1], [4]). So a question arises whether $L(x, D)$ is hypoelliptic at $x = 0$ for $\sigma \geq 2$ or not. Our main result is the following.

Theorem 1.3. ([3]) *Let $\sigma \geq 0$. Then $L(x, D)$ is hypoelliptic at $x = (0, 0)$ if and only if $(0 \leq) \sigma < 2$.*

Remark. In the above theorem $x = (0, 0)$ can be replaced by $x = (0, a)$ with $a \in \mathbb{R}$. Moreover, $P(x, D)$ is elliptic at $x = (x_1, x_2)$ with $x_1 \neq 0$ and, therefore, $P(x, D)$ is hypoelliptic at $x = (x_1, x_2)$ with $x_1 \neq 0$.

2 Preliminaries

First we shall give well-known facts on local solvability and hypoellipticity.

Lemma 2.1. *Let $x^0 \in \mathbb{R}^2$. Then P is locally solvable at x^0 if and only if there is a neighborhood ω of x^0 such that for any $s \in \mathbb{R}$ there are $t \in \mathbb{R}$ and $C > 0$ satisfying*

$$\|u\|_s \leq C(\|{}^t Pu\|_t + \|u\|_0) \quad \text{for } u \in C_0^\infty(\omega),$$

where $\| \cdot \|_s$ denotes the Sobolev norm of order s.

Lemma 2.2. *Assume that P is hypoelliptic at x^0. Then there is a neighborhood ω of x^0 such that for any nonvoid open subsets ω_i ($i = 1, 2$) of ω with $\omega_1 \subset\subset \omega_2 \subset \omega$ and any $p \in \mathbb{Z}_+$ there exist $q \in \mathbb{Z}_+$ and $C > 0$ satisfying*

$$\sup_{\substack{x \in \omega_1 \\ |\alpha| \leq p}} |D^\alpha u(x)| \leq C \{ \sup_{\substack{x \in \omega_2 \\ |\alpha| \leq q}} |D^\alpha Pu(x)| + \sup_{x \in \omega_2} |u(x)| \} \tag{2.1}$$

for any $u \in C^\infty(\overline{\omega_2})$. Here $\omega_1 \subset\subset \omega_2$ means that $\overline{\omega_1}$ is a compact subset of the interior $\overset{\circ}{\omega_2}$ of ω_2, and $C^\infty(\overline{\omega_2}) = \{ u \in C^0(\overline{\omega_2}); \text{ there is } U(x) \in C^\infty(\mathbb{R}^2) \text{ such that } U|_{\overline{\omega_2}} = u \}$.

Remark. If ${}^t P$ is locally solvable at x^0, then (2.1) holds for $u \in C_0^\infty(\omega_1)$.

In order to prove Theorem 1.3 we shall construct asymptotic solutions $u_\rho(x)$, which violate (2.1), in the form

$$u_\rho(x) = U_\rho(x_1) \exp[(4 \log \rho)^{2/\sigma} x_2]$$

when $\sigma \geq 2$. Write

$$L_\rho(x_1, \partial_1) U_\rho(x_1) = -\exp[-(4 \log \rho)^{2/\sigma} x_2] L(x, D) u_\rho(x),$$

where $\rho \geq 4$. Then we have

$$L_\rho(x_1, \partial_1) = f_\sigma(x_1)\partial_1^2 + (4 \log \rho)^{4/\sigma} x_1^4 - 1.$$

Asymptotic solutions will be constructed in two intervals $[t_\rho^-, t_\rho^+]$ and $[t_\rho, 1]$, respectively, where

$$t_\rho^\pm = (4 \log \rho)^{-1/\sigma} (1 \pm 2\rho^{-1}) \text{ and } t_\rho = (4 \log \rho)^{-1/\sigma} (1 + \rho^{-1}).$$

In order to estimate and connect these asymptotic solutions we need the following.

Lemma 2.3. *Let $\rho \geq 4$ and let $R(t; \rho)$ be a real-valued function defined for $\rho \geq 4$ and $t \in [t_\rho, 1]$ such that, with some $M \in \mathbb{R}$,*

$$|\partial_t^k R(t; \rho)| \leq C_k \rho^{-M + 3k/2}$$

for $\rho \geq 4$, $t \in [t_\rho, 1]$ and $k \in \mathbb{Z}_+$. Moreover, let $u(t; \rho)$ be a solution of the initial-value problem

$$\begin{cases} (\partial_t^2 + p(t; \rho))u(t; \rho) = R(t; \rho) & (t \in [t_\rho, 1]), \\ u(t_\rho; \rho) = \alpha(\rho), \quad (\partial_t u)(t_\rho; \rho) = \beta(\rho), \end{cases} \tag{2.2}$$

where $p(t; \rho) = f_\sigma(t)^{-1}((4 \log \rho)^{4/\sigma} t^4 - 1)$ and $\alpha(\rho)$ and $\beta(\rho)$ are real-valued functions of ρ (≥ 4). (i) Assume that $R(t; \rho) \equiv 0$. Then we have

$$|u(t; \rho)| \leq C(\rho^4 (\log \rho)^{2/\sigma} |\alpha(\rho)| + \rho^{5/2} (\log \rho)^{2/\sigma} |\beta(\rho)|) \tag{2.3}$$

for $t \in [t_\rho, 1]$. (ii) Assume that $\alpha(\rho) \equiv \beta(\rho) \equiv 0$. Then we have

$$|\partial_t^k u(t; \rho)| \leq C_k \rho^{-M-2+3k/2} (\log \rho)^{-1/(2\sigma)}, \tag{2.4}$$

for $t \in [t_\rho, t_\rho^+]$ and $k \in \mathbb{Z}_+$.

Proof. Put

$$U(t; \rho) = p(t; \rho)u(t; \rho)^2 + (\partial_t u(t; \rho))^2$$

for $t \in [t_\rho, 1]$. From (2.2) we have

$$2^{-1} \partial_t U(t; \rho) = 2^{-1} (\partial_t p(t; \rho))u(t; \rho)^2 + R(t; \rho)\partial_t u(t; \rho)$$

and, therefore,

$$2^{-1} U(t; \rho) - 2^{-1} U(t_\rho; \rho)$$
$$= 2^{-1} \int_{t_\rho}^t (\partial_s p(s; \rho))u(s; \rho)^2 \, ds + \int_{t_\rho}^t R(s; \rho)\partial_s u(s; \rho) \, ds.$$

Since

$$\partial_t p(t; \rho) = -\sigma t^{-\sigma-1} p(t; \rho) + 4(4 \log \rho)^{4/\sigma} t^3 \exp[t^{-\sigma}]$$
$$\leq 4(4 \log \rho)^{1/\sigma}((4 \log \rho)^{1/\sigma} t - 1)^{-1} p(t; \rho),$$
$$4(4 \log \rho)^{1/\sigma}((4 \log \rho)^{1/\sigma} t - 1)^{-1} \geq 4$$

for $t \in [t_\rho, 1]$, we have

$$U(t; \rho) \leq U(t_\rho; \rho) + \int_{t_\rho}^t R(s; \rho)^2 \, ds$$

$$+ 4 \int_{t_\rho}^t (4 \log \rho)^{1/\sigma}((4 \log \rho)^{1/\sigma} s - 1)^{-1} U(s; \rho) \, ds$$

for $t \in [t_\rho, 1]$. Putting $\tau = (4 \log \rho)^{1/\sigma} t - 1$, $V(\tau) = U(t; \rho)$ and $S(\tau) = R(t; \rho)$, we have

$$V(\tau) \leq V(1/\rho) + \int_{1/\rho}^{\tau} (4 \log \rho)^{-1/\sigma} S(s)^2 \, ds + 4 \int_{1/\rho}^{\tau} \frac{V(s)}{s} \, ds$$

for $\tau \in [1/\rho, (4 \log \rho)^{1/\sigma} - 1]$. Therefore, $F(\tau) \equiv \tau^{-4} \int_{1/\rho}^{\tau} V(s)/s \, ds$ satisfies

$$\tau^5 F'(\tau) \leq V(1/\rho) + \int_{1/\rho}^{\tau} (4 \log \rho)^{-1/\sigma} S(s)^2 \, ds.$$

This gives

$$F(\tau) \leq (\rho^4/4 - 1/(4\tau^4)) V(1/\rho)$$
$$+ \int_{1/\rho}^{\tau} (1/(4s^4) - 1/(4\tau^4))(4 \log \rho)^{-1/\sigma} S(s)^2 \, ds,$$

$$V(\tau) \leq \rho^4 \tau^4 V(1/\rho) + \int_{1/\rho}^{\tau} (\tau/s)^4 (4 \log \rho)^{-1/\sigma} S(s)^2 \, ds,$$

$$U(t; \rho) \leq \rho^4 ((4 \log \rho)^{1/\sigma} t - 1)^4 U(t_\rho; \rho)$$
$$+ \int_{t_\rho}^{t} ((4 \log \rho)^{1/\sigma} t - 1)^4 ((4 \log \rho)^{1/\sigma} s - 1)^{-4} R(s; \rho)^2 \, ds$$

for $t \in [t_\rho, 1]$.   (2.5)

(i) We first assume that $R(t; \rho) \equiv 0$. Since $p(t_\rho; \rho) \leq C\rho^3$ and $p(t; \rho)^{-1} \leq \rho/(4e)$ for $t \in [t_\rho, 1]$, (2.5) yields (2.3). (ii) Assume that $\alpha(\rho) \equiv \beta(\rho) \equiv 0$. From (2.5) we have

$$U(t; \rho) \leq C\rho^{-2M-1} (\log \rho)^{-1/\sigma} \quad \text{for } t \in [t_\rho, t_\rho^+].$$

Since $p(t; \rho)^{-1} \leq C\rho^{-3}$ for $t \in [t_\rho, t_\rho^+]$, this proves that (2.4) is valid for $k = 0, 1$. Note that

$$|\partial_t^k p(t; \rho)| \leq \begin{cases} C\rho^3 & (k = 0), \\ C_k \rho^4 (\log \rho)^{k/\sigma + k - 1} & (k \geq 1) \end{cases} \quad (2.6)$$

for $t \in [t_\rho, t_\rho^+]$. Now suppose that (2.4) is valid for $k \leq l$, where $l \geq 1$. Let $k = l + 1$. Then, from (2.2) and (2.6) we have

$$|\partial_t^k u(t; \rho)| \leq \sum_{j=0}^{k-2} \binom{k-2}{j} |\partial_t^j u(t; \rho)| |\partial_t^{k-2-j} p(t; \rho)| + |\partial_t^{k-2} R(t; \rho)|$$

$$\leq C_k \rho^{-M-2+3k/2} (\log \rho)^{-1/(2\sigma)} \quad \text{for } t \in [t_\rho, t_\rho^+],$$

which proves the assertion (ii).   □

3   Proof of Theorem 1.1

Let $x^0 \in \mathbb{R}^2$. If $a(x^0)b(x^0) \neq 0$, then $P(x, D)$ is elliptic, and $P(x, D)$ is locally solvable at x^0. Assume that $a(x^0)b(x^0) = 0$. Let U be a neighborhood of x^0, and put

$$Z := \{x \in U; \ a(x)b(x) = 0\},$$
$$Z_\varepsilon := \{x \in \mathbb{R}^2; \ |x - y| < \varepsilon \text{ for some } y \in Z\},$$

where $\varepsilon > 0$. We fix $s \in \mathbb{R}$ and define

$$P_s(x, D) := \langle D \rangle^{-s} P(x, D) \langle D \rangle^s.$$

Let U_j ($j = 0, 1, 2$) be neighborhoods of x^0 such that $U_0 \subset\subset U_1 \subset\subset U_2 \subset\subset U$. Choose $\chi_\varepsilon(x) \in C_0^\infty(U_2)$ so that $0 \leq \chi_\varepsilon(x) \leq 1$, $\chi_\varepsilon(x) = 1$ for $x \in U_1 \cap Z_{3\varepsilon}$ and supp $\chi_\varepsilon \subset\subset U_2 \cap Z_{4\varepsilon}$, where $\varepsilon > 0$. Since $P_s(x, D)^*$ is elliptic in $\mathbb{R}^2 \setminus Z_\varepsilon$, we have, with some $C_{s,\varepsilon} > 0$,

$$\|(1 - \chi_\varepsilon(x))u\|_2 \leq C_{s,\varepsilon}(\|P_s(x, D)^* u\|_0 + \|u\|_{-1}),$$
$$\|[\chi_\varepsilon, P_s(x, D)^*]u\|_0 \leq C_{s,\varepsilon}(\|P_s(x, D)^* u\|_0 + \|u\|_{-1}) \quad (3.1)$$

for $u \in C_0^\infty(U_0)$, where $[A, B] = AB - BA$ for operators A and B. Noting that

$$\chi_\varepsilon(x) P_s(x, D)^* u = P_s(x, D)^*(\chi_\varepsilon(x)u) + [\chi_\varepsilon, P_s(x, D)^*]u,$$

we see that for any $\nu > 0$ there is $C_{s,\varepsilon,\nu} > 0$ satisfying

$$\text{Re} \ (\chi_\varepsilon(x) P_s(x, D)^* u, \chi_\varepsilon(x)u)_{L^2} \geq \text{Re} \ (P_s(x, D)^* \chi_\varepsilon(x)u, \chi_\varepsilon(x)u)_{L^2}$$
$$-C_{s,\varepsilon,\nu}(\|P_s(x, D)^* u\|_0^2 + \|u\|_{-1}^2) - \nu \|\chi_\varepsilon(x)u\|_0^2 \quad (3.2)$$

for $u \in C_0^\infty(U_0)$, where $(u, v)_{L^2} = \int_{\mathbb{R}^2} u(x)\overline{v(x)} \, dx$. Let $u \in C_0^\infty(U_0)$, and put $v_\varepsilon = \chi_\varepsilon(x)u(x)$. Note that $v_\varepsilon \in C_0^\infty(U_0 \cap Z_{4\varepsilon})$ and

$$\text{Re} \ (P_s(x, D)^* v_\varepsilon(x), v_\varepsilon(x))_{L^2} = \text{Re} \ (P_s(x, D)v_\varepsilon(x), v_\varepsilon(x))_{L^2}. \quad (3.3)$$

We can write

$$P_s(x, D)u = D_1(a(x)D_1 u) + D_2(b(x)D_2 u) + i\gamma_s^1(x, D)u$$
$$+ (1 + \gamma_s^0(x, D))u + r_s(x, D)u,$$
$$\gamma_s^1(x, \xi) = c(x)\xi_1 + d(x)\xi_2$$
$$+ s\langle\xi\rangle^{-2}(a_{x_1}\xi_1^3 + b_{x_1}\xi_1\xi_2^2 + a_{x_2}\xi_1^2\xi_2 + b_{x_2}\xi_2^3) + a_{x_1}\xi_1 + b_{x_2}\xi_2,$$
$$\gamma_s^0(x, \xi) = -s\langle\xi\rangle^{-2}\{c_{x_1}\xi_1^2 + (c_{x_2} + d_{x_1})\xi_1\xi_2 + d_{x_2}\xi_2^2\}$$
$$- s(s + 2)\langle\xi\rangle^{-4}\{a_{x_1x_1}\xi_1^4 + 2a_{x_1x_2}\xi_1^3\xi_2 + (a_{x_2x_2} + b_{x_1x_1})\xi_1^2\xi_2^2$$
$$+ 2b_{x_1x_2}\xi_1\xi_2^3 + b_{x_2x_2}\xi_2^4\}/2$$
$$+ s\langle\xi\rangle^{-2}\{(a_{x_1x_1} + a_{x_2x_2})\xi_1^2 + (b_{x_1x_1} + b_{x_2x_2})\xi_2^2\}/2,$$
$$r_s(x, \xi) \in S_{1,0}^{-1}, \quad (3.4)$$

where $a_{x_j} = \partial_j a(x)$ and $a_{x_j x_k} = \partial_j \partial_k a(x)$. Then it follows from the condition (A) that

$$(\partial_x^\alpha \gamma_s^1)(x, \xi) = 0 \text{ and } \gamma_s^0(x, \xi) = 0 \qquad (3.5)$$

if $a(x)b(x) = 0$ and $|\alpha| = 1$. By (3.4) we have

$$\text{Re } (P_s(x, D)v_\varepsilon, v_\varepsilon)_{L^2} \geq \|v_\varepsilon\|_0^2 + \text{Re } (\gamma_s^0(x, D)v_\varepsilon, v_\varepsilon)_{L^2}$$
$$+ (\Gamma_s(x, D)v_\varepsilon, v_\varepsilon)_{L^2} - C_s \|v_\varepsilon\|_{-1/2}^2, \qquad (3.6)$$

where $\Gamma_s(x, D) = (i/2)(\gamma_s^1(x, D) - \gamma_s^1(x, D)^*)$. From (3.5) we can write

$$\Gamma_s(x, \xi) = \tilde{\gamma}_s^0(x, \xi) + \tilde{r}_s(x, \xi)$$

so that $\tilde{\gamma}_s^0(x, \xi) \in S_{1,0}^0$, $\tilde{r}_s(x, \xi) \in S_{1,0}^{-1}$ and $\tilde{\gamma}_s^0(x, \xi) = 0$ if $a(x)b(x) = 0$. This, together with (3.5), yields

$$1/2 + \chi_{2\varepsilon}(x)(\gamma_s^0(x, \xi) + \tilde{\gamma}_s^0(x, \xi)) \geq 0 \quad \text{if } \varepsilon \ll 1.$$

Therefore, using the sharp Gårding inequality or Fefferman–Phong's inequality we have

$$\text{Re } (\gamma_s^0(x, D)v_\varepsilon, v_\varepsilon)_{L^2} + (\Gamma_s(x, D)v_\varepsilon, v_\varepsilon)_{L^2}$$
$$\geq -\|v_\varepsilon\|_0^2/2 + \text{Re } (\{1/2 + \chi_{2\varepsilon}(x)(\gamma_s^0(x, D) + \tilde{\gamma}_s^0(x, D))\}v_\varepsilon, v_\varepsilon)_{L^2}$$
$$- C_{s,\varepsilon}\|v_\varepsilon\|_{-1/2}^2$$
$$\geq -\|v_\varepsilon\|_0^2/2 - C'_{s,\varepsilon}\|v_\varepsilon\|_{-1/2}^2 \qquad (3.7)$$

if $\varepsilon \ll 1$. From (3.2), (3.3), (3.6) and (3.7) we have

$$\|v_\varepsilon\|_0 \leq C_s(\|P_s(x, D)^* u\|_0 + \|u\|_{-1}).$$

This, together with (3.1), yields

$$\|u\|_0 \leq C_s(\|P_s(x, D)^* u\|_0 + \|u\|_{-1}) \quad \text{for } u \in C_0^\infty(U_0).$$

Let ω be a neighborhood of x^0 such that $\omega \subset\subset U_0$, and choose $\chi(x) \in C_0^\infty(U_0)$ so that $\chi(x) = 1$ near ω. Then we have

$$\|(1 - \chi(x))\langle D \rangle^s u\|_0 \leq C_{s,s'}\|u\|_{s'},$$
$$\|P_s(x, D)(1 - \chi(x))\langle D \rangle^s u\|_0 \leq C_s\|u\|_{s-1}$$

for $u \in C_0^\infty(\omega)$, where $s' \in \mathbb{R}$. Therefore, we have

$$\|u\|_s \leq \|\chi(x)\langle D \rangle^s u\|_0 + \|(1 - \chi(x))\langle D \rangle^s u\|_0$$
$$\leq C_s(\|P(x, D)^* u\|_s + \|u\|_{s-1})$$

for $u \in C_0^\infty(\omega)$. Since $P(x, D)^* u = \overline{{}^t P(x, D)\bar{u}}$, we have

$$\|u\|_s \leq C_s(\|{}^t P(x, D)u\|_s + \|u\|_{s-1}) \quad \text{for } u \in C_0^\infty(\omega),$$

which proves Theorem 1.1 due to Lemma 2.1.

4   Proof of Theorem 1.3

We proved Theorem 1.3 in [3]. We repeat here its proof.

In order to prove Theorem 1.3 it suffices to show that $L(x, D)$ is not hypoelliptic at $x = 0$ when $\sigma \geq 2$. Assume that $\sigma \geq 2$. As we stated in Section 2, we shall construct asymptotic solutions $u_\rho(x)$ in the form $u_\rho(x) = U_\rho(x_1) \exp[(4 \log \rho)^{2/\sigma} x_2]$. Note that

$$\sup_{|x_2| \leq 1} \exp[(4 \log \rho)^{2/\sigma} x_2] \leq \rho^4.$$

First we shall construct asymptotic solutions $U_\rho(x_1)$ satisfying $L_\rho(x_1, \partial_1) U_\rho(x_1) \sim 0$ in $[t_\rho^-, t_\rho^+]$. Putting $t = \rho\{(4 \log \rho)^{1/\sigma} x_1 - 1\}$ and $V_\rho(t) = U_\rho(x_1)$, we can write

$$\rho^{-2}(4 \log \rho)^{-2/\sigma} f_\sigma(x_1)^{-1} L_\rho(x_1, \partial_1) U_\rho(x_1) = \widetilde{L}_\rho(t, \partial_t) V_\rho(t)$$

for $t \in [-2, 2]$, where

$$\widetilde{L}_\rho(t, \partial_t) = \partial_t^2 + 4\rho(4 \log \rho)^{-2/\sigma} t + (4 \log \rho)^{-2/\sigma}(6t^2 + 4\rho^{-1}t^3 + \rho^{-2}t^4)$$
$$+ \sum_{j \geq k \geq 1} c_{j,k}\, \rho^{1-j} (\log \rho)^{-2/\sigma+k}\, t^{j+1}.$$

Indeed, we have

$$(1 + \rho^{-1}t)^{-\sigma} = 1 + \sum_{k=1}^{\infty} \binom{-\sigma}{k} \rho^{-k} t^k,$$

$$f_\sigma(x_1)^{-1} = \exp[(4 \log \rho)(1 + \rho^{-1}t)^{-\sigma}]$$
$$= \rho^4 \left(1 + \sum_{j \geq k \geq 1} c'_{j,k}\, \rho^{-j} (\log \rho)^k t^j\right),$$

if $\rho \geq 4$ and $|t| \leq 2$. Write

$$V_\rho(t) = \mathrm{Ai}(-c_\rho t) V_\rho^0(t) + \rho^{-1/6}(\log \rho)^{1/(3\sigma)}\, \mathrm{Ai}'(-c_\rho t) V_\rho^1(t), \qquad (4.1)$$

where $c_\rho = 4^{1/3}\rho^{1/3}(4 \log \rho)^{-2/(3\sigma)}$ and $\mathrm{Ai}(t)$ denotes the Airy function. The Airy function $\mathrm{Ai}(t)$ is defined, for example, by

$$\mathrm{Ai}(t) = \pi^{-1} \int_0^\infty \cos(s^3/3 + ts)\, ds$$

and satisfies $\mathrm{Ai}''(t) = t\, \mathrm{Ai}(t)$. A simple calculation gives

$$\widetilde{L}_\rho(t, \partial_t) V_\rho(t)$$
$$= \rho^{1/3}(\log \rho)^{-2/(3\sigma)}\, \mathrm{Ai}'(-c_\rho t)\Big\{-2 \cdot 4^{(1-2/\sigma)/3} \partial_t V_\rho^0(t)$$
$$+ \rho^{-1/2}(\log \rho)^{1/\sigma} \partial_t^2 V_\rho^1(t) + \sum_{\substack{j \geq k \geq 0 \\ j \geq 1}} c_{j,k}\, \rho^{1/2-j}(\log \rho)^{-1/\sigma+k}\, t^{j+1} V_\rho^1(t)\Big\}$$

$$+ \rho^{1/2}(\log\rho)^{-1/\sigma}\,\mathrm{Ai}(-c_\rho t)\Big\{2\cdot 4^{2(1-2/\sigma)/3}\,t\partial_t V_\rho^1(t) + 4^{2(1-2/\sigma)/3} V_\rho^1(t)$$

$$+ \rho^{-1/2}(\log\rho)^{1/\sigma}\partial_t^2 V_\rho^0(t) + \sum_{\substack{j\ge k\ge 0 \\ j\ge 1}} c_{j,k}\,\rho^{1/2-j}(\log\rho)^{-1/\sigma+k}\,t^{j+1}V_\rho^0(t)\Big\},$$

where $c_{1,0} = 6\cdot 4^{-2/\sigma}$, $c_{2,0} = 4^{1-2/\sigma}$, $c_{3,0} = 4^{-2/\sigma}$ and $c_{j,0} = 0$ for $j \ge 4$. Put

$$J(j) = \{-\mu/\sigma + l;\ \mu, l \in \mathbb{Z},\ |\mu| \le j \text{ and } 0 \le l \le j\}$$

for $j \in \mathbb{Z}_+$, and write $J(j) = \{v_{j,1}, v_{j,2}, \ldots, v_{j,r(j)}\}$, where $v_{j,1} > v_{j,2} > \cdots > v_{j,r(j)}$. Note that $J(0) = \{0\}$, $r(0) = 1$ and $v_{0,1} = 0$. We define $I(\cdot\,; j) : \mathbb{R} \ni \delta \longmapsto I(\delta; j) \in \{0, 1, \ldots, r(j)\}$ $(j \in \mathbb{Z}_+)$ by

$$I(\delta; j) = \begin{cases} k & \text{if } \delta \in J(j) \text{ and } \delta = v_{j,k}, \\ 0 & \text{if } \delta \notin J(j). \end{cases}$$

We also define $I(\delta; j) = 0$ if $j < 0$. Let us determine $V_\rho^i(t)$ $(i = 0, 1)$ in the form

$$V_\rho^i(t) \sim \sum_{j=0}^{\infty}\sum_{k=1}^{r(j)} \rho^{-j/2}(\log\rho)^{v_{j,k}} V_{j,k}^i(t) \tag{4.2}$$

so that $\widetilde{L}_\rho(t, \partial_t)V_\rho(t) \sim 0$, i.e.,

$$\Big|\partial_t^l \widetilde{L}_\rho(t, \partial_t) \sum_{j=0}^{N}\sum_{k=1}^{r(j)} \rho^{-j/2}(\log\rho)^{v_{j,k}}\{\mathrm{Ai}(-c_\rho t)V_{j,k}^0(t)$$

$$+ \rho^{-1/6}(\log\rho)^{1/(3\sigma)}\,\mathrm{Ai}'(-c_\rho t)V_{j,k}^1(t)\}\Big| \le C_{N,l}\,\rho^{-a_{N,l}}$$

for $\rho \ge 4$, $t \in [-2, 2]$ and $N, l \in \mathbb{Z}_+$, where $a_{N,l} \to \infty$ as $N \to \infty$. Then we have the transport equations

$$\begin{cases} -2\cdot 4^{(1-2/\sigma)/3}\partial_t V_{j,k}^0(t) + \partial_t^2 V_{j-1,I(v_{j,k}-1/\sigma;j-1)}^1(t) \\ \quad + \sum_{\mu\ge v\ge 0,\,\mu\ge 1} c_{\mu,v} t^{\mu+1} V_{j-2\mu+1,I(v_{j,k}+1/\sigma-v;j-2\mu+1)}^1(t) = 0, \\ 4^{2(1-2/\sigma)/3}(2t\partial_t + 1)V_{j,k}^1(t) + \partial_t^2 V_{j-1,I(v_{j,k}-1/\sigma;j-1)}^0(t) \\ \quad + \sum_{\mu\ge v\ge 0,\,\mu\ge 1} c_{\mu,v} t^{\mu+1} V_{j-2\mu+1,I(v_{j,k}+1/\sigma-v;j-2\mu+1)}^0(t) = 0 \end{cases} \tag{4.3}$$

for $j \in \mathbb{Z}_+$ and $1 \le k \le r(j)$, where $V_{j,0}^i \equiv 0$ for $j \in \mathbb{Z}_+$ and $i = 0, 1$. Let $V_{j,k}^i(t)$ $(i = 0, 1,\ j \in \mathbb{Z}_+$ and $1 \le k \le r(j))$ be solutions of (4.3) with the initial conditions

$$\begin{cases} V_{0,1}^0(0) = 1, \\ V_{j,k}^i(0) = 0 & \text{if } i = 0, 1,\ i + j \ge 1 \text{ and } 1 \le k \le r(j). \end{cases}$$

Then the $V^i_{j,k}(t)$ are determined inductively by

$$\begin{cases} V^0_{j,k}(t) = \delta_{j,0} + 2^{-1} \cdot 4^{-(1-2/\sigma)/3} \Big\{ \partial_t V^1_{j-1,I(\nu_{j,k}-1/\sigma;j-1)}(t) \\ \quad - (\partial_t V^1_{j-1,I(\nu_{j,k}-1/\sigma;j-1)})(0) \\ \quad + \sum_{\mu \geq \nu \geq 0,\, \mu \geq 1} c_{\mu,\nu} \int_0^t s^{\mu+1} V^1_{j-2\mu+1,I(\nu_{j,k}+1/\sigma-\nu;j-2\mu+1)}(s)\, ds \Big\}, \\ V^1_{j,k}(t) = -2^{-1} \cdot 4^{-2(1-2/\sigma)/3} \int_0^t t^{-1/2} s^{-1/2} \Big\{ \partial_s^2 V^0_{j-1,I(\nu_{j,k}-1/\sigma;j-1)}(s) \\ \quad + \sum_{\mu \geq \nu \geq 0,\, \mu \geq 1} c_{\mu,\nu} s^{\mu+1} V^0_{j-2\mu+1,I(\nu_{j,k}+1/\sigma-\nu;j-2\mu+1)}(s) \Big\}\, ds \end{cases}$$

$$(4.4)$$

($j \in \mathbb{Z}_+$ and $1 \leq k \leq r(j)$). Since $t^{-1/2}(t\theta)^{-1/2} = t^{-1}\theta^{-1/2}$ for $t \neq 0$ and $0 \leq \theta \leq 1$ and

$$\int_0^t t^{-1/2} s^{-1/2} f(s)\, ds = \int_0^1 \theta^{-1/2} f(t\theta)\, d\theta \in C^\infty([-2, 2])$$

for $f(t) \in C^\infty([-2, 2])$, we have $V^i_{j,k}(t) \in C^\infty([-2, 2])$ ($i = 0, 1$, $j \in \mathbb{Z}_+$ and $1 \leq k \leq r(j)$). Substituting (4.2) and (4.4) in (4.1), we have $\tilde{L}_\rho(t, \partial_t) V_\rho(t) \sim 0$. Indeed, we see that

$$\{1, 2, \ldots, r(j)\} \subset \{I(\nu_{j+1,k} - 1/\sigma; j)\,;\ 1 \leq k \leq r(j+1)\},$$
$$\{1, 2, \ldots, r(j)\} \subset \{I(\nu_{j+2\mu-1,k} + 1/\sigma - \nu; j)\,;\ 1 \leq k \leq r(j+2\mu-1)\}$$

if $j \in \mathbb{Z}_+$, $\mu \geq 1$ and $\mu \geq \nu \geq 0$. We have also used the estimates

$$\begin{cases} |\operatorname{Ai}^{(k)}(t)| \leq C_k (1+t)^{-1/4+k/2} \exp[-2t^{3/2}/3], \\ |\operatorname{Ai}^{(k)}(-t)| \leq C_k (1+t)^{-1/4+k/2} \end{cases}$$

$$(4.5)$$

for $t \geq 0$ (see, e.g., [5]). We note that $V^0_{2j-1,k}(t) \equiv V^1_{2j,l}(t) \equiv 0$ if $j \in \mathbb{N}$, $1 \leq k \leq r(2j-1)$ and $1 \leq l \leq r(2j)$. Put

$$U^N_\rho(x_1) = \sum_{j=0}^N \sum_{k=1}^{r(j)} \rho^{-j/2} (\log \rho)^{\nu_{j,k}} [\operatorname{Ai}(-c'_\rho(x_1 - s_\rho)) V^0_{j,k}(c''_\rho(x_1 - s_\rho))$$
$$+ \rho^{-1/6} (\log \rho)^{1/(3\sigma)} \operatorname{Ai}'(-c'_\rho(x_1 - s_\rho)) V^1_{j,k}(c''_\rho(x_1 - s_\rho))],$$
$$R^N_\rho(x_1) = \rho^{-2} (4 \log \rho)^{-2/\sigma} f_\sigma(x_1)^{-1} L_\rho(x_1, \partial_1) U^N_\rho(x_1)$$

$$(4.6)$$

for $x_1 \in [t^-_\rho, t^+_\rho]$ and $N \in \mathbb{Z}_+$, where $c'_\rho = 4^{1/3} \rho^{4/3} (4 \log \rho)^{1/(3\sigma)}$, $c''_\rho =$

$\rho(4\log\rho)^{1/\sigma}$ and $s_\rho = (4\log\rho)^{-1/\sigma}$. Then we have

$$R_\rho^N(x_1) = \mathrm{Ai}'(-c_\rho'(x_1 - s_\rho))\left[\sum_{k=1}^{r(N)} \rho^{-N/2-1/6}(\log\rho)^{1/(3\sigma)+\nu_{N,k}}\partial_t^2 V_{N,k}^1(t)\right.$$

$$+ \sum_{j=0}^{N}\sum_{\substack{2\mu \geq N+2-j \\ \mu \geq \nu \geq 0}}\sum_{k=1}^{r(j)} c_{\mu,\nu}\,\rho^{5/6-\mu-j/2}(\log\rho)^{-5/(3\sigma)+\nu+\nu_{j,k}}\,t^{\mu+1}V_{j,k}^1(t)\Bigg]$$

$$+ \mathrm{Ai}(-c_\rho'(x_1 - s_\rho))\left[\sum_{k=1}^{r(N)} \rho^{-N/2}(\log\rho)^{\nu_{N,k}}\partial_t^2 V_{N,k}^0(t)\right.$$

$$+ \sum_{j=0}^{N}\sum_{\substack{2\mu \geq N+2-j \\ \mu \geq \nu \geq 0}}\sum_{k=1}^{r(j)} c_{\mu,\nu}\,\rho^{1-\mu-j/2}(\log\rho)^{-2/\sigma+\nu+\nu_{j,k}}\,t^{\mu+1}V_{j,k}^0(t)\Bigg],$$

where $t = \rho\{(4\log\rho)^{1/\sigma}x_1 - 1\}$. From (4.5) we have

$$|\partial_1^k R_\rho^N(x_1)|$$

$$\leq \begin{cases} C_{N,0}\,\rho^{-N/2}(\log\rho)^{1-2/\sigma+(1+1/\sigma)N} & (k=0), \\ C_{N,k}\,\rho^{-1/12+3k/2-N/2}(\log\rho)^{1-11/(6\sigma)+(1+1/\sigma)N} & (k \geq 1) \end{cases} \quad (4.7)$$

for $x_1 \in [s_\rho, t_\rho^+]$, and

$$|\partial_1^k R_\rho^N(x_1)| \leq \begin{cases} C_{N,0}\,\rho^{-N/2}(\log\rho)^{1-2/\sigma+(1+1/\sigma)N} \\ \quad \times \exp[-2(c_\rho')^{3/2}(s_\rho - x_1)^{3/2}/3] & (k=0), \\ C_{N,k}\,\rho^{-1/12+3k/2-N/2}(\log\rho)^{1-11/(6\sigma)+(1+1/\sigma)N} \\ \quad \times \exp[-2(c_\rho')^{3/2}(s_\rho - x_1)^{3/2}/3] & (k \geq 1) \end{cases} \quad (4.8)$$

for $x_1 \in [t_\rho^-, s_\rho]$. Indeed, for example, we have

$$|\partial_1^k \mathrm{Ai}(-c_\rho'(x_1 - s_\rho))| \leq \begin{cases} C_0 & (k=0), \\ C_k\,\rho^{-1/12+3k/2}(\log\rho)^{1/(6\sigma)} & (k \geq 1), \end{cases}$$

$$|\partial_1^k \mathrm{Ai}'(-c_\rho'(x_1 - s_\rho))| \leq C_k\,\rho^{1/12+3k/2}(\log\rho)^{-1/(6\sigma)}$$

for $x_1 \in [s_\rho, t_\rho^+]$. Moreover, we have

$$\rho^{1-\mu-j/2}(\log\rho)^{\nu+\nu_{j,k}} \leq \rho^{-N/2}(\log\rho)^{1+(1+1/\sigma)N}$$

if $0 \leq j \leq N$, $1 \leq k \leq r(j)$, $2\mu \geq N+2-j$ and $\mu \geq \nu \geq 0$, and

$$(\log\rho)^{1/(3\sigma)+\nu_{N,k}} \leq (\log\rho)^{1-5/(3\sigma)+(1+1/\sigma)N}$$

if $1 \leq k \leq r(N)$. Similarly, we have

$$\begin{cases} |U_\rho^N(t_\rho)| \leq C_N\,\rho^{-1/12}(\log\rho)^{1/(6\sigma)}, \\ |(\partial_1 U_\rho^N)(t_\rho)| \leq C_N\,\rho^{17/12}(\log\rho)^{1/(6\sigma)}. \end{cases} \quad (4.9)$$

Let $\widetilde{U}_\rho^N(t)$ be a solution of (2.2) with $R(t;\rho) \equiv 0$, $\alpha(\rho) = U_\rho^N(t_\rho)$ and $\beta(\rho) = (\partial_t U_\rho^N)(t_\rho)$. We choose a function $\chi(t) \in C^\infty(\mathbb{R})$ so that $\chi(t) = 1$ for $t \leq 1$ and $\chi(t) = 0$ for $t \geq 2$, and put

$$u_\rho^N(x_1) = \{\chi_\rho^0(x_1)U_\rho^N(x_1) + \chi_\rho^1(x_1)\widetilde{U}_\rho^N(x_1)\}\exp[(4\log\rho)^{2/\sigma}x_2],$$

where $N \in \mathbb{Z}_+$, $\rho \geq 4$ and

$$\chi_\rho(x_1) = \chi(c_\rho''(x_1 - s_\rho)),$$
$$\chi_\rho^0(x_1) = \chi_\rho(x_1)\chi(c_\rho''(s_\rho - x_1)),$$
$$\chi_\rho^1(x_1) = 1 - \chi_\rho(x_1).$$

Lemma 4.1. (i) *For every* $k, N \in \mathbb{Z}_+$

$$|(\partial_1^k u_\rho^N)(s_\rho, 0)| = 4^{k/3}\rho^{4k/3}(4\log\rho)^{k/(3\sigma)}\left(|\,\mathrm{Ai}^{(k)}(0)| + o(1)\right) \tag{4.10}$$

as $\rho \to \infty$. *In particular, there are* $c_{N,k} > 0$ *and* $\rho_{N,k} \geq 4$ *such that*

$$|(\partial_1^{3k}u_\rho^N)(s_\rho, 0)| \geq c_{N,k}\rho^{4k}(\log\rho)^{k/\sigma} \tag{4.11}$$

if $\rho \geq \rho_{N,k}$. (ii) *For* $N \in \mathbb{Z}_+$ *and* $\rho \geq 4$,

$$\sup_{|x_2|\leq 1}|u_\rho^N(x)| \leq \begin{cases} C_N\,\rho^4\exp[-2(c_\rho')^{3/2}(s_\rho - x_1)^{3/2}/3] & \text{if } x_1 \leq s_\rho, \\ C_N\,\rho^4 & \text{if } x_1 \in [s_\rho, t_\rho], \\ C_N\,\rho^{95/12}(\log\rho)^{13/(6\sigma)} & \text{if } x_1 \in [t_\rho, 1]. \end{cases}$$

(iii) *For* $N \in \mathbb{Z}_+$, $\alpha = (\alpha_1, \alpha_2) \in (\mathbb{Z}_+)^2$ *and* $\rho \geq 4$,

$$\sup_{|x_2|\leq 1}|D^\alpha L(x, D)u_\rho^N(x)|$$

$$\leq \begin{cases} 0 & \text{if } x_1 \in (-\infty, t_\rho^-] \cup [t_\rho^+, 1], \\ C_{N,M,\alpha}\,\rho^{-M} & \text{if } x_1 \in [t_\rho^-, s_\rho^-] \text{ and } M \in \mathbb{Z}_+, \\ C_{N,\alpha}\,\rho^{2+3\alpha_1/2-N/2}(\log\rho)^{1+1/(6\sigma)+2\alpha_2/\sigma+(1+1/\sigma)N} \\ \qquad \text{if } x_1 \in [s_\rho^-, t_\rho], \\ C_{N,\alpha}\,\rho^{3+3\alpha_1/2-N/2}(\log\rho)^{1-1/(3\sigma)+2\alpha_2/\sigma+(1+1/\sigma)N} \\ \qquad \text{if } x_1 \in [t_\rho, t_\rho^+], \end{cases} \tag{4.12}$$

where $s_\rho^- = (4\log\rho)^{-1/\sigma}(1 - \rho^{-1})$.

Proof. (i) Note that $u_\rho^N(x_1, 0) = U_\rho^N(x_1)$ for $x_1 \in [s_\rho^-, t_\rho]$. This, together with (4.6), yields (4.10). Since $\mathrm{Ai}(0) = 3^{-2/3}\Gamma(2/3)^{-1}$ and $\mathrm{Ai}^{(3k)}(0) = 1 \cdot 4 \cdots \cdots (3k - 2)\,\mathrm{Ai}(0)\ (\neq 0)$ for $k \in \mathbb{N}$, we have (4.11). (ii) Note that

$$u_\rho^N(x) = \begin{cases} 0 & \text{if } x_1 \leq t_\rho^-, \\ \chi_\rho^0(x_1)U_\rho^N(x_1)\exp[(4\log\rho)^{2/\sigma}x_2] & \text{if } x_1 \leq t_\rho, \\ U_\rho^N(x_1)\exp[(4\log\rho)^{2/\sigma}x_2] & \text{if } x_1 \in [s_\rho^-, t_\rho], \\ \widetilde{U}_\rho^N(x_1)\exp[(4\log\rho)^{2/\sigma}x_2] & \text{if } x_1 \in [t_\rho^+, 1]. \end{cases} \tag{4.13}$$

From Lemma 2.3 and (4.9) it follows that

$$|\tilde{U}_\rho^N(x_1)| \le C_N \, \rho^{47/12}(\log \rho)^{13/(6\sigma)} \quad \text{for } x_1 \in [t_\rho, 1].$$

This, together with (4.5), (4.6) and (4.13), proves the assertion (ii). (iii) It is obvious that

$$|f_\sigma^{(k)}(x_1)| \le C_k \rho^{-4}(\log \rho)^{(1+1/\sigma)k}$$

for $x_1 \in [t_\rho^-, t_\rho^+]$. For $x_1 \in [t_\rho^-, s_\rho^-]$ we have

$$L(x, D)u_\rho^N(x) = -\exp[(4\log\rho)^{2/\sigma} x_2]\{[L_\rho, \chi_\rho^0(x_1)]U_\rho^N(x_1)$$
$$+ \rho^2(4\log\rho)^{2/\sigma} f_\sigma(x_1)R_\rho^N(x_1)\chi_\rho^0(x_1)\},$$
$$[L_\rho, \chi_\rho^0(x_1)] = 2f_\sigma(x_1)(\partial_1\chi_\rho^0(x_1))\partial_1 + f_\sigma(x_1)(\partial_1^2\chi_\rho^0(x_1)).$$

Therefore, by (4.5), (4.6) and (4.8) we can see that (4.12) is valid if $x_1 \in [t_\rho^-, s_\rho^-]$. For $x_1 \in [s_\rho^-, t_\rho]$ we have

$$L(x, D)u_\rho^N(x) = -\exp[(4\log\rho)^{2/\sigma} x_2]\rho^2(4\log\rho)^{2/\sigma} f_\sigma(x_1)R_\rho^N(x_1).$$

This, together with (4.7) and (4.8), shows that (4.12) is valid if $x_1 \in [s_\rho^-, t_\rho]$. For $x_1 \in [t_\rho, t_\rho^+]$ we have

$$L(x, D)u_\rho^N(x) = -\exp[(4\log\rho)^{2/\sigma} x_2]$$
$$\times L_\rho(x_1, \partial_1)\{\chi_\rho^0(x_1)(U_\rho^N(x_1) - \tilde{U}_\rho^N(x_1))\}.$$

Since

$$(\partial_1^2 + p(x_1; \rho))(U_\rho^N(x_1) - \tilde{U}_\rho^N(x_1)) = \rho^2(4\log\rho)^{2/\sigma} R_\rho^N(x_1),$$
$$U_\rho^N(t_\rho) - \tilde{U}_\rho^N(t_\rho) = 0, \quad (\partial_1 U_\rho^N)(t_\rho) - (\partial_1\tilde{U}_\rho^N)(t_\rho) = 0,$$

Lemma 2.3 and (4.7) give

$$|\partial_1^k(U_\rho^N(x_1) - \tilde{U}_\rho^N(x_1))| \le C_k \, \rho^{3k/2-N/2}(\log\rho)^{1-1/(3\sigma)+(1+1/\sigma)N}$$

for $x_1 \in [t_\rho, t_\rho^+]$. This proves that (4.12) is also valid for $x_1 \in [t_\rho, t_\rho^+]$. $\square$

By Lemma 4.1, for any neighborhoods ω_i ($i = 1, 2$) of $x = 0$ with $\omega_1 \subset\subset \omega_2 \subset \{x \in \mathbb{R}^2 ; |x_1| \le 1 \text{ and } |x_2| \le 1\}$ and any $N \in \mathbb{Z}_+$ there are $c > 0$, $\rho_0 \ge 4$, $C > 0$ and $C_j > 0$ ($j \in \mathbb{Z}_+$) such that

$$\sup_{x\in\omega_1, |\alpha|\le 6} |D^\alpha u_\rho^N(x)| \ge c\rho^8,$$
$$\sup_{x\in\omega_2, |\alpha|\le q} |D^\alpha L(x, D)u_\rho^N(x)| \le C_q \, \rho^{4+3q/2-N/2},$$
$$\sup_{x\in\omega_2} |u_\rho^N(x)| \le C\rho^{95/12}(\log\rho)^{13/(6\sigma)}$$

if $\rho \geq \rho_0$ and $q \in \mathbb{Z}_+$. This, together with Lemma 2.2, implies that $L(x, D)$ is not hypoelliptic at $x = 0$.

References

[1] M. Derridj, *Sur une classe d'opérateurs différentiels hypoelliptiques à coefficients analytiques*, Séminaire Goulaouic-Schwartz 1970–1971, Exposé No. 12, Ecole Polytechnique, Paris.

[2] N. Nakazawa, *On hypoellipticity for class of pseudo-differential operators*, Tsukuba J. Math. **24** (2000), 257–278.

[3] N. Nakazawa and S. Wakabayashi, *On hypoellipticity of the operator* $\exp[-|x_1|^{-\sigma}]D_1^2 + x_1^4 D_2^2 + 1$, Publ. RIMS, Kyoto Univ. **38** (2002), 135–146.

[4] O. A. Oleĭnik and E. V. Radkevič, *Second Order Equations with Nonnegative Characteristic Form*, Itogi Nauki, Ser. Mat., Mat. Anal., Moscow, 1971. English translation: Plenum Press, New York, London, 1973.

[5] F. W. J. Olver, *Asymptotics and Special Functions*, Academic Press, New York, London, 1974.

[6] F. Treves, *Topological Vector Spaces, Distributions and Kernels*, Academic Press, New York, 1967.

[7] A. Yoshikawa, *On the hypoellipticity of differential operators*, J. Fac. Sci. Univ. Tokyo Sect. IA, **14** (1967), 81–88.

Seiichiro Wakabayashi
Institute of Mathematics
University of Tsukuba
Tsukuba 305-8571, Japan
wkbysh@math.tsukuba.ac.jp

Nobuo Nakazawa
Institute of Mathematics
University of Tsukuba
Tsukuba 305-8571, Japan

Progress in Nonlinear Differential Equations and Their Applications

Editor
Haim Brezis
Département de Mathématiques
Université P. et M. Curie
4, Place Jussieu
75252 Paris Cedex 05
France
and
Department of Mathematics
Rutgers University
Piscataway, NJ 08854-8019
U.S.A.

Progress in Nonlinear Differential Equations and Their Applications is a book series that lies at the interface of pure and applied mathematics. Many differential equations are motivated by problems arising in such diversified fields as Mechanics, Physics, Differential Geometry, Engineering, Control Theory, Biology, and Economics. This series is open to both the theoretical and applied aspects, hopefully stimulating a fruitful interaction between the two sides. It will publish monographs, polished notes arising from lectures and seminars, graduate level texts, and proceedings of focused and refereed conferences.

We encourage preparation of manuscripts in some form of TₑX for delivery in camera-ready copy, which leads to rapid publication, or in electronic form for interfacing with laser printers or typesetters.

Proposals should be sent directly to the editor or to: Birkhäuser Boston, 675 Massachusetts Avenue, Cambridge, MA 02139

Printed in the United States
121447LV00003B/267/A